NEURAL PROSTHESES
FOR **RESTORATION**
OF **SENSORY** AND
MOTOR FUNCTION

METHODS & NEW FRONTIERS IN NEUROSCIENCE SERIES

Series Editors
Sidney A. Simon, Ph.D.
Miguel A.L. Nicolelis, M.D., Ph.D.

Published Titles

Apoptosis in Neurobiology
Yusuf A. Hannun, M.D. and Rose-Mary Boustany, M.D.

Methods of Behavior Analysis in Neuroscience
Jerry J. Buccafusco, Ph.D.

Methods for Neural Ensemble Recordings
Miguel A.L. Nicolelis, M.D., Ph.D.

NEURAL PROSTHESES FOR RESTORATION OF SENSORY AND MOTOR FUNCTION

Edited by
John K. Chapin
Karen A. Moxon

CRC Press
Taylor & Francis Group
Boca Raton London New York

CRC Press is an imprint of the
Taylor & Francis Group, an **informa** business

CRC Press
Taylor & Francis Group
6000 Broken Sound Parkway NW, Suite 300
Boca Raton, FL 33487-2742

First issued in paperback 2019

© 2001 by Taylor & Francis Group, LLC
CRC Press is an imprint of Taylor & Francis Group, an Informa business

No claim to original U.S. Government works

ISBN-13: 978-0-8493-2225-9 (hbk)
ISBN-13: 978-0-367-39808-8 (pbk)

Library of Congress Cataloging-in-Publication Data

Neural prostheses for restoration of sensory and motor function / edited by John K. Chapin, Karen A. Moxon.
    p. cm. -- (Methods and new frontiers of neuroscience)
Includes bibliographical references and index.
ISBN 0-8493-2225-1 (alk. paper)
    1. Neural stimulation. 2. Myoelectric prosthesis. 3. Cochlear implants. 4. Artificial vision. I. Chapin, John K., Ph.D. II. Moxon, Karen A. III. Series.

RC350.N48 N465 2000 616.8⊠
04645--dc21   00-010648

Visit the Taylor & Francis Web site at
http://www.taylorandfrancis.com

and the CRC Press Web site at
http://www.crcpress.com

Methods and New Frontiers in Neuroscience

Our goal in creating the Methods and New Frontiers in Neuroscience Series is to present the insights of experts on emerging experimental techniques and theoretical concepts that are, or will be, at the vanguard of neuroscience. Books in the series will cover topics ranging from methods to investigate apoptosis to modern techniques for neural ensemble recordings in behaving animals. The series will also cover new and exciting multidisciplinary areas of brain research, such as computational neuroscience and neuroengineering, and will describe breakthroughs in classical fields like behavioral neuroscience. We want these to be the books every neuroscientist will use in order to get acquainted with new methodologies in brain research. These books can be given to graduate students and postdoctoral fellows when they are looking for guidance to start a new line of research.

The series will consist of casebound books of approximately 250 pages. Each book will be edited by an expert and will consist of chapters written by leaders in a particular field. Books will be richly illustrated and contain comprehensive bibliographies. Each chapter will provide substantial background material relevant to the particular subject. Hence, these are not going to be only "methods books." They will contain detailed "tricks of the trade" and information as to where these methods can be safely applied. In addition, they will include information about where to buy equipment, web sites that will be helpful in solving both practical and theoretical problems, and special boxes in each chapter that will highlight topics that need to be emphasized along with relevant references.

We are working with these goals in mind and hope that as the volumes become available the effort put in by us, the publisher, the book editors, and individual authors will contribute to the further development of brain research. The extent that we achieve this goal will be determined by the utility of these books.

Sidney A. Simon, Ph.D.
Miguel A. L. Nicolelis, M.D., Ph.D.
Series Editors

The Editors

John K. Chapin is professor of physiology and pharmacology at the State University of New York Health Science Center in Brooklyn. Born in Denver, CO, he received his B.S. from Antioch College in Yellow Springs, OH, and his Ph.D. in physiology from the University of Rochester, NY. His dissertation work proved that somatosensory responses of single neurons in the brain are biased according to motor expectation. In 1980 he won the Donald Lindsley Prize for the best Ph.D. dissertation in behavioral neuroscience. As an assistant professor at the University of Texas Southwestern Health Science Center at Dallas, and later as an associate and full professor of neurobiology at MCP Hahnemann University in Philadelphia, he helped develop techniques for recording large numbers of neurons in the brains of awake behaving animals. This work enabled his recent demonstration that animals can learn to control robotic devices purely through brain derived commands, as decoded from large scale neuronal population recordings in the motor cortex and other regions.

Karen A. Moxon received her bachelor's degree from the University of Michigan, Ann Arbor in chemical engineering and her Ph.D. in systems engineering from the University of Colorado Department of Aerospace Engineering, Boulder CO. After completing postfellowships at University of Colorado Health Science Center and MCP Hahnemann University from 1995-1999, she became a research assistant professor in the Department of Neurobiology and Anatomy in 1999. In 2000, Dr. Moxon joined the faculty of the School of Biomedical Engineering, Science, and Health Systems at Drexel University where she is currently an assistant professor. Dr. Moxon also maintains an adjunct position as assistant professor at MCP Hahnemann School of Medicine.

Contributors

John K. Chapin, Ph.D.
Department of Neurobiology
State University of New York
Health Science Center at Brooklyn
Brooklyn, NY

Simon F. Giszter, Ph.D.
Department of Neurobiology and
  Anatomy
MCP Hahnemann University
Philadelphia, PA

Warren M. Grill, Ph.D.
Department of Biomedical Engineering
Case Western Reserve University
Cleveland, OH

Joaquín-Andrés Hoffer, Ph.D.
Neurokinesiology Laboratory
School of Kinesiology
Simon Fraser University
Burnaby, B.C., Canada

Klaus Kallesøe, Ph.D.
Neurokinesiology Laboratory
School of Kinesiology
Simon Fraser University
Burnaby, B.C., Canada

Michael W. Keith, M.D.
MetroHealth Medical Center
Department of Veterans Affairs
Case Western Reserve University
Cleveland, OH

Phillip R. Kennedy, M.D., Ph.D.
Department of Neurology
Emory University
Atlanta, GA

Kevin L. Kilgore, Ph.D.
MetroHealth Medical Center
Department of Veterans Affairs
Case Western Reserve University
Cleveland, OH

B. King
Neural Signals, Inc.
Duluth, GA

Michel A. Lemay, Ph.D.
Department of Biomedical Engineering
Case Western Reserve University
Cleveland, OH

Gerald E. Loeb, M.D.
Department of Biomedical Engineering
University of Southern California
Los Angeles, CA

Nandor Ludvig, Ph.D.
Department of Physiology and
  Pharmocology
State University of New York
Health Science Center at Brooklyn
Brooklyn, NY

James Morizio, Ph.D.
Department of Electrical Engineering
Duke University
Durham, NC

Karen A. Moxon, Ph.D.
School of Biomedical Engineering
Science and Health Systems
Drexel University
Philadelphia, PA

Vivian K. Mushahwar, Ph.D.
Division of Neuroscience
University of Alberta
Edmonton, Canada

Miguel A. L. Nicolelis, M.D., Ph.D.
Department of Neurobiology
Duke University Medical Center
Durham, NC

P. Hunter Peckham, Ph.D.
MetroHealth Medical Center
Department of Veterans Affairs
Case Western Reserve University
Cleveland, OH

Bryan E. Pfingst, Ph.D.
Kresge Hearing Research Institute
Department of Otolaryngology
University of Michigan
Ann Arbor, MI

Arthur Prochazka, Ph.D.
Division of Neuroscience
University of Alberta
Edmonton, Canada

Frances J.R. Richmond, Ph.D.
School of Pharmacy
University of Southern California
Los Angeles, CA

Patrick D. Wolf, Ph.D.
Department of Bioengineering
Duke University
Durham, NC

Table of Contents

Introduction

The possibility of interfacing the nervous system with electronic devices has long fascinated scientists, engineers and physicians. In general, an ability to expand the bandwidth of the communication between brain and machine would provide many interesting possibilities, ranging from faster human–computer interfaces to direct remote control (i.e., "telekinesis"). In medicine, the field of neuroprosthetics has grown rapidly to include a variety of devices for stimulating neural tissue. Chapters 1–5 of this volume outline the present usefulness and future promise of such devices in the central and the peripheral nervous systems and in muscle. One of the most interesting areas for further development of this field is the use of simultaneously recorded neurons in the brain to control robotic devices. Chapters 6–9 consider the possible scientific and clinical advantages of using implanted devices to *record* signals from the nervous system, and then employ them to restore neurological function. The combination of these two technologies could play a particularly important role in restoring lost motor function in paralysis patients. An injury to the spinal cord in the neck, for example, could interrupt the flow of "motor command" information from the brain to the arm and also the flow of sensory feedback from the arm to the brain.

As schematized in Figure 1, the emerging field of "neurorobotics" seeks to obtain motor command signals from motor control regions of the brain and transform them into electronic signals suitable for controlling a robotic device. The primary motor cortex (MI), in the precentral gyrus of the human cerebral cortex, has long been known to be important for the control of voluntary limb movements. It is therefore conceivable that one could record commands for arm movement in the MI cortex and use those signals to directly drive a robotic arm of similar configuration (see Chapters 7 and 8). We have recently demonstrated the feasibility of such neurorobotic control in rats, and similar studies are ongoing in monkeys.

"Neuroprosthetics" (Figure 1) constitute another approach to the problem of restoring movement in paralyzed patients through functional electrical stimulation (FES; also called functional neuromuscular stimulation, FNS) of muscles or muscle nerves. For example, devices currently in use can produce grasping movements in a paralyzed hand through FES in the forearm musculature. These devices can be controlled through movement of other body areas, such as the unparalyzed contralateral shoulder.

Notwithstanding these futuristic scenarios, neural prostheses are rapidly becoming viable therapies for a broad range of patients with injury or disease of the nervous system. For example, over 3000 auditory prostheses have been successfully implanted in deaf patients (see Chapter 1) and over 150 devices for FES that restore grasping have been implanted in patients who have suffered loss of function in their upper extremities (see Chapters 2 and 3). This volume will present several different types of neural prosthetic devices as well as recent advances in cutting-edge research

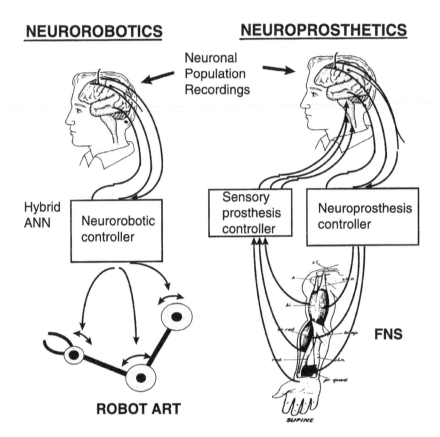

NEUROROBOTICS **NEUROPROSTHETICS**

Neuronal Population Recordings

Hybrid ANN | Neurorobotic controller

Sensory prosthesis controller | Neuroprosthesis controller

FNS

ROBOT ART

SUPINE

FIGURE 1 "Neurorobotics" refers to the use of brain-derived activity to control a robotic device. In the hypothetical case illustrated here, arrays of recording electrodes surgically implanted in the motor cortex and adjoining areas could be used to extract motor information from the brain. An online computational device (here, a neural network) could be used to transform the multichannel neural information into output signals appropriate for controlling a human-like robot arm. "Neuroprosthetics" generally refers to the use of electrical stimulation to artificially restore function of neural or muscle tissue. Chapters 2 and 3 describe existing methods for functional neuromuscular stimulation (FNS) to activate paralyzed muscles. As illustrated here, one might utilize brain-derived signals to provide command signals for such an FNS system to control the musculature of a paralyzed arm. As in the neurorobotic paradigm, a realtime computational device would be needed to transform the multichannel brain signals into FNS output signals sufficient to activate several arm muscles and coordinate their movements. Beyond this, it will ultimately be desirable to obtain sensory feedback from nerves in the arm (see Chapter 5) and feed that sensory information back into the brain. This would provide "closed-loop" control of the prosthetic device, allowing the brain to control the arm with much greater accuracy.

for novel devices to restore sensory and motor function in patients with neural damage. These chapters are intended to review the techniques underlying recently developed neural prosthetics that stimulate nerves and muscles to restore sensory or motor function. This area has shown rapid growth in the past decade. The realm of neural prosthetic devices spans stimulation of peripheral nerves for the restoration of motor function (FES), stimulating electrodes for repair of sensory systems (i.e., auditory), as well as the emerging field of devices implanted directly into the brain to control these prosthetic devices.

This book is divided into two sections. The first section provides details about some of the most successful sensory and motor prosthetic devices available. The second section reveals recent research into using brain signals to control a neural prosthetic device or an external device to restore sensory or motor function. In the first section, Chapter 1 presents an overview of the highly successful auditory prosthetic devices. These devices are now routinely implanted in patients with nerve damage and are widely successful in restoring hearing. Chapters 2 through 5 explore different approaches to the use of functional electrical stimulation for the restoration of motor control. Chapters 2 and 3 present devices developed and used in clinical settings for restoration of motor function in patients with spinal cord injury or disease. Chapter 2 describes a device that has been used to restore grasping in spinal-cord-injured patients. Chapter 3 presents the BION™ implant that performs FES to maintain muscle tone in patients who have lost the use of their limbs through spinal cord injury or stroke and its use for stimulating muscles to restore grasp. In Chapter 4 the latest advances in direct stimulation of the spinal cord for restoration of locomotion are examined. This chapter focuses on using FES of the spinal cord to control the limb movements during locomotion. Chapter 5 presents a nerve cuff electrode to record and modulate neural activity.

The next three chapters, making up the second part of the book, explore the growing field of brain-implantable devices to control artificial prosthetic devices or neural prosthetics described in Part 1 of this book. Chapter 6 introduces the electrodes and hardware that are traditionally used to record brain signals and the issues involved with creating a device for clinical applications. Some of the inherent problems with devices implanted directly into the brain are also discussed. Chapter 7 presents a case study of a successful control device implanted into the cortex of a severe shut-in patient. The patient has successfully used this device to control a cursor on a computer screen. Chapter 8 presents recent research data showing feasibility of brain-controlled neurorobotic devices. Finally, Chapter 9 provides a new perspective on combining neurochemical and neurophysiological information to create prosthetic control devices that restore chemical balance to the brain.

We would like to take this opportunity to thank the editors of the CRC Methods in Neuroscience Series, Drs. Nicolelis and Simon, for giving us the opportunity to put this book together. Without their enthusiasm and continued support, this project could not have been completed. Finally, we would like to thank the many people at CRC Press for their constant support and especially Barbara Norwitz, CRC Life Sciences Publisher, for her critical support in bringing this project to print.

John K. Chapin, Ph.D.
Department of Neurobiology
State University of New York
Health Science Center at Brooklyn
Brooklyn, NY

Karen A. Moxon, Ph.D.
School of Biomedical Engineering
Science and Health Systems
Drexel University
Philadelphia, PA

Part 1

Sensory and Motor Prostheses

1 Auditory Prostheses

Bryan E. Pfingst

CONTENTS

0-8493-2225-1/01/$0.00+$.50
© 2001 by CRC Press LLC

1.1 INTRODUCTION

Auditory prostheses, which function through electrical stimulation of the peripheral auditory nervous system, are the only commercially manufactured sensory-neural prostheses developed to date. Several companies around the world are currently manufacturing FDA-approved auditory prostheses for human use. As of June 1999, over 30,000 multichannel prostheses had been successfully implanted in deaf human patients ranging in age from 12 months to 80-plus years. Patients vary considerably in the amount of auditory information that they can understand using an auditory prosthesis. With the most recently implemented prosthesis designs, many patients can understand speech well enough in a familiar context to carry on a normal conversation without the aid of any visual cues. For the majority of patients, the prostheses provide enough information that normal conversation aided by lipreading is possible. A small number of patients are unable to understand speech even with the aid of lipreading but still find the implants useful for making them aware of their auditory environment: hearing an approaching automobile, hearing the doorbell ring, etc.

There are many reasons for the success of auditory prostheses. Prominent among these are:

1. **A strong basic-science foundation** — The physiological mechanisms by which auditory information is encoded in the central nervous system are well understood from many years of auditory, physiological and neuro-physiological research. In addition, many detailed studies in the area of auditory psychophysics and speech science have contributed to the design of processors for cochlear implants.

2. **A strong desire in the medical and industrial communities to bring this technology to deaf patients** — Several prominent otolaryngologists have aggressively pushed for the development and application of these devices. They, together with private companies, have shepherded the evaluation and approval of these devices by the U.S. FDA and other regulatory agencies and subsequently by medical insurance groups.

3. **Government support for ongoing research on these devices, as well as support of technology-transfer efforts that have led to the development of private companies that manufacture the devices** — Government support for development and implementation of this technology has been particularly strong in the United States and Australia.

4. **The redundant nature of acoustic speech signals and the ability of people to adapt to considerable distortions of, or deletions from, these signals** — An everyday example of this robustness is use of the ordinary telephone, which transmits acoustic information only between 300 Hz and 3.3 kHz. Although speech contains information in the range from about 80 Hz (the voice fundamental frequency for some male speakers) to about 8 kHz, the omissions imposed by the telephone go completely unnoticed by most users of this device.

1.2 HISTORY

The history of electrical stimulation of the auditory system can be traced to a number of early experiments in electricity and in neurophysiology (see Simmons,[1] Chapter 1). However, the event most frequently cited as inspiring the development of the auditory prosthesis was the experiment published in 1957 by Djourno and Eyries,[2] who implanted wires in the inner ear of a deaf patient and were able to produce sensations of sound by electrical stimulation. The experiment was subsequently tried by Doyle et al.[3] in Los Angeles and then by William House[4] and investigators at the House Ear Institute. It was House who was most enthusiastic about this procedure, and he eventually developed, and achieved FDA approval for, a single channel cochlear implant.[5] Blair Simmons, at Stanford University, first implemented the concept of a multichannel implant. He placed a set of four wires directly into the auditory nerve.[6] Shortly thereafter Robin Michelson at the University of California San Francisco implemented a multichannel implant placed in the scala tympani of the cochlea.[7] The UCSF implant, developed and subsequently modified by a research team under the leadership of Michael Merzenich, helped establish and test many of the design concepts and hypotheses that have led to the current generation of prosthesis.[8] The UCSF laboratories also began the development[9] of what is now the Clarion® prostheses, manufactured by Advanced Bionics® Corporation.[10,11]

In the meantime, a large and very productive research program was developing at the University of Melbourne in Australia under the leadership of Graeme Clark.[12] The implants resulting from this research program are manufactured by Cochlear Ltd., a subsidiary of Nucleus Pty. Ltd. in Sydney. Cochlear Ltd. currently has the largest population of implant users worldwide (over 23,000 patients).

Other implants have been brought to market as a result of research in London;[13] Salt Lake City, UT (the Ineraid prosthesis);[14,15] Vienna and Innsbruck, Austria (the Combi 40+ prosthesis manufactured by Med-El);[16-18] Paris (the Digisonic® prosthesis manufactured by MXM);[19] Antwerp, Belgium (the Laura prosthesis manufactured by Philips Hearing Technologies, B.V.);[20,21] and elsewhere.

1.3 BASIC DESIGN

The key elements in the auditory prosthesis are illustrated schematically in Figure 1.1. Acoustic signals (A) are picked up by a microphone (B) that converts them into electric signals and delivers them to an electronic signal processor (C). The processor typically divides the signal into multiple components using filters or other electronic processors and sends each component to a separate output channel as illustrated in Figure 1.2. The signal on each of these processor output channels is then converted into some sort of electrical waveform. Options for the electrical waveform include analog waveforms, which faithfully represent the temporal characteristics of the analog output from each channel of the processor; pulse trains, which can be fixed in rate or rate-modulated according to some temporal characteristic of the processor output channel; or amplitude-modulated pulse trains for which the modulation frequency is controlled by the processor channel output characteristics.

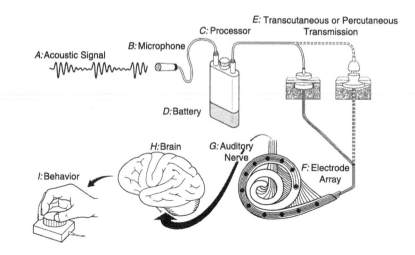

FIGURE 1.1 Schematic depiction of the components of an auditory prosthesis. A: Amplitude vs. time waveform of the acoustic speech signal. B: Microphone to capture the speech signal for delivery to the processor. C. Box containing the processor (See Figure 1.2), and in some cases the controlled-current stimulator. D: Batteries to power the processor and the stimulator. E: Transmission systems. The transcutaneous transmission system depicted on the left consists of an external antenna for transmitting signals and power across the skin and an implanted electronics package for receiving the signals and power and delivering controlled currents to the electrode array. The percutaneous connector depicted on the right can serve as an alternative for the transcutaneous transmission system. When the percutaneous connector is used, the controlled-current stimulator is housed with the processor. F: Electrode array implanted near the target neurons. In this depiction the electrode array is implanted in the scala tympani of the cochlea near the auditory nerve fibers (also see Figure 1.4). Possible alternative approaches include thin-film electrode arrays implanted directly into the auditory nerve bundle and surface or penetrating electrode arrays implanted on or within the cochlear nucleus in the auditory brainstem. G: Axons of the stimulated auditory neurons carrying nerve impulses to the central auditory system. This depiction illustrates the auditory nerve. H: The brain, including central auditory pathways, auditory cortex, and higher brain centers involved in speech recognition, speech production, and behavior. I: Detection, recognition, and discrimination of the electrical stimulation are indicated by subject responses, which can include oral and motor responses. This depiction shows a control knob that the subject can use to adjust the level of the signal during calibration sessions until it is just detectable or until it is at maximum comfortable loudness.

Examples of some stimulus waveforms are illustrated in Figure 1.3. The electrical signal must be compressed to conform to the dynamic range of electrical hearing. The signal must then be sent to the implanted electrodes.

Power for the electronics and for the stimulator is provided by batteries. Typically, a rechargeable battery pack is attached to the unit containing the processor (Figure 1.1D).

The most common scheme for delivering current to the electrodes is to implant a stimulator under the skin (Figure 1.1E: left drawing) behind the external ear and to lead wires from there to the electrode array that is implanted in or near auditory

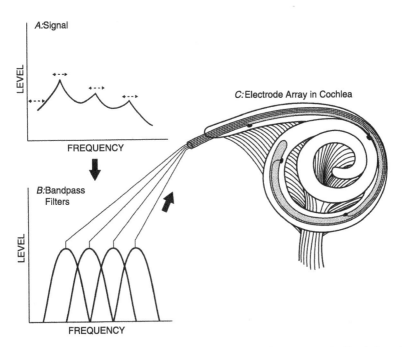

FIGURE 1.2 Schematic representation of a generic processing strategy. A: Depiction of the amplitude vs. frequency spectrum of a speech signal, in this case a vowel. For vowels, the spectrum is characterized by peaks called formants (F_1, F_2 and F_3). The horizontal arrows above the formant peaks and near the fundamental frequency (F_0) indicate that the frequencies of these formants change as a function of time during the speech utterance. B: Schematic representation of a processor. The processor extracts certain features from the acoustic signal and then these features are assigned to specific output "channels" based on their frequency content. In this illustration, bandpass filters are used to divide the acoustic signal into four specific frequency-limited bands and the output of each filter then comprises a channel. C: The output of each channel is sent to a specific place in or near the neural array, depending on its frequency. In a cochlear implant, channels that carry high frequency information are sent to the basal region of the cochlea and channels carrying low frequency information are sent to more apical regions, as illustrated in this example.

neurons. Instructions and power are then transmitted to the implanted receiver-stimulator by an antenna placed on the skin overlying the implanted receiver. Typically, the antenna is aligned and held in place by a magnet. The antenna outside the skin receives its information by cable from the battery-powered signal processor worn externally. The microphone is positioned near the external ear and is hardwired to the signal processor. Recently developed implanted electronics packages include the capability of transmitting information from the implanted electrodes in the cochlea, back out across the skin, to an external receiver so that electrical activity within the inner ear that is picked up by the implanted stimulating electrodes can be monitored, as can the electrode impedances. Some earlier designs of auditory prostheses for human patients and some current designs used for experimental studies

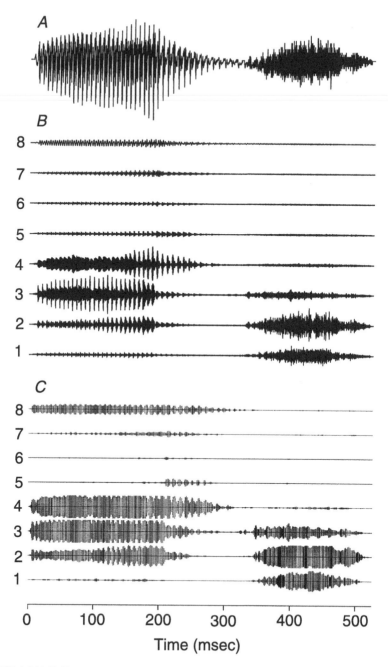

FIGURE 1.3(A,B,C) Illustration of various patterns of electrical stimulation used in auditory prostheses. **A:** Acoustic amplitude (ordinate) vs. time (abscissa) waveform for the word "ears." The time axis is shown below Figure 1.3C. **B:** Analog electrical waveforms for the word in Figure 1.3A following band-pass filtering. The waveform amplitudes would be compressed to fall within the subject's dynamic range for electrical hearing and then sent to sites in the

in animals use a percutaneous connector (Figure 1.1E: right hand drawing) instead of a transcutaneous transmission system. With a percutaneous connector, the stimulator can be connected directly to the internal electrodes, and direct monitoring of those electrodes is possible.

The output of the stimulator is sent to an array of electrodes that is implanted near neurons in the auditory pathway. Since most regions of the auditory pathway are tonotopically organized, similar processing strategies can be used regardless of the location of the electrode array.

The most common site for placement of the electrodes by far is in the inner ear, inside the snail-shaped cochlea (Figure 1.1F and Figure 1.4). These prostheses are commonly called *cochlear implants*. Typically, an array of electrodes is placed in the cochlea through the round window or through a fenestration made near the round window. Thus, barring an accidental deviation from the intended path, the electrode array will lie in the scala tympani and will occupy approximately the first 1½ turns of the cochlea. The human cochlea is approximately 2⅝ turns with the scala tympani having a length of approximately 35 mm, so the electrode array lies in the basal portion, which comprises the mid- to high-frequency region of a normal cochlea. The smaller, tighter turns of the apical region of the cochlea make it more difficult to place electrodes in that region without causing severe mechanical trauma to the remaining cochlear structures.

The scala tympani of the cochlea is an appropriate location for the electrode array because auditory nerve fibers are systematically arrayed along the longitudinal axis of the spiraling cochlea (Figure 1.4). In theory, specific groups of neurons can be stimulated by passing current between two electrodes placed near those neurons. Frequently, particularly in more recent implants, one or two extra-cochlear electrodes are implanted to serve as remote return electrodes for individual intracochlear electrodes. Stimulation in this configuration (monopolar stimulation) produces a larger current field and requires lower currents than bipolar configurations where both

---

FIGURE 1.3(A,B,C) (continued) electrode array, with the lowest frequency channel being sent to the most apical site and the highest frequency channel being sent to the most basal site. In this example, filter passbands (in Hz) for the eight rows (top to bottom) are: 187–437, 437–687, 687–1062, 1062–1562, 1562–2312, 2312–3437, 3437–5187, and 5187–7937. Analog waveforms such as these would be used in a compressed-analog (CA) or simultaneous-analog stimulation (SAS) strategy (Advanced Bionics Corporation). C: Fixed-channel amplitude-modulated interleaved pulse trains for the word in Figure 1.3A. Fixed-rate trains of symmetric biphasic pulses are amplitude modulated by the envelopes of filtered analog waveforms. These envelopes are extracted from filtered waveforms, such as those shown in Figure 1.3B, by half-wave rectification and low-pass filtering. The pulses are interleaved; that is, they are staggered in time so that no two channels are stimulated simultaneously. This interleaving is illustrated in Figure 1.3F. The pulse amplitudes would be compressed to fall within the subject's dynamic range for electrical hearing and then sent to sites in the electrode array, with the lowest frequency channel being sent to the most apical site and the highest frequency channel being sent to the most basal site. In this example, filter passbands are the same as those in Figure 1.3B, and pulse rate is 1200 pulses per second per channel. Amplitude-modulated pulse trains such as these are used in the continuous-interleaved-sampling (CIS) processing strategy.[66]

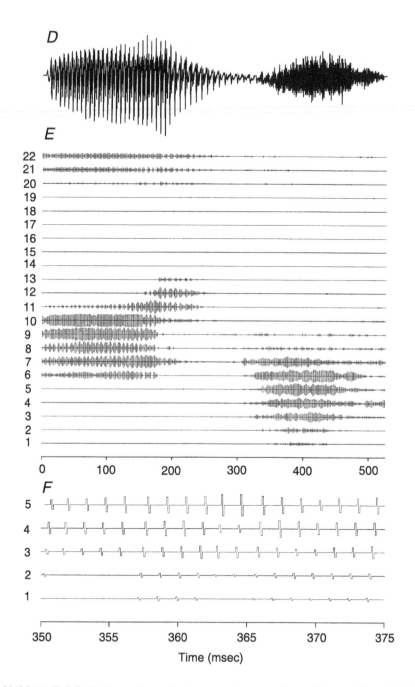

FIGURE 1.3(D,E,F) D: Acoustic amplitude versus time waveform of the word "ears" dupli-cated from Figure 1.3A. The time axis is shown below Figure 1.3E. **E:** Varied-channel ampli-tude-modulated interleaved pulse trains for the word in Figure 1.3D. Trains of symmetric biphasic pulses are amplitude modulated by the envelopes of the filtered analog waveforms in a subset of the number of available channels of stimulation. The subset of channels to be

electrodes are close together within the scala tympani. The monopolar configurations may have some additional advantages, as described later in this chapter.

The nerve fibers most likely stimulated by electrodes in the scala tympani are the auditory nerve fibers. In a normal cochlea these fibers have their cell bodies in the spiral ganglia within the cochlea and send peripheral processes along the basilar membrane, which forms the roof of the scala tympani (Figure 1.4). The central processes of these neurons exit the cochlea through the modiolus, which is located in the center of the snail-shaped cochlea. Following deafness, the peripheral processes of the neurons frequently degenerate, leaving the cell bodies and central process intact,[22,23] so the site of action potential initiation in many cases is probably at the central process.

Patients who are deaf due to bilateral auditory or vestibular nerve tumors (e.g., neurofibromatosis) may have the auditory nerves removed, in which case it is possible to place an array of electrodes on the cochlear nucleus (the first brainstem nucleus in the auditory pathway) where auditory nerve normally terminates. The current generation of *auditory brainstem implants* uses an array of electrodes that is placed on the surface of the cochlear nucleus.[24,25] However, very thin multicontact stimulating electrodes that penetrate the cochlear nucleus are currently under development.[26-28] These penetrating implants are manufactured using thin-film technology, and they potentially offer several significant advantages over surface electrode arrays. First, many more contact sites can be achieved because the electrodes penetrate through the layers of the nucleus and because the design technology allows a much more efficient and dense placement of stimulation sites on the electrode array.

FIGURE 1.3(D,E,F) (continued) stimulated on a given cycle is determined by the largest peaks in the outputs of the available filters. The pulses are interleaved; that is, they are staggered in time so that no two channels are stimulated simultaneously. This interleaving is illustrated in Figure 1.3F. The pulse amplitudes would be compressed to fall within the subject's dynamic range for electrical hearing and then sent to sites in the electrode array, with the lowest frequency channel being sent to the most apical site and the highest frequency channel being sent to the most basal site. In this example, the acoustic signal is passed through a bank of 22 band-pass filters spanning frequencies from 187 Hz to 7937 Hz. Between 0 and 9 channels, with 8 channels as the default average value, are stimulated at 900 pulses per second per channel. Varied-channel amplitude-modulated pulse trains such as these are used in the Advanced Combined Encoding (ACE) and SPEAK processing strategies (Cochlear Ltd.), and *n*-of-*m* processing strategy (Med-El). F: Expanded display pulse trains from Figure 1.3E, channels 1–5 in the time-range from 350 to 375 msec. Outputs from the active channels occur in round-robin fashion in rapid succession so that no two electrodes are stimulated simultaneously. Short pulse durations must be used so that the rate of the carrier pulse train on each electrode will be sufficiently high to faithfully carry the envelope waveform. In this example the pulse duration is 0.1 msec/phase, and the pulse rate on each stimulated electrode is normally 900 pulses/sec. Figures 1.3C, E, and F were constructed using the sCILab software program written by Lai and Dillier.[166] The pulse trains illustrated in Figures 1.3C, E, and F are fixed-rate pulse trains that are used as carriers for analog envelope waveforms. Another option, not illustrated here, is to modulate the rate of a pulse train based on some low-frequency feature of the signal, as is done in the early versions of the Nucleus processing strategies[12,54] and in the Laura phase-locked continuous interleaved strategy.[68]

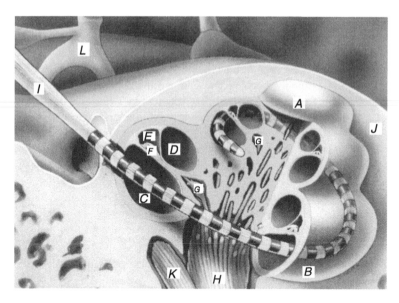

FIGURE 1.4 Artist's depiction of a dissected cochlea with a banded 22-electrode array inserted in the scala tympani. A wedge-shaped cut has been made in one side of the cochlea to reveal the internal structures. A: Outer bony shell of the apex of the cochlea. The apical region of the cochlea normally processes low-frequency sounds. B: Outer bony shell of the basal end of the cochlea. The basal region of the cochlea normally processes high-frequency sounds. C: Scala tympani. D: Scala vestibuli. E: Scala media, which contains the organ of Corti (F) in the normal cochlea. The organ of Corti normally contains about 3,000 inner hair cells that transduce mechanical vibrations caused by sounds into nerve impulses. In most profoundly deaf individuals, the hair cells have died. Implanted electrodes can be used in these cases to stimulate the auditory nerve. G: Parts of the spiral ganglion, which contains the cell bodies of the auditory nerve. Peripheral processes from these cell bodies project to hair cells in the organ of Corti (F) in an intact ear. Central processes from these cell bodies comprise the auditory nerve (H), which passes through the bony cone-shaped modiolus in the center of the cochlea and projects to the cochlear nucleus in the auditory brainstem. I: Multicontact electrode array inserted through a fenestration in the bone near the round window into the scala tympani and inserted for the first one and one-half turns of the cochlea. The Nucleus-22 electrode array depicted in this illustration consists of 22 platinum band electrodes supported by a silicone rubber carrier. Wires attached to the electrodes are located inside the silicone carrier and proceed to the receiver stimulator, which is implanted in the mastoid bone behind the external ear. Ten additional platinum bands (stiffening rings) at the basal end of the array are not connected to wires, but serve to improve the mechanical characteristics of the array to facilitate implantation. The cochlear structures overlying the implant are depicted as transparent so that the implant can be seen along its full length. J: Temporal bone surrounding the cochlea. This is the densest bone of the body. K: A portion of the vestibular nerve, which runs parallel to the auditory nerve and the facial nerve (hidden in this view) in the internal auditory meatus. L: The stapes, one of the middle ear ossicles that conduct sound from the ear drum to the fluids in a cochlear scalae in the normal ear. This drawing is reproduced with permission from Cochlear Corporation.

Second, the penetrating electrode arrays allow very close contact between the stimulation sites and the target neurons, resulting in lower current requirements and more specific activation of small populations of neurons. Third, the manufacturing process allows more flexibility in the design of the electrode arrays, batch fabrication of thousands of identical electrode arrays, and much lower production costs.

The potential advantages achieved with a penetrating cochlear nucleus implant could also apply to patients with an intact auditory nerve. A thin-film penetrating electrode could be placed in the auditory nerve in the modiolus. The surgical feasibility of such a *modiolar implant* was demonstrated very early in the history of cochlear implants by Simmons, who placed a bundle of wires into the auditory nerve. However, the amount of damage done by such a gross set of electrodes was, no doubt, considerably more than would be done by a thin-film penetrating electrode array. This modiolar implant is still in the early experimental stages.[29,30]

Thin-film technology can also be applied to scala tympani electrode arrays (cochlear implants). The technical requirements in this case are more difficult than with the penetrating brainstem or modiolar implants. The implant must be flexible so that it can follow the turns of the scala tympani, but it must be sufficiently stiff so that it can be inserted through the round window and pushed along the longitudinal axis of the scala tympani without buckling or sticking to the tissue surfaces. The current concept is to produce the electrode array and contact leads on a silicon substrate and then mount the thin-film array on other materials.[31] This would achieve the desired stiffness, curvature and bulk to facilitate insertion of the electrode array and to hold it in position once it is inserted. Currently most commercial scala tympani implants are manufactured by hand, which is a labor-intensive process. Thin-film technology offers the advantage of greatly reduced cost, as well as the potential for inserting a larger number of electrodes in a greater variety of spatial configurations.

1.4 RESEARCH ISSUES

Both theoretical and practical issues guide research on cochlear implants. A basic goal of most of the research is to understand the mechanisms underlying the function of the prosthesis. Knowledge of the mechanisms underlying the perception of electrical stimulation can be used to improve the design of the stimulus-processing strategies and to select optimal configurations for stimulus presentation (electrode configurations, stimulus waveforms, etc.). More specifically, this information can be used to improve the fitting of processing strategies and stimulus configurations to an individual patient operating in a specific environment. Both clinical and experimental data have shown that subjects vary enormously in their ability to use a particular auditory prosthesis.[32,33] Furthermore, recent studies have shown that when subjects are tested with several different processing strategies, they often prefer and/or perform better with one particular strategy. The optimal strategy varies from patient to patient.[34,35] For example, electrode configurations that are more effective for one patient may make no difference in another.[36] In addition to the individual variability among patients, the characteristics of the acoustic environment must also be considered. Thus, a patient who has good speech understanding with a particular processing strategy and stimulus configuration when listening in a quiet environment may require a different strategy in a noisy environment. Processors that provide a patient with good speech recognition may serve poorly when listening to music[37,38] and vice versa. Without knowing the mechanisms by which these various strategies, configurations and environments affect perception, it is difficult to know how to choose a strategy for a particular patient.

In the following sections we will review some of the practical and basic-science issues that have been addressed by researchers over the past 25 years.

In the early years of prosthesis research, safety was a major topic. Animal studies were used to determine effects of inserting electrode arrays and safe limits of electrical stimulation. While the basic questions for the existing generation of implants have been answered, new stimulating strategies raise new issues that must be addressed. Since safety considerations limit the range of prosthesis design and use, we will address the basic safety issues before proceeding with a discussion of prosthesis design and efficacy.

Within the safe limits of electrical stimulation, a number of options are available for improving prosthesis function. All of the stages represented in Figure 1.1 are potential sources for improving electrical hearing. In Section 1.6, we will review some of the practical and research issues associated with each of the stages of processing illustrated in Figure 1.1. In the process, we will review some of the improvements in prosthesis performance that have occurred over the last 25 years. Figure 1.5 shows some examples of the range of performance achieved with various versions of the auditory prosthesis.

While there has clearly been an improvement in the average speech-recognition scores achieved by auditory prosthesis users over years, there remains a great deal of individual variation in the benefit patients receive from these devices. Research directed at assessing and responding to these individual differences will be addressed in Section 1.7.

1.5   SAFETY

The number of medical problems (side effects) resulting from implantation and electrical stimulation of the auditory system has been remarkably small.[39,40] Problems primarily involve the skin overlying the implanted electronics, and these problems are rare.

The primary research issue regarding safety of the implant is with respect to safe limits of stimulation. These limits were documented in early studies in animals and were found to be higher than the currents required for operation of the implant, given the designs that were in use at that time.[41,42] However, new implant and processor designs have raised new questions about the safe limits of stimulation.

Two factors that affect the safe limits of stimulation are (1) the dissolution of toxic products at the electrode-tissue interface due to electrochemical reactions and (2) metabolic effects at high rates of electrically-induced neural activity. Factors that affect electrochemical reactions are the current level and the size of the stimulating electrodes. Electrode size is an issue because irreversible reactions can occur at the surface of the electrode based on *current density,* which increases as an inverse function of electrode size. The important variable here is the "real surface area," which can be measured electrically.[43] If the implant design calls for smaller electrode sites in order to increase the density of electrodes or to optimize their orientation, surface area becomes an issue. There is a relationship between real surface area and electrode materials, so it may be possible to decrease the acceptable electrode size by changing the composition of the electrodes.[44]

At high pulse rates, a major safety issue is neural fatigue. Neural damage can be produced by high rates of prolonged electrical stimulation. One indication that this injury is activity related is that the injury can be prevented by blocking the neural activity with a local anaesthetic.[45] In studies in animals, investigators have found that high rates of stimulation at high levels can produce a temporary elevation in the thresholds for evoking a response with electrical stimulation. The physiological mechanism for this temporary threshold elevation is not known, but it is assumed that stimuli that produce such changes should be avoided. Thus, animal models have been used to determine what combinations of stimulus rates and levels produce these changes.[46] Another potential concern with high rates of stimulation is that if pulses occur very close together in time, mechanisms for assuring charge balance may be ineffective. Charge balance is necessary to prevent charge buildup and to balance reversible reactions at the electrode-tissue interface, thus minimizing the deposition of toxic substances in the tissues.[47] If charge balance is achieved passively by shorting electrodes together and allowing any DC charge to equalize, sufficient time must be given for this equalization to occur.

1.6 OPPORTUNITIES FOR IMPROVING PROSTHESIS FUNCTION

1.6.1 CAPTURING THE ACOUSTIC SIGNAL

The first stage in auditory prosthesis signal processing involves capturing the acoustic signal that the patient wants to hear (see Figure 1.1). Most commonly, this is done with a small microphone that the subject wears near the ear. One of the most important problems that can be aided by the acoustic pick-up system is recognition of speech signals in a noisy environment. Noisy environments represent a particularly difficult situation for cochlear implant users, as well as for most individuals with any type of hearing impairment. The problem can be lessened by acoustic pick-up systems that focus on the source of the desired signal. A simple though not always practical approach is to put the pick-up close to the source of the signal. Thus, there are acoustic pick-ups that can be attached directly to a telephone, a television or a radio. In addition, where there is one-to-one contact between the speaker and the listener, the speaker can wear a microphone that transmits the signal to the subject's speech processor. Directional microphones worn by the listener can be advantageous. The best performance can be achieved by a beam former, which uses an array of microphones,[48,49] but this is not always practical for the wearer. Further development in this area is warranted.

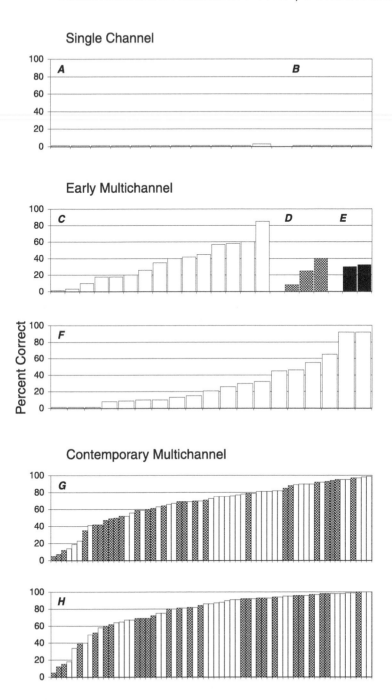

FIGURE 1.5 Speech recognition results for patients fitted with various types of cochlear prostheses. Each bar represents an individual patient. For each type of prosthesis, subjects are arranged in order based on their speech recognition scores. **A-F:** Results of a study by Gantz et al.[32] The Iowa Sentence Test was used for all configurations tested in this study. The

Another potential approach, still under development, is to extract the desired signal from background noise electronically. For example, signals with acoustic characteristics of speech might be extracted electronically from non-speech-like noise.[50] Such feature-extraction approaches are also sometimes used for converting acoustic signals to electrical signals as described in Section 1.6.2.

1.6.2 PROCESSING STRATEGIES

1.6.2.1 Basic Designs

The most plentiful and diverse opportunities for improving auditory prosthesis function lie in the stimulus-processing strategies. These strategies determine the way in which various features of the acoustic signal are selected and how those features are represented across the various channels of the prosthesis.

In designing processing strategies, some consideration is usually given to the way in which the normal auditory system analyzes and encodes acoustic signals. In the normal system, there are two primary codes: a temporal code and a spatial or

---

FIGURE 1.5 (continued) speech recognition results (percentage of key words correct) were obtained for presentation of the sentences in quiet with sound only (no lipreading). Prosthesis designs tested are identified by letters as follows. **A:** Eleven subjects using the House Ear Institute single-channel cochlear implant, manufactured by 3M Company.[51] **B:** Four subjects using a single channel implant developed at the University of Vienna.[52] Note that although subjects in this study who had single-channel processors showed little or no open-set speech recognition, the subjects did receive some benefit from the prosthesis in that it served as an aid to lipreading, made the subjects aware of environmental sounds, etc.[32] **C:** Fifteen subjects using an early version of the Nucleus multichannel cochlear implant developed at the University of Melbourne and manufactured by Cochlear Corporation.[12] These subjects were tested using an $F_0F_1F_2$ processing strategy where F_0 determined the pulse rate and F_1 and F_2 determined the place of stimulation. Pulsatile stimulation was used with only one electrode stimulated at a time. A longitudinal bipolar electrode configuration was used. **D (shaded bars):** Three subjects using an early Nucleus processing strategy in which the place of stimulation was determined by the second formant (F_2). **E (black bars):** Two subjects with a multichannel implant developed at the University of California San Francisco and manufactured by Storz.[9] **F:** Nineteen subjects using a four-channel implant developed at the University of Utah and manufactured by Symbion.[14] This implant used band-pass filters and sent the compressed analog output of each filter to an appropriate electrode. It used a monopolar electrode configuration and simultaneous stimulation of the four electrodes. This processing strategy is similar to that illustrated in Figure 1.2 and Figure 1.3B. **G and H:** Results from a study by Skinner et al.[33] comparing two more recent versions of the 22-electrode Nucleus prosthesis manufactured by Cochlear Corporation. This study used CUNY (clear bars) or SIT (shaded bars) sentences in quiet with no visual cues. Sixty-one subjects were tested using both MPEAK (Figure 1.5G) and SPEAK (Figure 1.5H) processing strategies. The MPEAK processing strategy is a feature extraction strategy, and the SPEAK processing strategy, is a filter-bank strategy. Twenty one of the 61 subjects showed significantly better speech recognition performance ($p < 0.05$) with the SPEAK strategy, and two showed significantly better performance with the MPEAK strategy. Even newer processors, currently in initial phases of testing, are showing promise of producing still better speech-recognition results.

place code. The temporal code consists of temporal patterns of neural discharges that follow some low-frequency aspect of the acoustic waveform. This code can be implemented in the auditory prosthesis because electrically stimulated neurons phase lock to low-frequency temporal patterns in the electrical stimulus.

A purely temporal-coding approach was used in the original single-channel implants that were developed by House and colleagues[51] and Hochmair-Desoyer and colleagues.[52] A few subjects were able to do quite well in understanding speech using just the temporal information provided on a single electrode.[53] In general, however, the speech recognition performance with these early single-channel implants was significantly inferior to that achieved with later multichannel implants (Figure 1.5).

Place codes in the normal auditory system result from mechanical analysis of the incoming acoustic signal by the basilar membrane, aided by an active process that depends on outer hair cells. The result is that different regions of the auditory nerve respond maximally to specific spectral components of the acoustic signal. This frequency-to-place code can be crudely duplicated in a cochlear prosthesis by assigning the place of electrical stimulation based on the energy in specific regions of the acoustic spectrum. This approach was taken in the early versions of the Cochlear Corporation "Nucleus-22" prosthesis. That prosthesis used a formant tracker to determine the frequencies of various formant peaks (Figure 1.2A) in the spectrum of the speech signal. In an early processing strategy (the F_0F_2 strategy), the second formant (F_2) was used.[12] In later processing strategies F_1 was added (the $F_0F_1F_2$ strategy), and then some additional higher-frequency spectral peaks were added (the multi-peak or MPEAK strategy).[54] The frequencies of these spectral peaks were used to determine the place of stimulation along a 22-electrode array implanted in the scala tympani. The signal used for stimulation at any given place was a train of pulses with a pulse rate determined by the fundamental frequency of the voice (F_0).

A third approach is to combine temporal and related spectral information. The strategy here is to use band-pass filters to divide the acoustic signal into various bands with a range of center frequencies (e.g., Figure 1.2B). The output of each filter then comprises a separate channel in the multichannel prosthesis and signals with specific center frequencies are sent to specific locations in the electrode array. This approach can be successfully implemented with only four or five channels, as was done in the early Ineraid implant.[14] The current Clarion prostheses use seven or eight channels.[11] Thus, reasonably good speech recognition can be obtained using only a few channels and largely temporal information.

1.6.2.2 Number of Channels

Dividing the information into at least four channels provides significant advantage over presenting all of the information on a single channel to a single pair of electrodes (Figure 1.5). This advantage of dividing the temporal information into channels can be demonstrated acoustically using *normal-hearing listeners*. The most recent experiments conducted using this acoustic-simulation approach were those of Shannon et al.[55] and Dorman et al.,[56] who extracted the temporal envelope from the speech signal and used band-pass filters to create one or more channels of information. The outputs of these filters were then used to amplitude modulate sinusoidal or narrow-band noise carriers so the output of each channel would stimulate different parts of the cochlea, depending on the spectrum of the carrier. This is similar to the early experiments with vocoders.[57-60] When only one channel was used, normal-hearing speakers with minimal practice performed rather poorly on speech recognition tasks but performance increased as a function of the number of channels. With four channels, performance was near 100% for simple sentences in a quiet background.[55] Under more difficult listening conditions, for example, in a noisy background, more channels are required in order for the subjects to achieve high speech recognition scores.[61] Young children also perform better with a greater number of channels.[62]

Several studies with *implanted patients* have indicated that speech recognition performance increases as a function of the number of channels up to 4–8 channels and then asymptotes, with no further increase for more than 8 channels in quiet[63] or in noise.[64] There may be interactions between the number of channels required and other variables. A recent study by Dorman et al.[65] showed an interaction between the number of channels required and the number of intensity steps available to the subject. Further research is needed to determine if implanted subjects can benefit from more that 8 channels and to define the conditions that limit the number of useful channels.

1.6.2.3   Current Interactions

In a multichannel prosthesis, each channel should carry information that is different from the information on adjacent channels. If current is output simultaneously to adjacent sets of electrodes in the stimulating electrode array, there are likely to be current interactions that will distort the waveform from each channel. A temporal solution to these spatial interactions of the electrical current fields is to stimulate one electrode at a time using pulsatile stimulation. However, a disadvantage is that the pulse rate is limited by the time it takes to cycle through the array of electrodes to be stimulated. For this reason, investigators are experimenting with processors that use simultaneous pulsatile stimuli to two or more electrodes.[11]

Pure analog stimulation, of course, requires simultaneous stimulation of multiple electrodes because the signals are continuous. However, it is possible to present much of the temporal information in a continuous analog signal by amplitude modulating pulse trains with the analog waveform. This strategy was successfully implemented by Wilson et al.[66] and now comprises the commercially available Continuous Interleaved Sampling (CIS) processing strategy. With this strategy the range of frequencies that can be encoded in the envelope of the amplitude-modulated pulse trains is, of course, limited by the pulse rate, which in turn is limited by the number of channels through which the processor must cycle. As discussed in Section 1.5, safety becomes a concern when very high rates of stimulation are used.

Many subjects show better speech recognition in the CIS strategy as compared to a compressed analog strategy in which all electrodes are stimulated simulta-neously.[66] However, recent studies have shown that some patients prefer and hear better with simultaneous analog stimulation (SAS) than with the CIS strategy.[34,35,67]

Several variables differ among these studies and could contribute to the differences in results. Additional research will be needed to sort out the effects of these variables.

Another approach to reducing current interactions is to space the electrode sites farther apart along the longitudinal axis of the scala tympani. Spacing the sites farther apart may also increase the discriminability of one electrode site from another. Since the electrode can be inserted only a given distance, spacing the electrodes farther apart reduces the number of total sites that can be implanted. However, since modern processors use only a few sites per cycle, it would be possible to impose algorithms that would keep adjacent sites farther apart. On the other hand, this may result in shifting place-pitch maps, which could degrade performance. This is an area for future research.

1.6.2.4   Some Contemporary Processing Strategies

Examples of some of the types of processing strategies currently in use are illustrated in Figure 1.3. An implementation of the analog strategy (Figure 1.3B) called simultaneous analog stimulation (SAS) is currently available as one of the options in the Clarion prosthesis.[11] With this strategy, the manufacturer attempts to reduce current interactions by using a bipolar stimulation configuration, which in theory has less current spread than a monopolar configuration. Another set of strategies available in the Clarion processor, and in the Nucleus, Combi 40+ and Laura processors, are the Continuous Interleaved Sampling (CIS) strategies (e.g., Figure 1.3C). These strategies, which vary in detail from processor to processor, are modeled after the CIS strategy developed by Wilson et al.[66] These strategies are similar to the simultaneous analog strategy in that the same channels are always stimulated with the outputs of fixed filters. However, instead of passing the compressed output of the filters directly to the electrodes, the envelopes of these filtered waveforms are used to amplitude modulate interleaved pulses, thus avoiding current interactions.

Two similar processing strategies used in current Nucleus (Cochlear Corporation) processors are the spectral peak (SPEAK) strategy and the advanced combined encoding (ACE) strategy (see Figure 1.3E). Like the CIS strategy, the SPEAK and ACE strategies use interleaved pulses and a fixed filter is assigned to each stimulation site, but unlike the CIS strategy the sites stimulated on any given cycle vary depending on the location of peaks in the signal amplitude spectrum. Also, because there are up to 22 stimulation sites, as opposed to 7 or 8 in the CIS and SAS strategies, the filter widths are narrower. With ACE the pulse rate is higher than with SPEAK. Up to 9 sites can be stimulated on a given cycle depending on the number and intensity of spectral peaks in the signal. The frequencies of the largest spectral peaks determine the places of stimulation on a given cycle. Since a high pulse rate is used in the ACE strategy, this strategy effectively represents the low-frequency temporal envelope of the signal in the amplitude envelope of the pulse train. The most prominent features of the stimulus spectrum are, of course, also represented in terms of place of stimulation along the 22-electrode scala tympani implant. The Med-EL Combi40+ uses a similar strategy, called "n of m," which stimulates the n (0 to 6) sites for which the filter outputs have the highest energy. However, the number of available stimulation sites ($m = 12$) is smaller than in the nucleus prostheses.

Some prostheses use a combination of strategies. For example, the Laura phase-locked CIS strategy uses a CIS type strategy for higher-frequency signals and pulse trains that are phase locked to lower-frequency components of the signal.[68]

1.6.3 Features of Electrical Stimulation

After decisions have been made about which features of the stimulus to send to which channels, the next step is to determine how the output of each channel will be presented to the designated site in the electrode array. That is, a given channel output can be presented to a given cochlear location using a variety of different stimulus waveforms, a variety of different electrode configurations, etc. Choice of the waveform, electrode configuration, etc. can make a large difference in the ability of the subject to recognize and discriminate the stimulus. In this section we will discuss some of the options available for stimulus presentation and their theoretical and practical implications. Some important features of electrical stimulation are listed in Table 1.1.

The various temporal, spatial, and level features of the electrical signal are controlled by the subject's processor (Figure 1.1C). The features of stimulation described in this section may be applied in a variety of the processing strategies described in the previous section. At the same time, many of these temporal, spatial and level features affect details of the spatial and temporal neural response patterns that are important for perception. Thus, from a conceptual point of view, they may be considered relevant to the electrical-neural interface depicted in Figure 1.1F and G. For example, the phase duration of the pulses used in any of the pulsatile processing strategies mentioned above may affect both the neural representation of temporal information in the pulse train and the spatial extent and distribution of neurons responding at a stimulus level that produces a comfortably loud sound. These effects on the neural response patterns may have large effects on perception of the signals presented in any of the processing strategies to which they are applicable.

1.6.3.1 Temporal Features

1.6.3.1.1 Electrical-stimulus waveforms

Temporal information in the auditory signal can be conveyed by one of three types of electrical waveforms: (1) analog waveforms in which the acoustic signal is filtered and compressed and conveyed as an analog electrical signal (Figure 1.3B); (2) pulse trains in which the frequency of the pulses corresponds to some frequency in the acoustic signal; or (3) amplitude-modulated pulse trains in which the rate of the pulse train is fixed and components of the acoustic signal are selected to modulate the amplitude of the pulse train (Figure 1.3C). In practice, good prosthesis performance has been obtained with all three types of waveforms, but comparative studies have shown differences among subjects in preference or performance based on the type of waveform used and on the listening context.[34,35,67] The compressed analog electrical waveform potentially carries the most information since all of the component frequencies passing through the filter will be represented in the electrical waveform. Some temporal detail is necessarily lost in the two pulsatile strategies. However, as noted in Section 1.6.2, in a multichannel prosthesis, pulsatile waveforms

---

TABLE 1.1
Some Features of Electrical Stimulation
Important for Electrical Hearing

Temporal Features
  Analog stimuli
    Frequency
    Phase
  Pulsatile stimuli
    Phase duration
    Pulse rate
    Pulse polarity
    Initial-phase polarity
    Pulse polarity order
  Analog or pulsatile stimuli
    Stimulus duration
    Duty cycle
Spatial Features
  Electrode location
    Longitudinal
    Medial-lateral
  Electrode orientation
    Longitudinal
    Radial
    Off-radial
  Electrode site separation
  Electrode configuration
    Multipolar
    Tripolar
      Quadrupolar
      Other tripolar
    Bipolar
    Monopolar
    Multiple monopoles
    Component electrode separation for configurations
      that use multiple electrodes
Stimulus Level
  Operating range
  Mean operating level
  Compression algorithm

---

offer the advantage that different channels can be stimulated at different times, thus avoiding problems of current interaction on adjacent channels.

In a multichannel prosthesis, the relative phase of the information presented on different channels is a potentially important variable. In a normal auditory system, the phase of the acoustically elicited traveling wave changes as a function of location along the basilar membrane. To date, only minimal attention has been given to this temporal feature of the electrical waveform. This is a potential source of improvement in future processing strategies.

1.6.3.1.2   Other temporal features

Some temporal features of the electrical stimulus are dependent upon the acoustic stimulus waveform,[69] but others are free parameters that can be set by the audiologist or the researcher. Which parameters are free depends on the processing strategy. In the older Nucleus processors, for example, the pulse *rate* was determined by the fundamental frequency of the voice (F_0). In more recent processing strategies, high-fixed-rate pulse trains are used and the temporal-envelope features of the acoustic signal are represented in a waveform that is used to amplitude modulate the pulses. Temporal stimulus features such as pulse *duration* (in both cases) and pulse rate (in the latter case) can be set by the experimenter independent of the acoustic signal. These free parameters can affect the physiological response patterns and hence the perception of the electrical signal. Effects of various temporal features of the electrical signal on both physiology and psychophysical perception have been studied in some detail. These features are listed in Table 1.1. Many of these stimulus features have been shown to affect detection of electrical signals,[70-74] and possibly they can affect speech recognition.[66,75]

Temporal features of the electrical stimulus may affect not only the temporal pattern of response of the auditory neurons, but in some cases, spatial features as well. For example, phase duration and pulse-polarity order may affect the site of action potential initiation.[76,77] Slopes of psychophysical strength-duration functions (detection threshold vs. pulse duration) are affected by pulse duration, stimulus duration and pulse-polarity order.[70,73,78] This may be a result, in part, of the effects of these variables on the site of action potential initiation.

Auditory neurons follow the temporal pattern of an electrical stimulus quite well up to frequencies as high as a few kilohertz.[79-81] Indeed the frequency-following may be better than that found in the normal acoustically-stimulated auditory system. This excellent frequency-following may be a drawback in that it is abnormal relative to the physiology of the intact auditory system. Investigators are now examining the use of refractory properties of the nerve to introduce more stochasticity in the neural response. In theory, this can be achieved by using high rates of pulsatile background stimulation.[82,83] Adding background noise to the signal may have additional advantages in enhancing detection of some spectral features of the signal.[84]

1.6.3.2   Spatial Features

1.6.3.2.1   Electrode configuration

For many years investigators have assumed that the best configuration for the electrodes in a multichannel electrode array would be one that would spatially restrict the current fields, thus maximizing the independence of each channel. Recent evidence, however, has indicated that configurations that are thought to produce large, broad current fields perform equally as well as, and in some cases better than, configurations that produce restricted current fields.[36,85,86] These observations call for a reexamination of the assumptions about the effects of electrode configuration on the size of the activated neural population, as well as assumptions about what is the optimal activation pattern for speech recognition. The relationship between electrode configuration and the size of the activated neural population is complex because in a functioning prosthesis the amount of current required to achieve a comfortable listening level varies as a function of electrode configuration and because the current level affects the number of neurons activated. The interactive effect of these two confounding variables (electrode configuration and stimulus level) on neural activation patterns is difficult to predict based on currently available models. Neural recording studies in behaviorally trained animals, in which thresholds and comfortably loud currents can be measured, are in progress to determine what neural activation patterns result when current levels for various configurations are set at equal-loudness levels.

In the normal auditory system the number of neurons activated by a specific stimulus, such as a pure tone, increases dramatically as a function of stimulus level. Some specificity remains in the discharge rate-place profile because neurons most sensitive to the stimulus frequency tend to fire at higher discharge rates than do neurons that are less sensitive to that frequency. However, rate saturation can result in a degradation of the rate-place profile at higher levels. Despite this degradation in the neural response pattern, the perception of the signals does not degrade. In fact, discrimination of one stimulus from another is typically better at high levels of stimulation than it is near the detection threshold.[87-92]

The optimal spatial pattern of neural excitation for perception is not known. Studies in deaf subjects with cochlear implants provide us with an opportunity to manipulate the spatial response patterns and thus test hypotheses about the optimal pattern. Recently, studies using acoustic stimuli that mimic the cochlear implant have also proven useful for testing these hypotheses.[93,94] Complete spatial overlap of the neural population excited by each channel of the prosthesis is not desirable because (as discussed earlier) it has been demonstrated that hearing is better when the acoustic signal is divided into restricted frequency bands, which are sent to multiple channels.[55] On the other hand, if the neural response pattern is too spatially restricted, this may have disadvantages. A spatial spread similar to that in the normal auditory system, i.e., fairly broad, may be desirable. It must be kept in mind, however, that predicting the optimal spatial pattern of neural excitation is complicated by the highly variable pattern of nerve loss across the population of deaf patients. The optimal spatial pattern may vary from patient to patient, requiring both tools for accurate estimation of the nerve survival pattern and rules for selecting an optimal electrode and stimulus configuration for various patterns.

1.6.3.2.2   Electrode orientation

For bipolar electrodes in the scala tympani, the most common orientation is longitudinal. That is, one electrode in the pair is apical to the other. In theory, a radial orientation in which one electrode in the pair is medial to the other would be more efficient because the principal current gradient would be aligned with the axis of the peripheral processes. However, there are some practical drawbacks to using the radial orientation. The principal drawback is that the diameter of the scala tympani is small, making it difficult to fit two electrodes in a radial orientation and have the electrode surface areas be sufficient to deliver the required levels of current safely.

Also, for the same reason, the electrodes must be close together, so the amount of current required to reach threshold is relatively high. This compounds the problem of safe stimulation given the small surface areas. Finally, one must consider that in many deaf cochleae, the peripheral processes may be degenerated so that the site of stimulation may be remote from the stimulating electrodes and the advantage of the radially-oriented current gradients may be lost.

1.6.3.2.3 Longitudinal electrode location and spacing

Current electrode designs and surgical procedures do not permit scala tympani electrode arrays to be inserted through all turns to reach the apex of the cochlea. This means that the electrode array is stimulating regions that in a normal ear carry higher frequencies than the actual frequencies that, in an implant processor, are determining the place of stimulation. Evidently, patients are able to adjust to this mismatch between the spectral content of the signal and the place of electrical stimulation, because many subjects are able to understand speech reasonably well despite the mismatch. On the other hand, there is some evidence that reducing the mismatch is advantageous. A study by Dorman et al.[94] using acoustic simulations of cochlear implants with amplitude-modulated sinusoidal carrier tones indicated an advantage of increasing insertion depth. Studies by Fu and Shannon have shown that matching spectral information with the place of stimulation that the subject has practiced using has significant advantages for speech recognition.[95]

Recently, investigators have been experimenting with surgical approaches that allow insertion of portions of the electrode array in fenestrations made in more apical regions of the cochlea.[96-98] Initial results from these experiments are promising. However, it should be noted that although the peripheral processes of the spiral ganglion cells project to all turns of the cochlea, the cell bodies are located near the modiolar wall of the scala tympani only through the first 1½ turns. Therefore, if the peripheral processes degenerate following deafness, efforts to extend the insertion of electrode array into the apical turns may be futile.

1.6.3.2.4 Radial electrode-site location

For a *cochlear* implant the cell bodies of the target neurons lie medial to the scala tympani (Figure 1.4). These target neurons send peripheral processes out through the habenula perforata (holes in the bony shelf near the dorsal-medial part of the scala) and under the basilar membrane, which forms the dorsal boundary of the scala tympani. The central processes of these neurons pass through the modiolus, which is medial to the scala tympani. In theory, the best location for the stimulating electrodes in an auditory prosthesis would be near the target neurons. In a cochlear prosthesis, this means that the electrodes should be near the medial or the dorsal-medial surface of the scala tympani. Recall, however, that survival of the peripheral processes is variable, and some deaf ears may lack peripheral processes altogether.[22,23,99]

If the cochlear electrode array is straight before it is inserted in the scala tympani, it is likely to lie against the outer wall of the scala when it is inserted, rather than at the presumably more desirable medial-wall location. Several attempts have been made to specifically design electrode arrays so that they will lie near the medial wall of the scala tympani. Experimental studies in animals have demonstrated that electrode arrays that lie near the modiolus require less current to elicit a threshold response than electrode arrays that lie laterally in the scala.[100] There are several advantages to using lower currents, including a greater margin of safety, as well as increased battery life. Increasing the proximity of the electrode array to the stimulated neurons offers other potential advantages, such as increased electrode discrimination and channel separation, but these potentials remain to be tested.

1.6.3.3   Stimulus Level

It is clear from neurophysiological experiments that stimulus level has large effects on the neural response pattern. Both the spread of excitation and the phase locking to temporal components of the signal increase significantly as a function of stimulus level.[101,102] Thus, stimulus level is one of the features of stimulation that can be useful in experiments aimed at understanding the relationship between neural response patterns and perception.

The level at which the electrical stimuli are presented, relative to the detection threshold, has significant effects on perception. The stimulus not only becomes louder as the level is increased, but it also becomes more discriminable in several respects. Psychophysical studies in animals and in humans have shown that several types of discrimination involving temporal cues improve as a function of level through at least the lower half of the dynamic range and, in some cases, through the entire dynamic range. Tasks that benefit from increases in level include sinusoidal-frequency discrimination,[88] pulse rate discrimination,[91] gap detection,[87] and temporal modulation transfer functions.[89,90] Discrimination of changes in stimulus level also improves as a function of the level of the stimulus above the detection threshold.[103,104] Discrimination of electrode place improves as a function of level in some cases, but not in others, and a decrease in electrode-place discrimination as a function of level is seen in some cases.[92,105] Given the general improvement in most types of discrimination as a function of stimulus level, one would expect that speech recognition would be better at higher stimulus levels in most subjects. This is more difficult to demonstrate because the speech signals span a range of levels that encompasses most of the dynamic range of electrical hearing. The input/output functions of stimulus processors can be set, however, and a common setting involves presentation of the majority of the signal in the upper regions of the dynamic range. Recent experiments have indicated that there is a fair amount of leeway in setting the parameters of these input/output functions.[106] However, there are demonstrated advantages for speech recognition to restricting stimuli to the upper regions of the electrical dynamic range.[107]

1.6.4   TRANSCUTANEOUS AND PERCUTANEOUS TRANSMISSION

Transcutaneous or percutaneous transmission systems (Figure 1.1E) should be able to accomplish three things. First, they must be able to deliver the appropriate spatial and temporal patterns of currents to the electrode array at appropriate levels under the guidance of the prosthesis processor. Second, they should provide a means for monitoring the integrity and impedance of the implanted electrode array. Third, they should be able to convey small electrical signals picked up by the implanted elec-trodes to external monitoring equipment. These latter potentials can be used to record the activity of the auditory nerve in response to electrical stimulation and, thus, may be helpful in evaluating the condition of the stimulated neurons and the effectiveness of the electrical stimulation.

Several additional features must be considered in the design of transmission systems. First, the system should be versatile so as to be adaptable to new processing strategies. At the current rate of prosthesis development, processing strategies are improving on a yearly basis. In some cases these more advanced strategies cannot be applied to older-model prostheses because the implanted electronics packages will not support them. In addition, the transmission system must be reliable since repair or replacement may require additional surgical intervention. Low power requirements are a significant practical advantage in a prosthesis because of the requirements of portability and use throughout the day. Direct percutaneous con-nections may result in significant power savings compared to transcutaneous sys-tems. Of course, the connection system must be safe and it must be cosmetically acceptable to the patient. Cost is also a consideration in many cases.

Some earlier versions of human auditory prostheses, as well as most experimen-tal implants in animals, have used a hard-wired percutaneous connector. This system provides maximum flexibility for delivering current, monitoring the condition of the electrodes, and recording evoked potentials from the cochlea using the implanted electrodes. Percutaneous connectors used in patients were typically made from a biocompatible ceramic. Connector plugs were mounted in a ceramic housing that was attached to the skull in the region just behind the external ear.[108,109] Complica-tions associated with these connectors included breakage and problems with infec-tion of the skin surrounding the connector housing. In contrast, titanium has been very successfully used in percutaneous skull-mounted devices and these problems have largely been avoided.[110,111] Titanium percutaneous connectors have been suc-cessfully adapted for use in nonhuman primates for experimental studies of auditory prostheses[112] and are being developed for human use.[113]

In current versions of *human* auditory prostheses, transcutaneous transmission systems are used almost exclusively to convey signals and power across the skin. Modern implanted electronics packages are equipped to both receive signals and power and to transmit signals recorded from the implanted electrodes back across the skin to a receiver on the outside.[18,114] As these transcutaneous transmission systems are becoming more transparent and more versatile, they are achieving some of the advantages previously unique to the percutaneous system. However, the percutaneous approach still provides the greatest flexibility and the lowest power consumption and cost. The technology is available to make the percutaneous system reliable and safe. Such a development would provide an excellent alternative to the systems currently used in patients.

1.6.5 Neural Survival Pattern

The most likely target for activation by electrical stimulation through a cochlear implant is the auditory nerve. Therefore, the condition of the auditory nerve is assumed to be an important variable determining the effectiveness of a cochlear implant. In a young, healthy, normally-hearing individual, the auditory nerve contains approximately 30,000 nerve fibers. However, profound deafness is almost always associated with partial degeneration of the auditory nerve. Post-mortem studies of temporal bones of patients who would have qualified audiometrically for a cochlear implant show nerve survival patterns ranging from near 0 to 76% of normal.[22] However, the nerve survival pattern that affects implant function is more complex than is reflected in simple counts of spiral ganglion cells. Spiral ganglion cells that are alive in a deaf individual can be modified in a variety of ways, including demyelination and partial degeneration of the peripheral process of the neuron.[115,116] It is likely that many neurons existing in this modified condition are capable of conducting action potentials, but the response properties of these neurons are probably not normal. Furthermore, the spatial-temporal pattern of activity elicited in a population of auditory nerve fibers depends on the location of each nerve fiber with respect to the stimulating electrodes. The spatial pattern of neuronal degeneration and/or modification following deafness can vary considerably from patient to patient. These complexities make it difficult to predict the nerve survival pattern in a deaf patient in any meaningful degree of detail or to develop noninvasive functional measures that can be used to estimate the nerve survival pattern in a deaf patient.

Studies in research animals, where both functional and histological data can be obtained from the same subject, have indicated that under certain conditions, thresholds for psychophysical detection of the stimulus or thresholds for brainstem auditory evoked responses show a moderately strong correlation with spiral ganglion cell counts in the region of the stimulated electrodes.[117-119] However, the mechanisms underlying these correlations are poorly understood and, as noted above, they are probably complex. A better understanding of the functional effects of each of the variables associated with neuronal damage and death is needed in order to measure and adjust for the individual differences among patients that may be due to differences in nerve survival patterns.

Recent studies have shown that degeneration of the auditory nerve following deafness can be retarded by stimulating the nerve either electrically[120-124] or with neurochemicals.[125-127] In those experiments, the total number of neurons remaining in the deafened cochlea was shown to be larger in animals that had been stimulated, compared to animals that received no stimulation. Interpretation of the histological results can be confounded by other effects of stimulation, such as changes in cell size[128] and bone growth.[129] Bone growth could decrease the size of Rosenthal's canal in which the spiral ganglion is housed and thus increase the density of nerve fibers without affecting their total number. However, the weight of evidence to date suggests that electrical and chemical stimulation do decrease the degeneration of auditory nerve fibers.

Effects of deafness and of subsequent stimulation on the nervous system are by no means limited to the auditory periphery. Neural changes throughout the central auditory pathways have been documented, both at the histological and neurophysiological levels. Following deafness due to cochlear destruction, changes in cell size,[130,131] synaptic structure,[132] neurotransmitter release,[133] and metabolic activity[134] have been seen. The spatiotopic organization of central nuclei is subject to marked rearrangement following deafness[135] and can be altered by patterned electrical stimulation of a cochlear prosthesis.[102] In addition, the temporal response properties of central neurons can be altered by periodic electrical stimulation of a deaf ear.[136]

An important direction for future research in this area is to determine the effects of these peripheral and central changes on perception. This type of study is currently possible using animal models in which both the psychophysical and the neural responses to peripheral electrical stimulation can be determined in the same subject. These experiments will be further enhanced by chronically implantable recording probes, which are currently under development.[137,138] Such studies could also contribute to our understanding of the mechanisms that influence intersubject differences in implant performance.

1.6.6. TRAINING

Cognitive ability has been shown to correlate with speech recognition performance.[139,140] Therefore, it stands to reason that training the patient to use the prosthesis[141] will result in improved performance. While a few programs have implemented this training, most centers do not consider it to be cost effective. Research is needed to determine the cost-benefit ratio of such training programs.

1.7 UNDERSTANDING INDIVIDUAL DIFFERENCES

One of the greatest challenges facing researchers is understanding the enormous interpatient variability in performance. Although the mean speech recognition scores of patients have increased steadily as new generations of processors have been introduced and patient candidacy requirements have been relaxed, patient scores for open set speech recognition in the absence of visual cues still range from near 0 to 100% (Figure 1.5).

A number of studies have attempted to identify and measure the variables that contribute to a patient's speech recognition performance. There are several uses for this information. First, the data might be useful for predicting the success of an implant during patient counseling sessions. The need for predicting the outcome of an implant during the initial counseling session is minimized somewhat by the fact that virtually all implanted patients receive some benefit from the implant. Nevertheless, pre-implant counseling regarding the probable degree of success of the implant is a consistent patient need. Second, in some cases data are needed to determine which ear to implant. Third, the data may be useful for determining which type of processor to prescribe for a particular patient in order to increase the probability of optimal performance. The number of processor options has increased dramatically in recent years, but the procedures for choosing which option is most appropriate for a given patient are still in their infancy. Finally, to the degree that the variables affecting speech recognition performance can be controlled, measurement of these variables can be a major contributor to the patient's rehabilitation.

A wide variety of procedures have potential for identifying and quantifying the underlying physiological and psychological variables that affect speech recognition with prosthetic electrical stimulation. These include assessment of historical factors, measurement of cognitive function, and measurement of electrophysiological and psychophysical variables that are thought to be related to the condition of the stimulated neural population.

Historical factors include age of onset at deafness and duration of deafness. For adults, if the onset of deafness occurred before the individual acquired language, the prognosis for speech recognition with the implant is much poorer than if the person became deaf postlingually. In addition, many studies have found negative correlations between duration of deafness and speech recognition ability in the postlingually deaf population.[140,142-144] Cognitive ability has also been correlated with speech recognition performance.[139]

Psychophysical and electrophysiological measures of the patient's response to electrical stimulation can be made prior to implantation by placing an electrode on the promontory bone near the round window of the cochlea. This is a relatively simple outpatient procedure that involves anesthetizing and then perforating the eardrum and inserting the electrode onto the exposed cochlear bone. There are some drawbacks to this procedure. A monopolar electrode placed on the promontory is clearly not going to provide the same detail of response that is provided by an intracochlear electrode array. In fact, in some cases, patients implanted despite a lack of response to extracochlear stimulation have shown normal responses to intra-cochlear stimulation.[145]

Once an electrode array has been implanted, a good deal of useful information can be obtained by recording psychophysical and electrophysiological responses to stimulation of the implanted electrodes. Threshold and maximum comfortable loud-ness levels are measured psychophysically at the time of the initial implant activation and periodically thereafter. These values are used to determine the levels of stimu-lation to be delivered in everyday use of the prosthesis. While these values have typically been determined psychophysically, investigators are currently exploring the possibility of using electrophysiological measures instead. In particular, com-pound action potentials generated by the auditory nerve, recorded using selected electrodes from the implanted scala tympani array, may prove useful for evaluating sensitivity and growth of loudness.[146] Potential advantages of electrophysiological over psychophysical procedures include shorter data acquisition times and the ability to obtain data from subjects, such as very young children, who are unable to perform the psychophysical procedures reliably.

In addition to their direct application in determining the upper and lower limits for current delivery, psychophysical and electrophysiological measures may be use-ful for indirect assessment of biophysical or physiological characteristics of the implanted region that may affect performance of the auditory prosthesis. For exam-ple, psychophysical or electrophysiological strength-duration functions may reflect the condition of the stimulated auditory neurons. These functions are similar to the classical physiological strength-duration function[147,148] in that they measure the threshold for eliciting a response as a function of the phase duration of the signal. In neurophysiological experiments and models these functions have been shown to reflect characteristics of the stimulated nerve, such as fiber diameter, degree of myelination, etc.[149] Models of the cochlea in which the auditory nerve is partially degenerated indicate that the slopes of these strength-duration functions should reflect such factors as the degree of demyelination or the length of the partially degenerated peripheral process of the nerve[76,150] or the number of nerve fibers activated by the electrical signal.[151] In psychophysical studies in animals, differences in slopes of psychophysical and electrophysiological strength-duration functions vary considerably from case to case.[70,152] These slopes have been shown to vary as a function of (1) degree of nerve loss;[153] (2) time after deafening,[152,154] which presumably is associated with degree of neural degeneration; and (3) electrode configuration,[155,156] which may affect the number of fibers activated and/or the site on the neuron where action potentials are initiated.

Another technique that is potentially useful for evaluating individual differences among patients is to measure discrimination of certain features of the electrical stimulus that serve as important components of the processed speech signal. Thus, for example, one could measure pulse-rate discrimination, modulation-frequency discrimination, or discrimination of one electrode from another. One concern with this approach is that the ability to discriminate particular features of the electrical stimulus is not always correlated with a subject's speech recognition performance. This difficulty also applies to measures that presumably reflect the physiological condition of the stimulated neurons, such as thresholds. For example, Zwolan et al.[157] found a poor correlation between electrode discrimination ability and speech recognition in a group of patients. On the other hand, they found that when they removed the most poorly discriminated electrodes, speech recognition improved. Other investigators have found moderate correlations between speech recognition and electrode discrimination[158] or pitch ranking based on electrode position.[159] Some investigators have found that detection threshold levels in human subjects were correlated with speech recognition,[78,160] while others have found poor correlations.[143] The variability from subject to subject and from study to study in the relationship between psychophysical data and speech recognition data probably reflects several factors, including the complexity and redundancy of the speech signal and the contribution of multiple variables to speech recognition ability. It may be necessary to use multiple measures, each with its own weighting, in order to accurately describe the variables that contribute to an individual patient's speech recognition ability with a given processing strategy.

1.8   FUTURE DIRECTIONS

1.8.1   Beyond the 40% Speech Recognition Criterion

When the first auditory prostheses were approved by the FDA, the criterion for implant candidacy was total deafness with no benefit from a hearing aid. In 1995, the NIH held a Consensus Conference to review the current status of cochlear implants and, as a result of that conference, the criterion for implant candidacy was changed to include severely hearing-impaired individuals who were achieving less than 30% speech recognition with hearing aids. This change reflected a significant improvement in the quality of speech recognition obtained with cochlear implants and suggested that some hearing-aid users could achieve better speech recognition with an implant. Further increases in speech recognition results have been achieved since that Consensus Conference and the criterion has now been raised to 40%. There is every reason to believe that further improvements are possible. Ongoing experiments in almost all of the areas listed in Section 1.6 are expected to yield benefits for future implantees.

Over the past 15 to 20 years, the major focus of auditory prosthesis research has been on improving speech recognition. This is a very reasonable goal, given the importance of speech communication in most daily activities. However, with the great success in this area and the likelihood of even further improvements in the near future, it would seem appropriate at this time to increase the emphasis on other aspects of hearing. For example, music perception with cochlear implants is still very crude.[37,38] Improving perception of music may require a different approach, or at least a much more detailed spectral representation of the acoustic signal, than is necessary for successful speech recognition. Thus, music perception represents perhaps one of the most significant challenges yet faced by designers of auditory prostheses.

1.8.2 VARIABLES VS. MECHANISMS

Research directed at understanding and improving the performance of auditory prostheses can be conducted at several levels. A common first approach is to manipulate some independent variable based on our current understanding of the mechanisms underlying prosthesis function and to test the effects of this variable on speech recognition or some other metric of prosthesis performance. Through this process, we either confirm or revise our theories about the underlying mechanisms. In addition, these theories can generate hypotheses than can be tested by other means, including psychophysical, neurophysiological and anatomical experiments. As our understanding of underlying mechanisms improves, so will our ability to make accurate predictions about modifications that will improve prosthesis function. For this reason, hypothesis-driven research aimed at understanding the mechanisms of underlying electrical perception is highly desirable. Much research on auditory prostheses currently uses that approach and should continue to do so in the future.

1.8.3 HARDWARE

Mechanical and electrical engineering technology plays a critical role in the development of improved auditory prostheses. Processing strategies in use today could not have been implemented 10 years ago because of hardware limitations. Current research suggests the need for even faster processors, more complex electrode arrays and better transmission systems. Technical developments in all of these areas are in progress and will be essential for the next generation of auditory prostheses.

1.8.4 BINAURAL HEARING

Normally-hearing individuals have two ears and the two ears receive slightly different information from a given sound source. The differences in the inputs received by the two ears are precisely analyzed in the central nervous system, allowing the person to accurately identify the location of the sound source and to selectively listen to a particular sound source in a noisy environment. Thus, an individual with two normal ears has a significant advantage over a person who is deaf or hearing impaired in one ear, particularly for understanding speech in a noisy environment. Some of the advantages of having two ears can be achieved by using two microphones, as noted in Section 1.6.1. However, much of the processing done by the brain in comparing the levels and precise timing of inputs from the two ears can only be achieved if both the left and right ears are stimulated in the proper relationship. To date, cochlear implants have, for the most part, been placed in only one ear and only in patients who have significant hearing impairment in both ears. Recently, however, investigators have begun to experiment with bilateral implants.[161] Designing these implants to take advantage of the precise timing and intensity comparisons that the central nervous system is capable of achieving will no doubt require some effort, but the results are potentially quite rewarding. In addition, as auditory prostheses improve, it may be possible to restore bilateral advantages to persons who are deaf in only one ear.

1.8.5 TISSUE ENGINEERING

A long-standing hypothesis is that performance with auditory prostheses depends in part on the nerve survival pattern near the implanted electrode array. Techniques are now becoming available to improve the nerve survival pattern in animal subjects (see Section 1.6.5 above). In addition, techniques for delivery of neurochemicals that may enhance nerve survival and/or regrowth of neurons near the electrode arrays are under development, including implantable mini-osmotic pumps,[162] gene transfer,[163,164] and stimulating electrode arrays with built-in mechanisms for drug delivery.[165] Studies in animals to assess the functional effects of these tissue engineering procedures, and the application of the procedures to human subjects, seem to be a promising future direction in this field.

ACKNOWLEDGMENTS

The author would like to express appreciation to Kevin Franck, Li Xu, Amy Miller and Mabel Holland for assistance in preparing this chapter.

REFERENCES

1. Simmons, F. B., Electrical stimulation of the auditory nerve in man, *Archives of Otolaryngology — Head and Neck Surgery,* 84, 24, 1966.
2. Djourno, A. and Eyries, C. H., Prothese auditive par excitation electrique a distance du nerf sensoriel a l'aide d'un bobinage inclus a demeure, *La Presse Medicale,* 65, 1417, 1957.
3. Doyle, J. H., Doyle, J. B., and Turnbull, F. M., Electrical stimulation of eighth cranial nerve, *Archives of Otolaryngology — Head and Neck Surgery,* 80, 388, 1964.
4. House, W. F. and Urban, J., Long term results of electrode implantation and electronic stimulation of the cochlea in man, *Annals of Otology, Rhinology and Laryngology,* 82, 504, 1973.

5. House, W. F. and Berliner, K. I., Safety and efficacy of the House/3M cochlear implant in profoundly deaf adults, *Otolaryngologic Clinics of North America*, 19, 275, 1986.

6. Simmons, F. B., Epley, J. M., Lummis, R. C., Guttman, N., Frishkopf, L. S., Harmon, L. D., and Zwicker, E., Auditory nerve: Electrical stimulation in man, *Science*, 148, 104, 1965.

7. Michelson, R. P., Electrical stimulation of the human cochlea: A preliminary report, *Archives of Otolaryngology — Head and Neck Surgery*, 93, 317, 1971.

8. Merzenich, M. M., Schindler, D. N., and White, M. W., Feasibility of multichannel scala tympani stimulation, *Laryngoscope*, 84, 1887, 1974.

9. Schindler, R. S., Kessler, D. K., and Rebscher, S. J., The University of California, San Francisco/Storz Cochlear Implant Program, *Otolaryngology Clinics of North America*, 19, 287, 1986.

10. Schindler, R. A. and Kessler, D. K., Clarion cochlear implant: Phase I investigational result, *American Journal of Otology*, 14, 263, 1993.

11. Kessler, D. K., The CLARION® Multi-Strategy cochlear implant, *Annals of Otology, Rhinology and Laryngology*, 108, 8, 1999.

12. Clark, G. M., Blamey, P. J., Brown, A. M., Gusby, P. A., Dowell, R. C., Franz, B. K., Pyman, B. C., Shepherd, R. K., Tong, Y. C., Webb, R. L., Hirshorn, M. S., Kuzma, J., Mecklenburg, M. J., Money, D. K., Patrick, J. F., and Seligman, P. M., The University of Melbourne — Nucleus multi-electrode cochlear implant, *Advances in Oto-Rhino-Laryngology*, 38, 1, 1987.

13. Douek, E., Fourcin, A. J., Moore, B. C. J., and Clarke, G. P., A new approach to the cochlear implant, *Proceedings of the Royal Society of London Series B: Biological Sciences*, 70, 379, 1977.

14. Eddington, D. K., Dobelle, W. H., Brackmann, D. E., Mladejovsky, M. G., and Parkin, J. L., Auditory prosthesis research with multiple channel intracochlear stimulation in man, *Annals of Otology, Rhinology and Laryngology*, 87, 1, 1978.

15. Parkin, J. L., Randolph, L. J., and Parkin, B. D., Multichannel (Ineraid®) cochlear implant update, *Laryngoscope*, 103, 835, 1993.

16. Hochmair-Desoyer, I. J., Results from better postlingual adult users of the MED-EL devices, in *Advances in Cochlear Implants*, Hochmair-Desoyer, I. J. and Hochmair, E. S., Eds., Manz, Vienna, 1994, 363.

17. Helms, J., Muller, J., Schon, F., Moser, L., Arnold, W., Janssen, T., Ramsden, R., vonIlberg, C., Kiefer, J., Pfennigdorf, T., Gstottner, W., Baumgarter, W., Ehrenberger, K., Skarzynski, H., Ribari, O., Thumfart, W., Stephan, K., Mann, W., Heinemann, M., Zorowka, P., Lippert, K. L., Zenner, H. P., Bohndorf, M., Huttenbrink, K., and Muller-Ascoff, E., Evaluation of performance with the COMBI 40 cochlear implant in adults. A multicentric clinical study, *ORL-Journal for Oto-Rhino-Laryngology and Its Related Specialties*, 59, 23, 1997.

18. Zierhofer, C. M., Hochmair, I. J., and Hochmair, E. S., The advanced Combi 40+ cochlear implant, *American Journal of Otology*, 18, S37, 1997.

19. Chouard, C. H., The Digisonic from MXM, *Advances in Otorhinolaryngology*, 52, 258, 1997.

20. Peeters, S., Marquet, J., Offeciers, F. E., Bosiers, W., Kinsbergen, J., and Van Durme, M., Cochlear implants: the Laura prosthesis, *Journal of Medical Engineering and Technology*, 13, 76, 1989.

21. Peeters, S., Offeciers, E., Kinsbergen, J., van Durme, M., van Enis, P., Dijkmans, P., and Bouchataoui, I., A digital speech processor and various speech encoding strategies for cochlear implants, *Progress in Brain Research*, 97, 283, 1993.

22. Hinojosa, R. and Marion, M., Histopathology of profound sensorineural deafness, *Annals of the New York Academy of Sciences,* 405, 459, 1983.

23. Suzuka, Y. and Schuknecht, H. F., Retrograde cochlear neuronal degeneration in human subjects, *Acta Oto-Laryngologica* (Stockholm), S450, 1, 1988.

24. Brackmann, D. E., Hitselberger, W. E., Nelson, R. A., Moore, J., Waring, M. D., Portillo, F., Shannon, R. V., and Telischi, F. F., Auditory brainstem implant. I. Issues in surgical implantation, *Otolaryngology — Head and Neck Surgery,* 108, 624, 1993.

25. Shannon, R. V., Fayad, J., Moore, J., Lo, W. W. M., Otto, S., Nelson, R. A., and O'Leary, M., Auditory brain-stem implant. II. Postsurgical issues and performance, *Otolaryngology — Head and Neck Surgery,* 108, 634, 1993.

26. Anderson, D. J., Najafi, K., Tanghe, S. J., Evans, D. A., Levy, K. L., Hetke, J. F., Xue, X., Zappia, J. J., and Wise, K. D., Batch-fabricated thin-film electrodes for stimulation of the central auditory system, *IEEE Transactions on Biomedical Engineering,* BME-36, 1206, 1989.

27. Niparko, J. K., Altschuler, R. A., Xue, X., Wiler, J. A., and Anderson, D. J., Surgical implantation and biocompatibility of central nervous system auditory prostheses, *Annals of Otology, Rhinology and Laryngology,* 98, 965, 1989.

28. Evans, D. E., Niparko, J. K., Miller, J. M., Jyung, R. W., and Anderson, D. J., Multiple channel stimulation of the cochlear nucleus, *Otolaryngology — Head and Neck Surgery,* 101, 651, 1989.

29. Zappia, J. J., Hetke, J. F., Altschuler, R. A., and Niparko, J. K., Evaluation of a silicon-substrate modiolar eighth nerve implant in a guinea pig, *Otolaryngology — Head and Neck Surgery,* 103, 575, 1990.

30. Jones, D., Arts, H., Bierer, S., Hetke, J., and Anderson, D., Development of an implantable auditory nerve prosthesis, *Otolaryngology — Head and Neck Surgery,* Submitted.

31. Bell, T., Wise, K., and Anderson, D., A flexible micromachined electrode array for a cochlear prosthesis, *Sensors and Actuators A: Physical,* 66, 63, 1998.

32. Gantz, B. J., Tyler, R. S., Knutson, J. F., Woodworth, G., Abbas, P., McCabe, B. F., Hinrichs, J., Tye-Murray, N., Lansing, C., Kuk, F., and Brown, C., Evaluation of five different cochlear implant designs: audiologic assessment and predictors of performance, *Laryngoscope,* 98, 1100, 1988.

33. Skinner, M. W., Clark, G. M., Whitford, L. A., Seligman, P. M., Staller, S. J., Shipp, D. B., Shallop, J. K., Everingham, C., Menapace, C. M., Arndt, P. L., Antogenelli, T., Brimacombe, J. A., Pijl, S., Daniels, P., George, C. R., McDermott, H. J., and Beirer, A. L., Evaluation of a new spectral peak coding strategy for the Nucleus 22 channel cochlear implant system, *American Journal of Otology,* 15, 15, 1994.

34. Battmer, R., P, H., Zilberman, Y., and Lenarz, T., Simultaneous analog stimulation (SAS)-continuous interleaved sampler (CIS) pilot comparison study in Europe, *Annals of Otology, Rhinology and Laryngology,* 108, 69, 1999.

35. Osberger, M., SAS-CIS preference study in postlingually deafened adults implanted with the Clarion cochlear implant, *Annals of Otology, Rhinology and Laryngology,* 108, 74, 1999.

36. Pfingst, B. E., Zwolan, T. A., and Holloway, L. A., Effects of stimulus configuration on psychophysical operating levels and on speech recognition with cochlear implants, *Hearing Research,* 112, 247, 1997.

37. Gfeller, K. and Lansing, C. R., Melodic, rhythmic, and timbral perception of adult cochlear implant users, *Journal of Speech and Hearing Research,* 34, 916, 1991.

38. Fujita, S. and Ito, J., Ability of Nucleus cochlear implantees to recognize music, *Annals of Otology, Rhinology and Laryngology,* 108, 634, 1999.

39. Cohen, N. L., Hoffman, R. A., and Stroschein, M., Medical or surgical complications related to the Nucleus multichannel cochlear implant, *Annals of Otology, Rhinology and Laryngology,* 97, 8, 1988.

40. Hoffman, R. A. and Cohen, N. L., Complications of cochlear implant surgery, *Annals of Otology, Rhinology & Laryngology,* 104 (Supp 166), 420, 1995.

41. Shepherd, R. K., Clark, G. M., and Black, R. C., Chronic electrical stimulation of the auditory nerve in cats. Physiological and histopathological results, *Acta Oto-Laryngologica* (Stockholm), S399, 19, 1983.

42. Leake, P. A., Rebscher, S. J., and Aird, D. W., Histopathology of cochlear implants: safety considerations., in *Cochlear Implants*, Schindler, R. A. and Merzenich, M. M., Eds., Raven Press, New York, 1985, 55.

43. Brummer, S. B. and Turner, M. J., Electrochemical considerations for safe electrical stimulation of the nervous system with platinum electrodes, *IEEE Transactions on Biomedical Engineering,* 24, 59, 1977.

44. Robblee, L. S. and Rose, T. L., Electrochemical guidelines for selection of protocols and electrode materials for neural stimulation, in *Neural Prostheses: Fundamental Studies*, Agnew, W. F. and McCreery, D. B., Eds., Prentice Hall, Englewood Cliffs, NJ, 1990,

45. Agnew, W. F., McCreery, D. B., Yuen, T. G. H., and Bullara, L. A., Local anaesthetic block protects against electrically-induced damage in peripheral nerve, *Journal of Biomedical Engineering,* 12, 301, 1990.

46. Huang, C. Q. and Shepherd, R. K., Reduction in excitability of the auditory nerve following electrical stimulation at high stimulus rates. IV. Effects of stimulus intensity, *Hearing Research,* 132, 60, 1999.

47. Donaldson, N. D. N. and Donaldson, P. E. K., When are actively balanced biphasic ('Lilly') stimulating pulses necessary in a neurological prosthesis? I. Historical background; Pt resting potential; Q studies, *Medical and Biological Engineering and Computing,* 24, 41, 1986.

48. Kompis, M. and Dillier, N., Noise reduction for hearing aids: combining directional microphones with an adaptive beamformer, *Journal of the Acoustical Society of America,* 96, 1910, 1994.

49. Hamacher, V., Doering, W. H., Mauer, G., Fleischmann, H., and Hennecke, J., Evaluation of noise reduction systems for cochlear implant users in different acoustic environment, *American Journal of Otology,* 18, 46, 1997.

50. Elberling, C., Ludvigsen, C., and Keidser, G., The design and testing of a noise reduction algorithm based on spectral subtraction, *Scandinavian Audiology Supplementum,* 38, 39, 1993.

51. Fretz, R. and Fravel, R., A physical and electrical description of the 3M/House cochlear implant system, *Ear and Hearing,* 6, 145, 1985.

52. Hochmair-Desoyer, I. J., Hochmair, E. S., Burian, K., and Fischer, R. E., Four years of experience with cochlear prostheses, *Medical Progress Through Technology,* 8, 107, 1981.

53. Tyler, R. S., Open-set word recognition with the 3M/Vienna single-channel cochlear implant, *Archives of Otolaryngology — Head and Neck Surgery,* 114, 1123, 1988.

54. Skinner, M. W., Holden, L. K., Holden, T. A., Dowell, R. C., Seligman, P. M., Brimacombe, J. A., and Beiter, A. L., Performance of postlinguistically deaf adults with the wearable speech processor (WSP III) and mini speech processor (MSP) of the Nucleus multi-electrode cochlear implant, *Ear and Hearing,* 12, 3, 1991.

55. Shannon, R. V., Zeng, F.-G., Kamath, V., Wygonski, J., and Ekelid, M., Speech recognition with primarily temporal cues, *Science,* 270, 303, 1995.

56. Dorman, M. F., Loizou, P. C., and Rainey, D., Speech intelligibility as a function of the number of channels of stimulation for signal processors using sine-wave and noise-band outputs, *Journal of the Acoustical Society of America,* 102, 2403, 1997.

57. Dudley, H., The vocoder, *Bell Labs Record,* 17, 122, 1939.

58. Hill, F. J., McRae, L. P., and McClellan, R. P., Speech recognition as a function of channel capacity in a discrete set of channels, *Journal of the Acoustical Society of America,* 44, 13, 1968.

59. Villchur, E., Electronic models to simulate the effect of sensory distortions on speech perception by the deaf, *Journal of the Acoustical Society of America,* 62, 665, 1977.

60. Zollner, V. M., Intelligiblity of the speech of a simple vocoder, *Acustica,* 43, 271, 1979.

61. Dorman, M. F., Loizou, P. C., Fitzke, J., and Tu, Z., The recognition of sentences in noise by normal-hearing listeners using simulations of cochlear-implant signal processors with 6–20 channels, *Journal of the Acoustical Society of America,* 104, 3583, 1998.

62. Eisenberg, L. S., Shannon, R. V., Martinez, A. S., and Wygonski, J., Do children require a greater number of spectral channels than adults to understand speech, in *Association for Research in Otolaryngology Abstracts,* Popelka, G. R., Ed. Association for Research in Otolaryngology, Mt. Royal, NJ, 1999, 71.

63. Fishman, K. E., Shannon, R. V., and Slattery, W. H., Speech recognition as a function of the number of electrodes used in the SPEAK cochlear implant speech processor, *Journal of Speech, Language, and Hearing Research,* 40, 1201, 1997.

64. Friesen, L. M., Shannon, R. V., and Slattery, W. H., Speech recognition in noise as a function of the number of electrodes used in the SPEAK, SAS and CIS speech processors, *Conference on Implantable Auditory Prostheses,* Pacific Grove, CA, 119, 1999.

65. Dorman, M. F., Loizou, P. F., Poroy, O., Spahr, T., and Maloff, E., The effect of intensity quantizing on the intelligibility of speech transmitted by signal processors using a small number of output channels, *Conference on Implantable Auditory Prostheses,* Pacific Grove, CA, 117, 1999.

66. Wilson, B. S., Finley, C. C., Lawson, D. T., Wolford, R. D., Eddington, D. K., and Rabinowitz, W. M., Better speech recognition with cochlear implants, *Nature,* 352, 236, 1991.

67. Kompis, M., Vischer, M. W., and Hausler, R., Performance of Compressed Analogue (CA) and Continuous Interleaved Sampling (CIS) coding strategies for cochlear implants in quiet and noise, *Acta Oto Laryngologica,* 119, 659, 1999.

68. Peeters, S., Offeciers, F. E., Joris, P., and Moeneclaey, L., The Laura cochlear implant programmed with the continuous interleaved and phase-locked continuous interleaved strategies, *Advances in Oto-Rhino-Laryngology,* 48, 261, 1993.

69. Rosen, S., Temporal information in speech: acoustic, auditory and linguistic aspects, *Philosophical Transactions of the Royal Society of London,* 336, 367, 1992.

70. Pfingst, B. E., De Haan, D. R., and Holloway, L. A., Stimulus features affecting psychophysical detection thresholds for electrical stimulation of the cochlea. I: Phase duration and stimulus duration, *Journal of the Acoustical Society of America,* 90, 1857, 1991.

71. Pfingst, B. E. and Morris, D. J., Stimulus features affecting psychophysical detection thresholds for electrical stimulation of the cochlea. II: Frequency and interpulse interval, *Journal of the Acoustical Society of America,* 94, 1287, 1993.

72. Pfingst, B. E., Holloway, L. A., and Razzaque, S. A., Effects of pulse separation on detection thresholds for electrical stimulation of the human cochlea, *Hearing Research,* 98, 77, 1996.

73. Coste, R. L. and Pfingst, B. E., Stimulus features affecting psychophysical detection thresholds for electrical stimulation of the cochlea. III: Pulse polarity, *Journal of the Acoustical Society of America,* 99, 3099, 1996.

74. Miller, A. L., Morris, D. J., and Pfingst, B. E., Interactions between pulse separation and pulse polarity order in cochlear implants, *Hearing Research,* 109, 21, 1997.

75. Zwolan, T. A., Kileny, P. R., and Boerst, A. K., Programming the Nucleus 20+2L cochlear implant: using the apical electrode as active, in *Association for Research in Otolaryngology Abstracts,* Popelka, G. R., Ed., Vol. 21, Association for Research in Otolaryngology, Mt. Royal, NJ, 1998, 220.

76. van den Honert, C. and Stypulkowski, P. H., Physiological properties of the electrically stimulated auditory nerve. II. Single fiber recordings, *Hearing Research,* 14, 225, 1984.

77. Parkins, C. W., Temporal response patterns of auditory nerve fibers to electrical stimulation in deafened squirrel monkeys, *Hearing Research,* 41, 137, 1989.

78. Moon, A. K., Zwolan, T. A., and Pfingst, B. E., Effects of phase duration on detection of electrical stimulation of the human cochlea, *Hearing Research,* 67, 166, 1993.

79. Glass, I., Phase-locked responses of cochlear nucleus units to electrical stimulation through a cochlear implant, *Experimental Brain Research,* 55, 386, 1984.

80. Javel, E., Tong, Y. C., Shepherd, R. K., and Clark, G. M., Responses of cat auditory nerve fibers to biphasic electrical current pulses, *Annals of Otology, Rhinology and Laryngology,* 96, 26, 1987.

81. Dynes, S. B. C. and Delgutte, B., Phase-locking of auditory-nerve discharges to sinusoidal electric stimulation of the cochlea, *Hearing Research,* 58, 79, 1992.

82. Morse, R. P. and Evans, E. F., Additive noise can enhance temporal coding in a computational model of analogue cochlear implant stimulation, *Hearing Research,* 133, 107, 1999.

83. Rubinstein, J. T., Pseudospontaneous activity: stochastic independence of auditory nerve fibers with electrical stimulation, *Hearing Research,* 127, 108, 1999.

84. Morse, R. P. and Evans, E. F., Preferential and non-preferential transmission of formant information by an analogue cochlear implant using noise: the role of the nerve threshold, *Hearing Research,* 133, 120, 1999.

85. Lehnhardt, E., Gnadeberg, D., Battner, R. D., and von Wallenberg, E., Experience with the cochlear miniature speech processor in adults and children together with a comparison of unipolar and bipolar modes, *ORL-Journal of Oto-Rhino-Laryngology and its Related Specialties,* 54, 308, 1992.

86. Zwolan, T. A., Kileny, P. R., Ashbaugh, C., and Telian, S. A., Patient performance with the Cochlear Corporation "20+2" implant: bipolar versus monopolar activation, *American Journal of Otology,* 17, 717, 1996.

87. Shannon, R. V., Detection of gaps in sinusoids and pulse trains by patients with cochlear implants, *Journal of the Acoustical Society of America,* 85, 2587, 1989.

88. Pfingst, B. E. and Rai, D. T., Effects of level on nonspectral frequency difference limens for electrical and acoustic stimuli, *Hearing Research,* 50, 43, 1990.

89. Shannon, R. V., Temporal modulation transfer functions in patients with cochlear implants, *Journal of the Acoustical Society of America,* 91, 2156, 1992.

90. Shannon, R. V., Reply to: "comment on 'temporal modulation transfer functions in patients with cochlear implants' "(J. Acoust. Soc. Am., 93, 1649, 1993), *Journal of the Acoustical Society of America,* 93, 1651, 1993.

91. Pfingst, B. E., Holloway, L. A., Poopat, N., Subramanya, A. R., Warren, M. F., and Zwolan, T. A., Effects of stimulus level on nonspectral frequency discrimination by human subjects, *Hearing Research,* 78, 197, 1994.

92. Pfingst, B. E., Holloway, L. A., Zwolan, T. A., and Collins, L. M., Effects of stimulus level on electrode-place discrimination in human subjects with cochlear implants, *Hearing Research*, 134, 105, 1999.

93. Shannon, R. V., Zeng, F.-G., and Wygonski, J., Speech recognition with altered spectral distribution of envelope cues, *Journal of the Acoustical Society of America*, 104, 2467, 1998.

94. Dorman, M. F., Loizou, P. C., and Rainey, D., Simulating the effect of cochlear-implant electrode insertion depth on speech understanding, *Journal of Acoustal Society of America*, 102, 2993, 1997.

95. Fu, Q.-J. and Shannon, R. V., Recognition of spectrally degraded and frequency-shifted vowels in acoustic and electric hearing, *Journal of Acoustical Society of America*, 105, 1889, 1999.

96. Lenarz, T., Battmer, R. D., Lesinski, A., and Parker, J., Nucleus double electrode array: a new approach for ossified cochleae, *American Journal of Otology*, 18, 39, 1997.

97. Bredberg, G., Lindström, B., Löppönen, H., Skarzynski, H., Hyodo, M., and Sato, H., Electrodes for ossified cochleas, *American Journal of Otology*, 18, 42, 1997.

98. Colletti, V., Fiorino, F. G., Saccetto, L., Giarbini, N., and Carner, M., Improved auditory performance of cochlear implant patients using the middle fossa approach, *Audiology*, 38, 225, 1999.

99. Fayad, J., Linthicum, F. H., Otto, S. R., Galey, F. R., and House, W. F., Cochlear implants-histopathologic findings related to performance in 16 human temporal bones, *Annals of Otology, Rhinology and Laryngology*, 100, 807, 1991.

100. Shepherd, R. K., Hatsushika, S., and Clark, G. M., Electrical stimulation of the auditory nerve: the effect of electrode position on neural excitation, *Hearing Research*, 66, 108, 1993.

101. Kral, A., Hartmann, R., Mortazavi, D., and Klinke, R., Spatial resolution of cochlear implants: the electrical field and excitation of auditory afferents, *Hearing Research*, 121, 11, 1998.

102. Snyder, R. L., Rebscher, S. J., Cao, K., Leake, P. A., and Kelly, K., Chronic intracochlear electrical stimulation in the neonatally deafened cat. I. Expansion of central representation, *Hearing Research*, 50, 7, 1990.

103. Pfingst, B. E., Burnett, P. A., and Sutton, D., Intensity discrimination with cochlear implants, *Journal of the Acoustical Society of America*, 73, 1283, 1983.

104. Shannon, R. V., Multichannel electrical stimulation of the auditory nerve in man. I. Basic psychophysics, *Hearing Research*, 11, 157, 1983.

105. McKay, C. M., O'Brien, A., and James, C. J., Effect of current level on electrode discrimination in electrical stimulation, *Hearing Research*, 136, 159, 1999.

106. Zeng, F. G. and Galvin, J. J., Amplitude mapping and phoneme recognition in cochlear implant listeners, *Ear and Hearing*, 20, 60, 1999.

107. Skinner, M. W., Holden, L. K., Holden, T. A., and Demorest, M. E., Comparison of two methods for selecting minimum stimulation levels used in programming the Nucleus 22 cochlear implant, *Journal of Speech, Language and Hearing Research*, 42, 814, 1999.

108. Parkin, J. L., Percutaneous pedestal in cochlear implantation, *Annals of Otology, Rhinology and Laryngology*, 99, 796, 1990.

109. Parkin, J. L. and Parkin, M. J., Multichannel cochlear implantation with percutaneous pedestal, *Ear, Nose and Throat Journal*, 73, 156, 1994.

110. Tjellström, A., Rosenhall, U., Lindström, J., Hallén, O., Albreksson, T., and Brånemark, P.-I., Five year experience with skin-penetrating bone-anchored implants in the temporal bone, *Acta Oto-Laryngologica* (Stockholm), 95, 568, 1983.

111. Håkansson, B., Liden, G., Tjellström, A., Ringdahl, A., Jacobsson, M., Carrlsson, P., and Erlandson, B.-E., Ten years of experience with the Swedish bone-anchored hearing system, *Annals of Otology, Rhinology and Laryngology,* 99, 1, 1990.

112. Pfingst, B. E., Albrektsson, T., Tjellström, A., Miller, J. M., Zappia, J., Xue, X. L., and Weiser, F., Chronic skull-anchored percutaneous implants in non-human primates, *Journal of Neuroscience Methods,* 29, 207, 1989.

113. Downing, M., Johansson, U., Carisson, L., Walliker, J. R., Spraggs, P. D., Dodson, H., Hochmair-Desoyer, I. J., and Albrektsson, T., A bone-anchored percutaneous connector system for neural prosthetic applications, *Ear, Nose and Throat Journal,* 76, 328, 1997.

114. Abbas, P. J., Brown, C. J., Shallop, J. K., Firszt, J. B., Hughes, M. L., Hong, S. H., and Staller, S. J., Summary of results using the Nucleus CI24M implant to record the electrically evoked compound action potential, *Ear and Hearing,* 20, 45, 1999.

115. Johnsson, L.-G. and Hawkins, J., Sensory and neural degeneration with aging, as seen in microdissections of the human ear, *Annals of Otology, Rhinology and Laryngology,* 81, 179, 1972.

116. Goycoolea, M. V., Stypulkowski, P., and Muchow, D. C., Ultrastructural studies of the peripheral extensions (dendrites) of type I ganglion cells in the cat, *Laryngoscope,* 100, 19, 1990.

117. Pfingst, B. E., Glass, I., Spelman, F. A., and Sutton, D., Psychophysical studies of cochlear implants in monkeys: clinical implications, in *Cochlear Implants,* Schindler, R. A. and Merzenich, M. M., Eds., Raven Press, New York, 1985, 305.

118. Miller, C. A., Abbas, P. J., and Robinson, B. K., The use of long duration current pulses to assess nerve survival, *Hearing Research,* 78, 11, 1994.

119. Hall, R. D., Estimation of surviving spiral ganglion cells in the deaf rat using the electrically evoked auditory brainstem response, *Hearing Research,* 45, 123, 1990.

120. Lousteau, R. J., Increased spiral ganglion cell survival in electrically stimulated, deafened guinea pig cochleae, *Laryngoscope,* 97, 836, 1987.

121. Hartshorn, D. O., Miller, J. M., and Altschuler, R. A., Protective effect of electrical stimulation in the deafened guinea pig cochlea, *Otolaryngology — Head and Neck Surgery,* 104, 311, 1991.

122. Leake, P. A., Hradek, G. T., Rebscher, S. J., and Snyder, R. L., Chronic intracochlear electrical stimulation induces selective survival of spiral ganglion neurons in neonatally deafened cats, *Hearing Research,* 54, 251, 1991.

123. Mitchell, A., Miller, J. M., Finger, P. A., and Heller, J. W., Effects of chronic high-rate electrical stimulation on the cochlea and eighth nerve in the deafened guinea pig, *Hearing Research,* 105, 30, 1997.

124. Leake, P. A., Hradek, G. T., and Snyder, R. L., Chronic electrical stimulation by a cochlear implant promotes survival of spiral ganglion neurons after neonatal deafness, *Journal of Comparative Neurology,* 412, 543, 1999.

125. Staecker, H., Kopke, R., Malgrange, B., Lefebvre, P., and Van de Water, T. R., NT-3 and/or BDNF therapy prevents loss of auditory neurons following loss of hair cells, *NeuroReport,* 7, 889, 1996.

126. Miller, A. L., Yamasoba, T., and Altschuler, R. A., Influence of growth factors on hair cell and spiral ganglion neuron preservation and regeneration, *Current Opinion in Otolaryngology and Head and Neck Surgery,* 6, 301, 1998.

127. Marzella, P. L. and Clark, G. M., Growth factors, auditory neurones and cochlear implants: A review, *Acta Oto-Laryngologica* (Stockholm), 119, 407, 1999.

128. Araki, S., Kawano, A., Seldon, H., Shepherd, R., Funasaka, S., and Clark, G., Effects of chronic electrical stimulation on spiral ganglion neuron survival and size in deafened kittens, *Laryngoscope,* 108, 687, 1998.

129. Li, L., Parkins, C. W., and Webster, D. B., Does electrical stimulation of deaf cochleae prevent spiral ganglion degeneration?, *Hearing Research,* 133, 27, 1999.

130. Powell, T. P. S. and Erulkar, S. D., Transneuronal cell degeneration in the auditory relay nuclei of the cat, *Journal of Anatomy,* 96, 249, 1962.

131. Webster, D. B., Auditory neuronal sizes after a unilateral conductive hearing loss, *Experimental Neurology,* 79, 130, 1983.

132. Miller, J. M., Altschuler, R. A., Hartshorn, D. O., Helfert, R. H., and Moore, J. K., Deafness-induced changes in the central nervous system: reversibility and prevention, in *Noise-Induced Hearing Loss*, Marshall, D., Ed. Mosby Year Book, Inc., St. Louis, 1992, 130.

133. Bledsoe, S. C., Nagase, S., Miller, J. M., and Altschuler, R. A., Deafness-induced plasticity in the mature central auditory system, *NeuroReport,* 7, 225, 1995.

134. El-Kashlan, H. K., Norrily, A. D., Niparko, J. K., and Miller, J. M., Metabolic-activity of the central auditory structures following prolonged deafferentation, *Laryngoscope,* 103, 399, 1993.

135. Irvine, D. R. F., and Rajan, R., Injury- and use-related plasticity in the primary sensory cortex of adult mammals: possible relationship to perceptual learning, *Clinical and Experimental Pharmacology and Physiology,* 23, 939, 1996.

136. Snyder, R. L., Rebscher, S. J., Leake, P. A., and Kelly, K., Chronic intracochlear electrical stimulation in the neonatally deafened cat. II. Temporal properties of neurons in the inferior colliculus, *Hearing Research,* 56, 246, 1991.

137. Najafi, K., Wise, K. D., and Mochizuki, T., A high-yield IC-compatible multichannel recording array, *IEEE Transactions on Electronic Devices,* ED-32, 1206, 1985.

138. Hetke, J. F., Lund, J. L., Najafi, K., Wise, K. D., and Anderson, D. J., Silicon ribbon cables for chronically implantable microelectrode arrays, *IEEE Transactions on Biomedical Engineering,* 41, 314, 1994.

139. Knutson, J. F., Hinrichs, J. V., Tyler, R. S., Gantz, B. J., Schartz, H. A., and Woodworth, G., Psychological predictors of audiological outcomes of multichannel cochlear implants: preliminary findings, *Annals of Otology, Rhinology and Laryngology,* 100, 817, 1991.

140. Gantz, B. J., Woodworth, G. G., Knutson, J. F., Abbas, P. J., and Tyler, R. S., Multivariate predictors of audiological success with multichannel cochlear implants, *Annals of Otology, Rhinology and Laryngology,* 102, 909, 1993.

141. Tye-Murray, N., *Foundations of Aural Rehabilitation*, Singular Publishing Group, San Diego, CA, 1998.

142. Kileny, P. R., Zimmerman-Phillips, S., Kemink, J. L., and Schmaltz, S. P., Effects of preoperative electrical stimulability and historical factors on performance with multichannel cochlear implant, *Annals of Otology, Rhinology and Laryngology,* 100, 563, 1991.

143. Blamey, P. J., Pyman, B. C., Gordon, M., Clark, G. M., Brown, A. M., Dowell, R. C., and Hollow, R. D., Factors predicting postoperative sentence scores in postlinguistically deaf adult cochlear implant patients, *Annals of Otology, Rhinology and Laryngology,* 101, 342, 1992.

144. Rubinstein, J. T., Parkinson, W. S., Tyler, R. S., and Gantz, B. J., Residual speech recognition and cochlear implant performance: effects of implantation criteria, *American Journal of Otology,* 20, 445, 1999.

145. Ito, J., Tsuji, J., and Sakakihara, J., Reliability of the promontory stimulation test for the preoperative evaluation of cochlear implants: a comparison with the round window stimulation test, *Auris Nasus Larynx,* 21, 13, 1994.

146. Brown, C. J., Abbas, P. J., Borland, J., and Bertschy, M. R., Electrically evoked whole nerve action potentials in Ineraid cochlear implant users: Responses to different stimulating electrode configurations and comparison to psychophysical responses, *Journal of Speech and Hearing Research,* 39, 453, 1996.

147. Hill, A. V., The strength-duration relation for electric excitation of medullated nerve, *Proceedings of the Royal Society of London Series B: Biological Sciences,* 119, 440, 1936.

148. Frankenhaeuser, B. and Huxley, A. F., The action potential in the myelinated nerve fibre of *Xenopus laevis* as computed on the basis of voltage clamp data, *Journal of Physiology* (London), 171, 302, 1964.

149. Bostock, H., The strength-duration relationship for excitation of myelinated nerve: computed dependence on membrane parameters, *Journal of Physiology* (London), 341, 59, 1983.

150. Colombo, J. and Parkins, C. W., A model of electrical excitation of the mammalian auditory-nerve neuron, *Hearing Research,* 31, 287, 1987.

151. Bruce, I. C., White, M. W., Irlicht, L. S., O'Leary, S. J., Dynes, S., Javel, E., and Clark, G. M., A stochastic model of the electrically stimulated auditory nerve: single-pulse response, *IEEE Transactions on Biomedical Engineering,* 46, 617, 1999.

152. Miller, C. A., Faulkner, M. J., and Pfingst, B. E., Functional responses from guinea pigs with cochlear implants. II. Changes in electrophysiological and psychophysical measures over time, *Hearing Research,* 92, 100, 1995.

153. Pfingst, B. E. and Sutton, D., Relation of cochlear implant function to histopathology in monkeys, *Annals of the New York Academy of Sciences,* 405, 224, 1983.

154. Pfingst, B. E., Changes over time in thresholds for electrical stimulation of the cochlea, *Hearing Research,* 50, 225, 1990.

155. Pfingst, B. E., Morris, D. J., and Miller, A. L., Effects of electrode configuration on threshold functions for electrical stimulation of the cochlea, *Hearing Research,* 85, 76, 1995.

156. Smith, D. W. and Finley, C. C., Effects of electrode configuration on psychophysical strength-duration functions for single biphasic electrical stimuli in cats, *Journal of the Acoustical Society of America,* 102, 2228, 1997.

157. Zwolan, T. A., Collins, L. M., and Wakefield, G. H., Electrode discrimination and speech recognition in postlingually deafened adult cochlear implant subjects, *Journal of the Acoustical Society of America,* 102, 3673, 1997.

158. Henry, B., McKay, C., McDermott, H., and Clark, G., Speech cues for cochlear implantees: spectral discrimination, *Sydney '97: XVI World Congress of Otolaryngology Head and Neck Surgery,* 89, 1997.

159. Nelson, D. A., VanTasell, D. J., Schroder, A. C., Soli, S., and Levine, S., Electrode ranking of "place pitch" and speech recognition in electrical hearing, *Journal of the Acoustical Society of America,* 98, 1987, 1995.

160. Shiroma, M., Honda, K., Yukawa, K., Yamanaka, N., Kumakawa, K., Kawano, J., and Funasaka, S., Factors contributing to phoneme recognition ability of users of the 22-channel cochlear implant system, *Annals of Otology, Rhinology and Laryngology,* 101, 32, 1992.

161. Lawson, D. T., Wilson, B. S., Zerbi, M., van den Honert, C., Finley, C. C., Farmer, J. C., McElveen, J. T., and Roush, P. A., Bilateral cochlear implants controlled by a single speech processor, *American Journal of Otology,* 19, 758, 1998.

162. Brown, J. N., Miller, J. M., Altschuler, R. A., and Nuttall, A. L., Osmotic pump implant for chronic infusion of drugs into the inner ear, *Hearing Research,* 70, 167, 1993.

163. Raphael, Y. and Yagi, M., Gene transfer in the inner ear, *Current Opinion in Otolaryngology and Head and Neck Surgery,* 6, 311, 1998.

164. Yagi, M., Magal, E., Sheng, Z., Ang, K. A., and Raphael, Y., Hair cell protection from aminoglycoside ototoxicity by adenovirus-mediated overexpression of glial cell line-derived neurotrophic factor, *Human Gene Therapy,* 10, 813, 1999.

165. Mensinger, A. F., Anderson, D. J., Buchko, C. J., Johnson, M. A., Martin, D. C., Tresco, P. A., Silver, R. B., and Highstein, S. M., Chronic recording of regenerating VIIIth nerve axons animals with a sieve electrode, *Journal of Neurophysiology,* 83, 611, 2000.

166. Lai, W. K. and Dillier, N., sCiLab — Swiss Cochlear Implant Laboratory. ENT-Department, University Hospital of Zurich, Zurich, 1997.

2 Advances in Upper Extremity Functional Restoration Employing Neuroprostheses

Kevin L. Kilgore, P. Hunter Peckham, and Michael W. Keith

CONTENTS

0-8493-2225-1/01/$0.00+$.50
© 2001 by CRC Press LLC

2.1   THE ROLE OF NEURAL PROSTHESES IN SPINAL CORD INJURY

Clinical care in spinal cord injury has advanced to provide individuals who have sustained spinal cord injury with a nearly normal life span by minimizing the leading causes of death: bladder infections and respiratory infections. While these advances have caused a dramatic reduction of morbidity and mortality, advances leading to functional restoration of the individual have been considerably less progressive. In point, conventional restoration of upper extremity function still utilizes orthoses or, more recently (since the mid-1960s), tendon transfers. Conventional mobility is focused on the wheelchair or standard orthotics, with some advancement in reciprocal gait orthoses. Conventional bladder management employs external collection devices or indwelling catheters, and defecation uses suppositories with the accompanying hours of time for bowel management. The evolving dissemination of neural prostheses is dramatically changing this environment.

Neural prostheses are devices that utilize electrical stimulation to activate the damaged or disabled nervous system to restore function. Neural prostheses employ the technique of functional electrical (or neuromuscular) stimulation (FES or FNS) to generate action potentials selectively and produce subsequent contraction of muscles. Recently, advances of bioengineering have reached wider clinical circulation, with approval by the FDA of the first generation neural prosthetic systems for hand control and for bladder/bowel management. Standing systems are nearing clinical trials. Deployment of these systems to address the multisystem needs of the individual with spinal cord injury promises to provide a dramatic change in rehabilitation outcome.

Clinical utilization and acceptance of neural prostheses requires that the neural tissue be activated reliably and specifically over the lifetime of the user without degradation of the tissue. The basic fundamentals of safe, precise, and repeatable electrical excitation of neural tissue and the technology to deliver stimulation are now well known and thoroughly reviewed in the literature.[1] The principles of this work have enabled fundamental technologies to be developed and clinically applied. In this chapter, we will focus on the application of neuroprostheses for the restoration of upper extremity function in spinal cord injury.

Restoration of hand function is a high priority for individuals with mid-cervical level spinal cord injury. Although orthotics are available, the function is limited and the rejection is high.[2,3] For lower cervical injuries (C7,C8), surgical alterations to transfer muscles under voluntary control are possible.[4,5]

Neuroprostheses utilizing FES for restoration of hand function have been clinically implemented and investigated by four independent research groups worldwide, including our center in Cleveland (see Chapter 3 for another example). Two of these groups use systems employing surface electrodes, and a third uses a percutaneous system.[6] In Cleveland, we have employed both of these types of systems over the past 25 years[7-11] but have focused most extensively on implantable systems. Although we believe that both surface and percutaneous stimulation have potential application in muscle conditioning and in short-term research applications, it has been our clinical experience that surface stimulation systems have a high rejection rate by patients in functional use, and that the maintenance associated with percutaneous systems is prohibitive for long-term clinical use.

A first generation implanted neuroprosthesis has been implemented successfully in over 150 patients, has been transferred to industry, and has received premarket approval from the FDA. More than two dozen sites internationally have been trained in its deployment. This clinical success is an indication of the impact of this technology on restoring function, and it is currently being investigated for use in restoring upper and lower limb function in individuals with stroke and cerebral palsy as well.

A second generation technology has also been developed and has begun clinical deployment. This technology provides bidirectional telemetry, which allows the use of implanted control sensors, thus freeing the user of all externally donned components except for a single transmitting coil. Additional stimulation channels are also provided for further restoration of function, including better hand control and adding arm control. The current status of the clinical implementation of upper extremity neuroprostheses for spinal cord injury is reviewed below.

2.2   NEUROPROSTHESES FOR MOTOR FUNCTION

All neuroprostheses utilizing electrical stimulation to provide motor function share several common elements. First, the user of a neuroprosthesis must have a way to communicate his or her intent to the device in order to control the resulting limb movement. This *command input* can take any number of forms, from simple switch closures and timer settings to more complicated sequences of EMG activity from muscles still under volitional control[12-16] or from brain-derived signals (see Chapters 7 and 8 for examples). Once the command is delivered, the device must process the input, which could be interpreted differently depending on the prior history of stimulation, the current state of device, and/or the status of the limb. After the *command processor* recognizes the intent of the user, the neuroprosthesis must take action. The *control processor* selects the appropriate channels (corresponding to the nerves or muscles required to effect the intended motion) as well as the relative timing, intensity, and frequency of stimulation. These parameters are then used by the *stimulus delivery* subsystem to create the stimulation wave forms and deliver current to the neural structures. These components of the system include cables, leads, and electrodes that interface with the biological system. The user can be made aware of what the system is doing through *cognitive feedback* of the state of the device via displays, warning lights, audio tones, or electrotactile feedback. The majority of clinically applied neuroprostheses operate *open loop*, that is, they are unresponsive to the environment and do not automatically correct for errors that arise between the intended and actual motions of the limbs. Sensors have been employed experimentally to feed the state of the biological system (joint angles, contact forces, etc.) back to the controller. Such *closed loop* systems require sensors to provide output feedback and a *sensor processor* to monitor the actions of the limbs and allow the control processor to adjust stimulus levels automatically without conscious input from the user. Finally, the user can be informed of the orientation and state of his own body (rather than the state of the device) through *substitute sensory feedback*. In this scheme, the closed loop sensor signals are used to modulate tactile stimulation to sensate areas or provide other indications of the status of the limbs and joints and their interaction with the environment.[17,18]

Various portions of the neuroprosthesis can be made implantable. Neuroprostheses can be completely external, in which case no foreign material is introduced into the body and only the stimulating current crosses the skin boundary. When subsystems are implanted (for example, the electrodes and/or stimulus delivery circuitry), communication must be maintained with those parts of the system remaining outside of the body. This can be done by direct percutaneous connection or via a radiofrequency (RF) transmission (see Chapter 6 for details). In the latter case, nothing crosses the skin except electromagnetic energy, reducing the likelihood of infection present with percutaneous connections and improving the convenience of donning and doffing over completely external systems. Implanting components of the system requires additional circuitry (RF transmitters and receivers) to complete the communication pathways and may increase the complexity of the design. In spite of the required surgery, implantable systems offer the advantages of placing the stimulating electrodes in close proximity to neural structures, greatly increasing the selectivity and efficiency of activation while simultaneously reducing the required current.

2.3 FIRST GENERATION NEUROPROSTHESES FOR UPPER EXTREMITY MOTOR FUNCTION IN SPINAL CORD INJURY

Functional electrical stimulation (FES) has been used experimentally to provide grasp and release for individuals with a spinal cord injury at the cervical level for many years.[9,19-24] The objectives of these systems are to reduce the need of individuals to rely on assistance from others, reduce the need for adaptive equipment, reduce the need to wear braces or other orthotic devices, reduce the time it takes to perform tasks, and enable individuals to perform tasks more normally. Neuroprostheses using FES make use of the individual's own paralyzed musculature to provide the power for grasp and the individual's voluntary musculature to control the grasp. Stimulation is applied to the muscles of the forearm and hand via implanted electrodes to provide grasp and release. Grasp movement is controlled using voluntary movements generated by the quadriplegic individual. A controller/stimulator is responsible for converting the command signal obtained from the individual into the appropriate stimulus levels for each electrode so that the desired grasp pattern is achieved. Typically, subjects use the neuroprosthesis for such tasks as eating, personal hygiene, writing, and office tasks. These systems have moved from the experimental process and are now available clinically. Users have demonstrated improved independence and daily home use of the neuroprosthesis.[11,25,26]

The first generation implanted neuroprosthesis has been developed at Case Western Reserve University/Cleveland Veterans Affairs Medical Center.[9-11,27] This system, shown in Figure 2.1, consists of an eight-channel implant receiver-stimulator, eight epimysial electrodes, leads, and connectors.[27] Electrodes are surgically placed on the muscles of the forearm and hand. A radio frequency inductive link provides the communication and power to the implant receiver-stimulator. The external components of the neuroprosthesis are an external control unit, a transmitting coil, and a shoulder control unit.[28] The external control unit performs the signal processing of the control inputs and generates the output signal (modulated radio frequency) to the implant receiver-stimulator. The radio frequency transmitting coil is taped to the individual's chest directly over the implant receiver-stimulator in order to make the inductive communication link. The shoulder control unit consists of a joystick mounted to the chest and a logic switch.

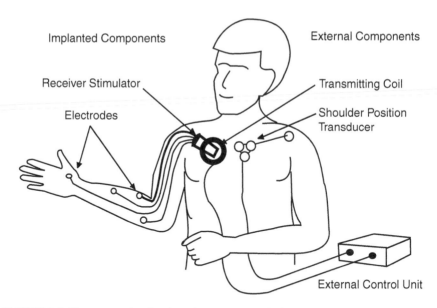

FIGURE 2.1 First generation implanted neuroprosthesis for hand control. This system is currently available commercially as the Freehand® system (NeuroControl Corporation, Cleveland, Ohio).

2.3.1 CANDIDATE SELECTION

2.3.1.1 Injury Classification

The most common upper extremity FES application has been to provide grasp and release for individuals with C5 and C6 motor level complete spinal cord injury. Damage at the fifth and sixth cervical vertebrae is the most common cervical level injury, resulting in reduced shoulder and elbow function, with complete loss of movement of the fingers and thumb. Elbow flexion is typically strong and can be near normal, but elbow extension is lost. Motor function at the C6 level includes wrist extension and forearm pronation. For these individuals, FES provides finger extension and flexion and thumb extension, flexion, and abduction. For injuries at the C4 level, control of elbow flexion must be provided by stimulation of the biceps and/or brachialis, or by a mechanical or surgical means. FES has been applied to a limited extent to these individuals, but there are no clinically deployed systems for this population to date.[29,30] For individuals with strong C6 or C7 level function, there are other surgical options such as tendon transfers to provide function, and FES is usually not indicated at the present time.[4,5,31]

2.3.1.2 Physiological Profile

The primary physiological characteristic necessary for the application of FES for motor control is the presence of intact lower motor neurons innervating the muscles to be stimulated. This is because the low levels of electrical stimulation used in FES result in excitation of intact motor neurons rather than direct excitation of muscle fibers.[1] When a spinal cord injury occurs, there may be damage to the anterior horn cells and/or spinal root evulsion at the injury site extending one or more levels in the cord. The extent of lower motor neuron damage varies from person to person and is determined by the precise nature of the injury and the location of the relevant motoneuron pools. Extensive damage to the motoneuron pool of the entire upper extremity is rare. Peckham and Keith[10] found that between 80 and 100% of the muscles necessary for grasp had sufficient innervation intact to generate functional levels of force. In many cases, other paralyzed muscles can be used to substitute for the function that is not available, although this may require surgical intervention.[32]

Neurological stability is necessary for implanted systems; therefore, implant surgery is not typically performed until at least one year after injury. Joint contractures must be corrected, or grasp functions may be limited. Spasticity must be under control. Neither age nor time post-injury appear to be major factors when considering neuroprosthetic applications.

2.3.2 OPERATING PRINCIPLES OF UPPER EXTREMITY NEUROPROSTHESES

The neuroprosthesis works in the following manner: the user depresses a switch on his or her chest which activates the system, and the user's hand opens in the lateral pinch. Graded elevation of the user's contralateral shoulder results in graded grasp closure. A quick jerk of the shoulder "locks" the hand so that it remains closed at the desired degree of closure until another quick jerk of the shoulder releases the lock command. Depressing the chest switch briefly causes the system to switch to the palmar grasp. Depressing the switch for three seconds turns the system off.

2.3.2.1 Implant Components

The implant device is designed to generate stimulus pulses of 0–200 μS pulse duration at four constant current levels (2–20 mA) through eight separate output channels. The implant receiver-stimulator consists of a titanium capsule housing the stimulator circuitry with an external stainless steel coil encased in epoxy.[27] The entire implant is coated in medical grade silicone for strain relief and to round the edges of the capsule. The electrode leads are multi-stranded helically wound stainless steel wires coated in Teflon. Two multi-stranded wires are wound in tandem and inserted into a silicone tube. Distal electrode leads are connected to the proximal implant leads using pins that insert into a center spring.[33] A silicone sleeve is placed over the connector and sutured at each end.

Epimysial electrodes are used which have a stimulating disk made of platinum-iridium with a surface area of 10 mm². The disk is embedded in a silicone elastomer backing that allows the electrode to be sutured to the muscle epimysium. Akers et al.[1] showed that the tissue response of these electrodes was acceptable, typically consisting of a thin fibrous encapsulation approximately 0.5 mm thick.

2.3.2.2 Stimulation Parameters

Balanced biphasic constant current stimulus pulses are used to maintain the integrity of both the electrodes and tissue.[1] Stimulation frequency is constant at either 12 or

16 Hz and the amplitude is constant at 20 mA. Muscle force is graded by varying the duration of the stimulus pulse from 0 to 200 μsec. The recharge phase was amplitude limited to 0.5 mA. The implant capsule serves as the anode for the return current from all eight electrodes.

2.3.2.3   Synthesis of Grasp and Release

Two basic grasp patterns are generally provided for functional activities: lateral pinch and palmar prehension.[34] The *lateral pinch* is used for holding small utensils such as a fork, spoon, or pencil. In the open phase of this grasp, the fingers and thumb are extended. The fingers are then fully flexed at all joints while the thumb remains extended. The thumb is then flexed against the lateral aspect of the index finger to produce pinch. Strong pinch forces can be achieved, typically up to 30 Newtons. *Palmar prehension* is used for acquiring large objects. The fingers are extended and the thumb is posted in full abduction. The fingers then flex against the thumb, ideally resulting in contact between the tip of the thumb and the tips of the index and long fingers. Thumb flexion can be added to increase grasp force if needed.

Nearly every muscle in the forearm and hand has been used in upper extremity neuroprostheses. At least five muscles are necessary to provide a palmar and lateral grasp. The most frequently used muscles are: adductor pollicis (AdP), extensor pollicis longus (EPL), abductor pollicis brevis (AbPB), extensor digitorum communis (EDC), and flexor digitorum superficialis (FDS). Typically, more force is required in flexion, so a second thumb and finger flexor are utilized, usually flexor digitorum profundus (FDP) and flexor pollicis longus (FPL). Wrist extensors, forearm pronators, and the finger intrinsic muscles have also been used.[35]

Grasp patterns are customized for each subject by analyzing the gradation of muscle force with stimulation (i.e., muscle recruitment properties) for each electrode. The pulse duration of the stimulus delivered to each electrode is used to modulate the muscle force. A series of "stimulus maps" are generated, which relate the single proportional command input signal to the stimulus output to each electrode.[34] In this way, coordinated control of hand movement and force of either grasp is achieved by the single control movement.

2.3.2.4   Mechanisms for Control of Grasp

Proportional control of grasp opening and closing is achieved using either shoulder or wrist motion.[13,36] Joint motion is obtained from a joystick taped to the chest or dorsum of the wrist. The subjects use a switch, mounted on the chest, to turn the system on and off and to switch between grasp patterns. When shoulder movement is used to control grasp, a quick jerk of the shoulder initiates a lock command. This causes the hand to remain at a constant level of stimulation regardless of shoulder position. Another quick jerk of the shoulder initiates an unlock command so that proportional control can be regained.

Control of grasp by wrist motion works in coordination with the tenodesis grasp that subjects are already trained to use. Wrist extension closes the grasp, wrist flexion (by gravity) opens the grasp. If a lock is necessary, it is usually provided by a switch located on the subject's wheelchair. Many subjects prefer to use wrist control rather than shoulder control because it seems more natural to them and they can perform activities faster.

2.3.2.5 Feedback Mechanisms

The implanted neuroprosthesis can provide a single channel of electrotactile feedback.[1] This is accomplished by placing a stimulating electrode in a region of normal sensation such as the shoulder. Feedback regarding the level of proportional control was provided using five levels of frequency from four to 55 Hz. Each level corresponded to 20% of the command range, with the lowest frequency indicating the range from 0 to 20%, when the grasp was open. In addition, the sensory feedback was turned off when the system was locked or when the system was turned off. In general, the sensory feedback provided subjects with information about the state of the system that would have been difficult to obtain visually.

2.3.3 AUGMENTATIVE SURGICAL PROCEDURES

The implanted neuroprosthesis is nearly always implanted in conjunction with augmentative surgical procedures of the hand and arm. These procedures provide the subject with improved function both with and without stimulation and enhance the capabilities provided by the neuroprosthesis. The procedures are: tendon transfers, tendon synchronizations, Zancolli-lasso of the flexor digitorum superficialis (FDS), and joint stabilizations.

2.3.3.1 Tendon Transfers of Voluntary and Paralyzed Muscles

Tendon transfers involve detaching the insertion of the donor tendon and suturing it into the insertion of another muscle so that the donor muscle performs a new function. Tendon transfers of muscles under voluntary control are performed to counteract the lack of voluntary wrist extension and/or elbow extension.[5] By far, the two most common procedures for the C5- and C6-level individuals are the transfer of brachioradialis to extensor carpi radialis brevis for wrist extension and the transfer of the posterior aspect of the deltoid to the triceps for elbow extension.

Tendon transfers can also be performed with paralyzed muscles.[32] This type of transfer can be used to compensate for lower motor neuron damage to critical muscle groups. For example, a paralyzed but innervated palmaris longus muscle could be transferred to provide thumb extension if the EPL has lower motor neuron damage. Also, stimulated muscles can be used to assist voluntary function, such as a transfer of a paralyzed extensor carpi ulnaris to extensor carpi radialis brevis. The specific procedures used with each subject are dependent upon the muscle groups that are weak or absent and upon the muscles that are sufficiently strong for transfer.

2.3.3.2 Finger Flexor and Extensor Synchronizations

The finger flexor tendons and finger extensor tendons can be sutured together side-to-side. The objective is (1) to provide more uniform motion of the fingers when a single electrode is used to activate the muscle, (2) to compensate for denervation (particularly of the FDP of the index finger), and (3) to provide uniform force distribution in flexion across the digits.

2.3.3.3 FDS Zancolli-Lasso Procedure

This procedure provides flexion of the metacarpal-phalangeal (MP) joints of digits II–V with motoring of the FDS by electrical stimulation. The procedure is performed as originally described[31] and involves releasing the FDS tendon from its insertion at the base of the middle phalanx and suturing the tendon around the A1 pulley.

2.3.3.4 Joint Stabilizations

The thumb interphalangeal (IP) joint is often too lax to provide good functional lateral pinch. Stabilizing the thumb IP joint is particularly important because it allows the FPL to be motored with electrical stimulation and used in functional grasp. Stimulation of the FPL without stabilization results in flexion of the thumb IP joint into the palm, a non-functional position. In the first few implantation procedures, IP stabilization was accomplished with an arthrodesis of the joint. More recently, a split transfer of part of the distal FPL tendon to the EPL insertion has been performed.[37] This procedure allows the thumb to be more mobile passively, but the joint is stiffened by FPL stimulation.

2.3.4 IMPLEMENTATION PROCESS

2.3.4.1 Pre-Surgical Conditioning

Subjects undergo an initial period of muscle conditioning using standard surface stimulation techniques. The goals of this conditioning are to build muscle strength and fatigue resistance prior to surgery. It is important to have conditioned muscles during surgery so that appropriate decisions can be made regarding electrode placement and any necessary muscle substitution patterns. Surface stimulation is applied to the radial, median and ulnar nerves for at least one month prior to implant surgery.

2.3.4.2 Surgical Procedure

The surgical procedure involves the implantation of the stimulator receiver unit and placement of electrodes as well as augmentative surgical procedures.[18,19,22] Intraoperatively, conventional hand surgical procedures are used, except that no anesthetic neuromuscular blocking agents are used and the tourniquet is released after the incisions are complete. This allows blood flow to the muscles so that they can be electrically stimulated for the duration of the procedure. Optimal placement of electrodes is determined using a mapping electrode that can be moved over the surface of the muscle. When the stimulated response shows the desired characteristics, an epimysial electrode is sutured on the muscle. Leads are tunneled from the electrodes up the arm to a connector in the mid-humeral area. The receiver-stimulator unit is implanted in the pectoral region with the leads tunneled subcutaneously to the humeral connector site, where a spring connector is used to connect the proximal implant lead to the distal electrode lead. The sensory electrode, when used, is placed rostral to the clavicle, with the stimulating portion facing the skin surface and sutured in place.

2.3.4.3   Post-operative Stabilization and Conditioning

Post-operatively the subject is placed in a long arm cast for three weeks for electrode stabilization. Elective modifications to the post-operative care are made if the subject received additional surgical alterations to the upper extremity, such as tendon transfers. Muscle conditioning using the neuroprosthesis is used to rebuild the muscle strength after the casts are removed and continues for at least a month prior to programming of a functional system.

2.3.4.4   Rehabilitation

Subjects are trained and evaluated in the use of their neuroprosthesis as an outpatient or during a three-week hospital stay. During this time, the grasp patterns, control parameters, and sensory feedback levels are determined; the subject is trained to use the neuroprosthesis; and a series of functional evaluations are performed. The subject is then discharged to use the device at home. Follow-up evaluations are performed after six months and one year.

2.3.4.5   Grasp and Control Programming

Grasp patterns and control parameters are determined by the treating clinician (typically an occupational therapist) with the guidance of a computer program specific to the implanted upper extremity neuroprosthesis. Grasp patterns are developed in two steps.[34] First, an electrode "profile" is performed in which each electrode is examined individually to determine the minimum (threshold) and maximum stimulus levels. These values are determined by slowly increasing the stimulus from zero until the first sign of muscle contraction is observed. This value is recorded by the computer as the threshold. Then the stimulus is increased until a second muscle begins to contract. This occurs because the stimulus field spreads over a larger area as the stimulus increases, thus recruiting more and more muscles. The stimulus value at which recruitment of a non-synergistic muscle occurs is recorded by the computer as the maximum stimulus level. If only a single muscle is recruited, then the maximum stimulus level is set to 200 μsec. When all the electrodes have been profiled in this manner, the computer program develops an initial palmar and lateral grasp pattern using a pre-determined rule-based algorithm.[34] These grasp patterns can then be tested, and the clinician can make on-line adjustments if necessary. Frequently the grasp pattern will be refined over the first few days of functional use.

The control parameters are also customized for each subject. First, the subject's range of motion is recorded by computer using the shoulder-mounted joystick controller. The computer then chooses a portion of the total shoulder range that will be used to control grasp opening and closing. Typically 30–50% of the total range is used. The specific parameters for the lock threshold (speed and size of shoulder movement) are also determined by computer by recording shoulder movement as the subject makes a series of repeated shoulder jerks. In all cases, the clinician can adjust the parameters determined by computer to optimize the functional outcome.

The entire system is typically programmed within a few hours, and subjects are able to immediately begin functional activities. Frequently, however, further modifications to both the grasp and control parameters take place during the first week of use as the subject attempts new activities.

2.3.4.6 Functional Evaluations

Functional assessment is an important component in the development and evaluation of rehabilitation technology such as FES neuroprostheses. The impact of a rehabilitation device must be assessed over a wide range of life impact measures.[38] Measures of *impairment* refer to the specific physiological limitation. For C5 and C6 injury levels, the impairment is the inability to move the fingers and thumb and generate grasp forces. However, providing these movements alone does not ensure that the subject can perform useful activities with the hand. Therefore, *functional limitation* must be measured. Functional limitation refers to the lack of ability to perform an activity in the manner considered normal. For C5 and C6 individuals, functional limitation manifests itself as an inability to perform activities necessary for daily living without adaptive equipment or assistance from others. Finally, the *disability*, or societal impact, must be measured. This measure refers to the limitation in performing socially defined activities within a social and physical environment. This may manifest itself in the inability to return to work or school or the inability to engage in various recreational activities. By evaluating the effect of a device over the spectrum of life impact measures, the benefits of the device can be truly demonstrated.

2.3.4.6.1 *Impairment*

Grasp force, range of motion and grasp-release ability are measured to demonstrate a reduction in impairment. Grasp and pinch force is measured using a modified pinch meter (B & L Engineering, Santa Fe Springs, CA) with metal bars added to extend and enlarge the grasping surfaces of the meter. Range of motion is measured during stimulation using a plastic goniometer. A six-object grasp-release test is used to evaluate the subject's ability to pick up and release objects.[39] The subject is instructed to move as many objects as possible within a thirty-second trial, without making mistakes. Three of the objects require the use of palmar grasp, and three require the use of lateral grasp. The objects are of varying size and weight, are manipulated using one hand only, and include a small juice can (210 grams), a paperweight (264 grams), a video tape (334 grams), and a simulated fork task (requiring 440 grams to depress the handle). This test is designed to demonstrate the ability to grasp and release objects while minimizing the influence of other variables, such as upper arm strength or postural control.

2.3.4.6.2 *Functional limitation*

The reduction in functional limitation for each subject is measured by performance in a variety of tasks encountered in everyday life, referred to as an Activities of Daily Living Test.[11,40] Each subject is tested in at least four different activities of daily living. The purpose of this test is to compare a subject's ability to perform tasks of daily living with the neuroprosthesis to their ability without using the neuroprosthesis. Tested tasks include eating, drinking, writing, and brushing teeth. In addition to these common tasks, subjects also have been tested in tasks of particular interest to them, including applying makeup, brushing hair, getting money out of a wallet, rolling dice, and painting. Individual phases of each task are identified, such as acquire, release and performance. The subject's level of independence is rated for each phase. The rating categories were needs physical assistance, needs adaptive equipment, needs an orthotic assist (braces), needs self assistance (second hand, mouth), or is independent. The subject is asked whether he or she prefers to perform each task with, or without, the neuroprosthesis.

2.3.4.6.3 Disability

Disability is measured using a satisfaction survey administered at least one year after the implantation of the neuroprosthesis.[39] These surveys are mailed to the subjects and then the answers are retrieved by phone. A five-point Likert scale (strongly agree, agree, ..., strongly disagree) is used to indicate the responses to most of the questions.

2.3.5 CLINICAL RESULTS OF A FIRST GENERATION IMPLANTED NEUROPROSTHESIS

2.3.5.1 Study Status and Demographics

There are currently over 150 subjects who have been implanted with this device at multiple sites in North America, Europe, Australia, and Asia.[25] A multicenter clinical trial of the implant device was completed in 1997 and pre-market approval was received from the FDA in August 1997. The subject demographics follow the typical demographics for spinal cord injury: 80% male and 66% between the ages of 20 and 39. The range in age was 16 to 57. The time between injury and the implantation of the neuroprosthesis ranged from 8 months to 32 years with 48% being less than four years post-injury. Three-quarters of the subjects had C5-level injuries and the remainder had C6-level injuries. At the time of this writing, the longest time post-implant is 13 years.[9]

2.3.5.2 Safety and Longevity of Components

The durability, reliability and stability of the implanted components of the first generation neuroprosthesis have been assessed over the course of a multicenter study. The incidents involving the implanted components have been few and can be divided into four categories: mechanical durability, electronic operation, device-tissue interface, and infection/rejection.

2.3.5.2.1 Mechanical durability

Over five-hundred electrodes have been implanted for at least two years, and over 200 have been implanted for more than four years. There have been a total of four electrode-lead failures, and in all cases the failure occurred at or near the electrode-lead junction. There have been no failures of the connector.

Only two of the four electrode failures appear to be related to use (fatigue fracture). In one case, a sensory lead broke secondary to an implant device rotating within the body. The rotation of the device caused the sensory lead to become wound around the remaining leads. The lead was eventually pulled apart in tension and broke near the electrode termination. In a second case, a local infection developed in a suture near a sensory electrode. When the electrode was removed, a portion of the stimulating disk had pulled away from the silastic backing. This probably occurred during the process of electrode removal and is not a fracture of any component. Both of the two remaining electrodes were located in the palm of the hand (1-AbPB muscle, 1-AdP muscle). One electrode broke at the electrode-lead weld after 2.0 years of normal operation, probably caused by repeated pressure over the electrode. This electrode is near the skin surface and is susceptible to repeated forces. We now place AbPB electrodes in a more protected location or use an intramuscular type of electrode. The fourth electrode failed two months after implantation. This failure occurred subsequent to a prolonged episode of constant use and an infection associated with the healing incision in the palm. The exact mechanism of failure is unknown.

2.3.5.2.2　Electronic operation

There have been no cases of device failure due to breakage or water ingress into the capsule. Of the 85 devices implanted to date, one is unable to generate stimulus pulses out of one of the eight channels. The source of the problem has not been determined; it may be fracture in the proximal lead near the implant device or it may be a fault within the device. Since the remaining seven channels are still operational and the patient gains beneficial function from this device, it remains implanted.

A number of devices have exhibited a susceptibility to altered function resulting from an electrostatic discharge, probably at the time of implant. In this altered state, the implant device is no longer able to provide a constant current pulse, but operates in a constant voltage mode. Function and safety are preserved in this altered state, so these devices remain implanted and in regular functional use.

2.3.5.2.3　Device-tissue interface

The implant, leads and electrodes directly interface with living tissue. In particular, the electrodes and the implant case (the anode) must maintain good electrical coupling with living tissue in order to operate properly. There have been no cases where this coupling was compromised by a buildup of encapsulation tissue. In the few cases where an implant device or electrode has been moved or upgraded, the encapsulation around the implanted components is typically 1mm or less thick and well-formed. Electrode threshold measurements taken on the group of eight electrodes implanted in the first subject have been recorded over the past 13 years and show no upward trend which might indicate increased encapsulation or migration. There has been no evidence that any of the electrodes implanted have moved, either during the time of initial encapsulation or over long-term usage.

Of the first 488 electrodes that were implanted, 16 (3.3%) were repositioned in a surgical revision procedure in order to try to obtain improved electrode recruitment. These procedures are often performed in conjunction with other procedures and are rarely the primary cause for a revision surgery.

Two implant devices rotated within the body in two of the early patients in this study. In both cases the implant was unwound and sutured down to the underlying pectoral fascia. We now routinely suture the device securely to the fascia, and no further cases of device rotation have been reported.

2.3.5.2.4   Infection/rejection

There have been five cases of infection and no cases of device rejection. Less than 2% of all electrodes implanted have been removed due to infection. One patient required the removal of all eight electrodes and the implant device at 1.4 years post-implant. This individual developed an open pressure sore in the arm near some of the leads. The individual allowed the area to remain untreated for a period of months. When medical intervention was sought, it was determined that it was prudent to remove the entire device to ensure that the infection could not track up the leads to the device itself.

The remaining four cases of infection all involved a localized infection of a single electrode. In all cases the infections arose at the incision site, usually in conjunction with expulsion of a suture. These cases happened at 0.2, 0.3, 0.6 and 1.0 years post-implant. In one case the entire lead was removed back to the connector site, and in the remaining three cases the distal lead was cut proximal to the infection site and the distal end of the electrode was removed. In all cases the infection was resolved without further incident.

In summary, these results indicate that this device is biologically and electrically safe within the body. The leads and electrodes demonstrate excellent mechanical stability. The device-tissue interface consists of minimal encapsulation that is stable over time. Overall, 97.9% of all electrodes (666/680) remain functional at the time of this writing.

2.3.5.3   Surgical Outcomes

All (100%) of the subjects who have received an implant stimulator have also had at least one additional augmentative surgery to their hand or arm. The most frequently performed procedures were finger synchronizations (87%), tendon transfer of posterior deltoid to triceps (77%), and tendon transfer of brachioradialis to extensor carpi radialis brevis (ECRB). Typical FES-related tendon transfers included extensor carpi ulnaris (ECU) to ECRB, ECU to EDC, palmaris longus to EPL, and flexor carpi ulnaris to FDP. Keith et al.[32] reported that 93% (38/41) of the FES-related procedures performed on 11 subjects resulted in a successful outcome. Finger synchronization and thumb stabilization procedures were the most successful procedures. The primary complication encountered was tendon adhesions, especially following tendon transfer surgery.

2.3.5.4   Functional Outcomes

2.3.5.4.1   Pinch force

Lateral and palmar grasp force increased in every subject (n = 43). The median lateral pinch force with stimulation was 12 Newtons, with some pinch forces greater than 25 Newtons. Without stimulation, subjects were able to use their tenodesis grasp to achieve a median of 1.6 Newtons. Similarly, subjects achieved a median palmar grip force of 6.4 Newtons with stimulation, but only a median of 0.7 Newtons voluntarily. These are statistically significant increases.

2.3.5.4.2   Range of motion

All subjects had a complete absence of voluntary movement of both the fingers and thumb, and all subjects demonstrated movement of all five digits with stimulation (n = 43). Range of motion with stimulation was, however, less than normal in most subjects, with a median total range of motion (sum of the range of all three joints for a single digit) of approximately 80 degrees for each digit including the thumb. Frequently, full normal range of motion cannot be achieved because of the presence of joint contractures, particularly at the proximal interphalangeal (PIP) joint.

2.3.5.4.3   Grasp-release test

The neuroprosthesis provided 98% (42/43) of the subjects with the ability to manipulate at least one more object than they could manipulate without the neuroprosthesis, and 86% could manipulate at least three additional objects or could manipulate all six objects in the test. Typically, subjects could grasp the two lightest objects both with and without the neuroprosthesis. The major difference in performance was exhibited in grasping heavier objects. For example, none of the subjects could manipulate the weight object without their neuroprosthesis, but 93% were successful with their neuroprosthesis. In addition, the number of completions was statistically significantly higher for all objects except the block and peg.

2.3.5.4.4   Activities of daily living

Subjects demonstrated improved independence using the neuroprosthesis in various activities. Subjects use the neuroprosthesis to perform activities such as using a utensil to eat, drinking from a glass, writing, using a phone, shaving, brushing teeth, brushing hair, putting on make-up, and painting. Twenty-five subjects have been tested on at least four different tasks each, and in every case (100%), subjects demonstrate improved performance in at least two tasks, as shown in Figure 2.2. Twenty-four of these subjects required physical assistance or adaptive equipment to accomplish at least one of the tasks tested, and in every case they were more independent using their neuroprosthesis in at least one of those tasks. The tasks in which subjects showed the most consistent improvement were eating with a fork, brushing teeth, and writing.

The neuroprosthesis reduces the subject's need for assistance and enables him or her to use non-adapted objects. For example, without stimulation, subjects generally lack the ability to stab with a fork. In order to accomplish this task, they use a fork with a built-up handle or they place a fork in a cuff strapped to their hand. They may or may not be able to strap the cuff on their hand themselves. They may also need help getting the fork out of the cuff, making it difficult to switch to another utensil. However, with the strong lateral pinch provided by the neuroprosthesis, subjects can often pick up and use an unmodified fork for eating, and they can easily put the fork down and pick up a spoon or knife.

The neuroprosthesis also provides a more secure grasp. Many subjects can pick up a glass without the neuroprosthesis by squeezing the glass between their two palms

Increased Independence Using the Neuroprosthesis

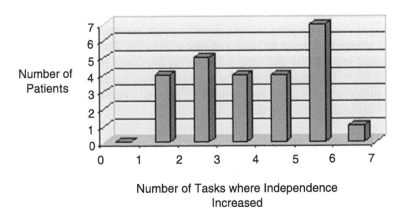

FIGURE 2.2 Change of performance of subjects using neuroprosthesis in activities of daily living. All subjects increase independence in at least two and as many as six activities.

and bringing it to their mouth. However, subjects report a much more secure grip on the glass using the palmar grasp in their neuroprosthesis. In addition, subjects prefer being able to pick up a glass with just one hand because it frees their other hand for other tasks (such as balance support), and because it looks more normal. In general, the neuroprosthesis improves subjects' independence in performing a task, improves the quality of the task, and enables subjects to perform tasks more normally.

2.3.5.4.5   Satisfaction survey

Twenty-three subjects have been surveyed at least one year after implantation of their neuroprosthesis regarding their satisfaction and utilization of the neuroprosthesis. Ninety-one percent reported overall satisfaction with the neuroprosthesis. Eighty-seven percent said that it was easier to perform activities of daily living using their neuroprosthesis and that they could perform more activities using the neuroprosthesis. Ninety-five percent said that they would recommend the neuroprosthesis to other individuals. Each subject was asked to identify the number of days per week that they typically donned their neuroprosthesis and used it for functional tasks. Eighty-three percent reported that they used the neuroprosthesis at least three days per week, with more than half of those subjects indicating that they used the neuroprosthesis every day.

2.3.5.4.6   Functional outcome summary

In summary, the neuroprosthesis reduces the impairment by providing grasp opening, pinch force, and grasp-release ability to every subject. The neuroprosthesis reduces functional limitation by enabling every subject to perform many activities of daily living more independently. The survey results indicate that disability is reduced for nearly all subjects, and the majority of subjects become regular users of the neuroprosthesis.

2.4   SECOND GENERATION NEUROPROSTHESES FOR UPPER EXTREMITY SPINAL CORD INJURY

A second generation implantable neuroprosthesis is now beginning clinical deployment.[41] This neuroprosthesis enables control using an implanted sensor (replacing the external shoulder controller) and provides better posture of the fingers in grasp and the ability to control forearm and elbow position. These features have been investigated individually and were subsequently integrated into the neuroprosthesis design.[20-23,42,43] The second generation neuroprosthesis has been implemented fully in two subjects, and without the implanted joint angle transducer (IJAT) controller in two additional subjects to date.

2.4.1   System Description

The second generation neuroprosthesis includes both implanted and external components, as shown in Figure 2.3. The implanted components are a ten-channel stimulator-telemeter (IST), an IJAT, and ten electrodes. The external components are the external control unit, which provides the power and intelligence for the system, a transmit/receive coil that is placed on the user's chest over the IST device, and an on/off switch typically placed on the user's wheelchair. The entire system is portable and, once donned, can be operated independently by the user throughout the day.

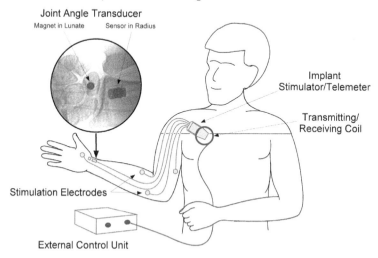

FES Hand Grasp System with
Implanted Joint Angle Transducer

Joint Angle Transducer
Magnet in Lunate   Sensor in Radius

Implant
Stimulator/Telemeter

Transmitting/
Receiving Coil

Stimulation Electrodes

External Control Unit

FIGURE 2.3 Second generation neuroprosthesis for upper extremity control. This advanced system adds to the basic hand control of the first generation system additional stimulation channels to be used for control of both the digits and proximal limb muscles. It also provides an implanted sensor for ipsilateral control of the limb through a sensor placed in the wrist.

The operation of the second generation neuroprosthesis is based on the operation of the first generation neuroprosthesis, in which the stimulus to the paralyzed muscles is controlled through user-initiated movements. Wrist motion is used to control grasp opening and closing. When the wrist is extended, the hand is closed, and when the wrist is flexed, the hand opens.

Communication and powering of the implanted components is accomplished by electromagnetic induction through radio frequency waves. To communicate with the implanted components, the external coil is placed over the implant and the system turned on. This transmits the radio frequency signal internally to power the IST and IJAT. Movements of the wrist are sensed and transmitted externally over the RF.[23,47] The transducer signal is used by the external control unit to determine the appropriate stimulus levels for each electrode, and this information is transmitted back to the IST device. Figure 2.4 shows a user with this system.

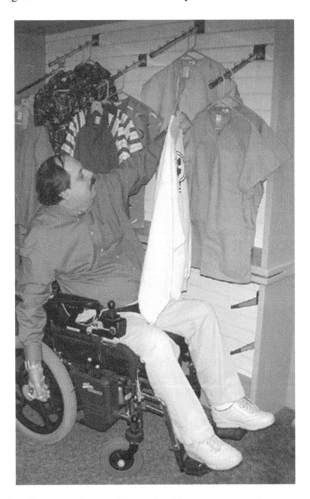

FIGURE 2.4 User demonstrating reaching task with second generation neuroprosthesis. The control device is implanted in the ipsilateral wrist, and stimulated elbow extension is provided.

2.4.2   ENHANCED MOTOR FUNCTION

The second generation neuroprosthesis provides two additional stimulus channels for added motor functions. The potential functional gains provided by adding stimulation of the finger intrinsic muscles, the triceps, and the pronator quadratus have been studied and are reviewed in the following paragraphs.

2.4.2.1   Intrinsic Muscles

Upper extremity paralysis often results in flexion tightness at the PIP joint. Stimulation of the EDC results in limited extension of the PIP and distal interphalangeal (DIP) joints. In order to improve grasp opening and hand posture for grasp, it is necessary to provide an MP flexion moment coupled with an IP extension moment. We have investigated surgical alternatives to supply these joint moments and have utilized tendon transfers and electrical stimulation of paralyzed and transferred muscles.[32] However, we have found that the most successful method for providing this force is to electrically stimulate the dorsal and palmar interossei muscles.[42] Although there are eleven finger intrinsic muscles, we have found that two electrodes are sufficient to provide forces on the index, long and ring fingers. Intramuscular electrodes are used.[48] One electrode is placed between the second and third metacarpal, stimulating the second dorsal and the first palmar interosseous muscles. The second electrode is placed between the third and fourth metacarpal, recruiting the third dorsal interosseous and the second palmar interosseous. This combination of electrodes has been used successfully with four patients.

2.4.2.2   Elbow Extension

We developed control techniques to provide elbow extension in persons with C5 tetraplegia and assessed the change in workspace that was provided.[44] In these studies, two persons with C5 tetraplegia were provided with control of the triceps, using percutaneous electrodes for stimulation. Control was derived from an accelerometer, used as a gravity sensor, mounted on the upper arm near the elbow to detect the gravity vector. As the arm was raised above horizontal, a signal was generated to regulate the stimulus applied to the triceps and extend the elbow. Elbow angle was controlled by voluntary activation of the elbow flexors in opposition to the triceps. We demonstrated that both subjects had increased ability to grasp objects over a greater workspace using this approach. However, there are technical limitations of this approach, in that it requires a second sensor (the accelerometer) and only provided control above the horizontal level of the shoulder. Crago et al.,[45] advanced this technique to eliminate the second sensor and demonstrated the stability provided by co-activation of the elbow flexors and extensors.

2.4.2.3   Supination/Pronation

The feasibility of providing forearm pronation/supination control to persons with C5/C6 tetraplegia was investigated using an electrically activated pronator and voluntary supination.[43] Two muscles are primarily responsible for generation of pronation moments, the pronator quadratus and pronator teres. We determined that the pronator quadratus (PQ) was easier to electrically activate than the pronator teres. By measuring the voluntary and electrically produced pronation/supination moments of tetraplegic subjects, we determined that the moments produced by electrical activation of the pronator were larger than 30 N-cm in all cases, and the voluntary supination moment was larger than the maximal PQ pronation moment in two of the three forearms tested. In all patients we tested, they were able to voluntarily supinate against the electrically activated pronator, producing intermediate angles between full pronation and full supination. This experiment demonstrated the feasibility of using "voluntary antagonist control" in neuroprosthetic systems.

2.4.3 BIMANUAL CONTROL

Bimanual implementation of neuroprostheses in both arms potentially provides the user with the capability of performing more complex tasks than can be accomplished unilaterally and increasing the working space for object manipulation. Implementation requires that the command control task not compromise activities in the contralateral arm, as this would be anticipated to impair operation of the neuroprosthesis. The first bilateral neuroprosthesis was implemented in a subject who has C6 function in his left arm and C5 function in his right. In the initial procedure, performed in August 1986, a first generation system was implanted in the left arm. In September 1997, the first generation system was upgraded to a second generation system with an IJAT. In March 1997, a second first-generation system was implanted in his right arm, after an extensive period of study of his right arm using percutaneous electrodes to model the function of the implantable system. The procedure performed included implantation of the stimulator, a revision of a previous tendon transfer of the brachioradialis to the extensor carpi radialis brevis, transfer of extensor carpi ulnaris to the flexor pollicis longus, Zancolli-lasso transfers of the flexor digitorum superficialis of the long, ring, and small fingers, and implantation of epimysial electrodes on the abductor pollicis brevis and flexor digitorum superficialis. Intramuscular electrodes were used for stimulation of the extensor digitorum communis, extensor pollicis longus, flexor digitorum profundus, flexor digitorum superficialis of the index, extensor carpi ulnaris, and triceps. Following post-op immobilization, gradual exercise was begun to condition the muscles in an exercise pattern. The subject is currently using this system with an external shoulder control source.

Activities of daily living (ADL) tasks have been evaluated with this subject. Usage was assessed in tasks where it was anticipated that bilateral control would provide a benefit, including opening a jar, answering a phone and writing messages, cutting food, pouring from a bottle, and placing a letter in an envelope. The quality of the task was improved in two of the four tasks and the subject preferred the bilateral system in three of the four tasks, indicating that it was easier to perform the tasks bilaterally. There was no change in the amount of assistance that was needed across the four ADL tasks evaluated. These early results indicate the potential for benefit provided by the bilateral neuroprostheses. Bilateral performance was limited by the nonoptimal grasp output in the subject's weaker right hand due to denervation of some key muscles.

2.5 FUTURE NEUROPROSTHETIC SYSTEMS FOR UPPER EXTREMITY SPINAL CORD INJURY

Future advances in upper extremity neuroprosthetic systems are likely to include advances in the areas of control, sensory feedback, and new methods of muscle recruitment. It is anticipated that upper extremity systems will be combined with bladder and lower extremity systems to provide even more functions for spinal cord injured individuals. Finally, neuroprosthetic systems will be applied to other CNS diseases.

2.5.1 FUTURE ADVANCES IN CONTROL

2.5.1.1 Myoelectric Control

Control signals for neuroprosthetic systems can be derived from the myoelectric signal (MES) generated by muscles under voluntary control. MES from the sternocleidomastoid muscle (SCM) has been used to control grasp opening and closing through a gated ramp algorithm.[8] The potential for bilateral control using the SCM muscles has also been investigated.[49] MES derived from voluntary wrist muscles can also be used for control of grasp and functions similarly to control of grasp using wrist position.[36] This type of control is amenable to implantable systems, and we are presently developing an implanted neuroprosthesis based on the IST device that will be capable of MES telemetry from two muscles.[50]

2.5.1.2 Cortical Control

The concept of cortical control is to utilize signals generated by cortical neurons to provide the command input to the neuroprosthesis, thus eliminating the need for physical sensors (e.g., joysticks, myoelectric signals, joint angle sensors) to artificially represent the user's command control intention. Considerable basic research is ongoing which is focused on two different approaches of acquiring cortical signals. The first is to record selectively from groups of neurons with indwelling electrodes. The second is to record and process the EEG signal and to use some of its properties to provide a command control input.[51] Using the latter method, we have shown that the frontal beta signal has adequate control information to enable simple control tasks in a neuroprosthesis and is minimally affected by proximal arm movements or stimulus artifacts.[52] A subject using an implanted neuroprosthesis accomplished pick and place tasks (e.g., pegs, weights) and activities of daily living (e.g., writing, drinking) in a laboratory environment.[53] Presently, any practical implementation of this system is limited by the extensive hardware required and its uncosmetic appearance. Nevertheless, this experiment, utilizing minimal control information that provided only simple logical commands, demonstrated the feasibility of producing functional control of a neuroprosthesis using cortical signals.

2.5.1.3 Closed Loop Systems

Closed loop control systems are a means for improving functional output in the presence of perturbations. Closed loop control systems use information about the system or hand to adjust the inputs. It has been demonstrated that closed loop control can improve grasp output.[54] Closed loop feedback would also have a substantial role in the control of proximal muscles, particularly at the shoulder joint where volitional control of individual muscles will be impractical. Closed loop systems require a source of information that is to be fed back to the controller for system regulation. Crago et.al. have provided performance specifications for sensors to be used in the upper extremity. Clinical acceptance of these sensors probably requires that the sensor system must be implanted. However, a principal difficulty in implementing systems employing feedback in the clinical environment is the lack of suitable sensors.[18] Small, implantable sensors for pinch force or finger position are not yet available. An alternative approach is to use nerve signal recordings from the intact sensory afferent fibers in the hand. Such applications use a nerve cuff to house the recording electrodes, as developed in Aalborg and Vancouver. [55] The Aalborg group has demonstrated in humans that, under restricted laboratory conditions, such sensors could be used to stop slipping of objects being held in lateral grasp.[56]

2.5.2 FUTURE ADVANCES IN SENSORY FEEDBACK

Although the focus for spinal cord injured individuals has been the lack of motor function, the sensory system is equally disabled in a spinal cord injury. Sensory feedback would augment the information provided by the visual system and presumably reduce the user's conscious effort that is directed toward neuroprosthesis operation. Methods of delivering such feedback conceivably include any sensory modality. For practical reasons, methods that have been used include skin senses (electrotactile and vibrotactile) and auditory. While the latter has been found disruptive, the others are more intimate with the user and have received more attention in applications related to neuroprostheses. In particular, electrotactile stimulation has been investigated most thoroughly, perhaps because its modality of delivery is compatible with the motor activation. Current methods of electrotactile sensory stimulation provide an encoded form of the information to be delivered in the feedback coding.[57] For example, increased stimulus frequency is used to indicate increased pinch force. Increases in grasp opening can be indicated by an increase in the spatial distribution of stimulation sites. Feedback of other modalities, such as temperature, may also be important to provide, as these could provide protective sensation in the case of prolonged force generation which could damage the skin, or to prevent burns. The limitation, as with closed loop control, is the lack of appropriate sensors which are required to generate the information to be delivered to the user. Additionally, the optimal means of delivery has not been developed.

2.5.3 FUTURE ADVANCES IN NEURAL ACTIVATION
AND MUSCLE RECRUITMENT

Basic research continues in the area of motor control using FES in a search for improved methods of stimulation and control. New electrode designs under research study have centered around direct nerve stimulation.[58] The appeal of nerve stimulation is that multiple muscles may be recruited by the same electrode, possibly reducing the number of electrode leads, surgery time, and number of incisions. Also, nerves do not move as much as muscles, possibly resulting in increased lead life and better recruitment properties. One type of electrode, a nerve cuff electrode, has been used for years in FES applications involving respiration[59] and bladder function,[60] but not in upper extremity applications. This is because stimulation of whole peripheral nerves in the arms results in mass contraction of multiple muscles, eliminating the ability to control specific grasp patterns. Grill and Mortimer[61] have developed a method for "steering" current to different locations within a nerve cuff electrode, demonstrating the ability to differentiate the stimulation of separate muscles. Yoshida and Horsch[62] have developed methods for placing fine wire electrodes within the nerve as a method for stimulating portions of the nerve. Another option is an electrode that slowly penetrates into the interfascicular space, as described by Tyler and Durand.[63]

Another approach to neural activation allows electrically generated action potentials to be propagated unidirectionally.[64] This could be important for several clinical conditions in which abnormal control of the nervous system creates unspecific activation of muscles (spasms or spasticity). In this situation, electrical impulses could be generated on the nerve to arrive along the nerve in coincidence with the undesired activity, with the resulting collision of impulses creating a conduction block of those impulses.

Another attempt to reduce the number of leads in a neuroprosthesis is the development of a microstimulator.[65] The microstimulator is a small, self-contained single channel unit which can be injected into the muscle with a hypodermic needle. This technique is being commercialized by Advanced Bionics (Sylmar, CA). Each stimulator is individually addressable and controlled by an externally imposed electromagnetic field. Multiple microstimulators that are included within the physical extent of the transmitting coil can be controlled through the external RF link. An advantage of the microstimulator is that it provides a means for reducing the extent of surgery needed to implant the electrodes.

2.5.4 FUTURE ADVANCES IN MULTIPLE LIMB AND ORGAN SYSTEMS

As advances to the control of individual systems become more prevalent, the need for system integration becomes a focus. For example, upper limb, lower limb, and urinary systems each are individually deployed at present, while some subjects have received both upper limb and bladder systems. To date, these systems have been treated as entirely separate elements, with unintegrated implanted and external components. In the future other system concepts will emerge. These might include a single implanted system and single external controller with multiple applications, or a fully distributed system, such as is possible with the micro-stimulator, or variations in between such as multiple implants controlled by a single external controller. The former approach is employed by Neopraxis, Ltd. (Sydney, N.S.W.) for lower extremity and bladder applications. To date, the technology has not been realized to implement multiple system applications, except in the simplest manner. Thus, while the control achieved with individual systems is quite remarkable at this time, realizing a fully integrated approach to functional restoration employing neuroprostheses awaits new technological concepts.

2.5.5 APPLICATION OF NEUROPROSTHETICS TO OTHER CNS DISEASES

Application to spinal cord injury has been by far the most popular application for functional upper extremity systems. However, FES is also applicable to other disabilities such as head injury,[11] stroke,[66] and cerebral palsy.[6] These disabilities all have certain characteristics that are similar to spinal cord injury, in that they all have electrically excitable lower motor neurons. However, differences between these populations and spinal cord injury are that these individuals frequently (1) have retained sensation in the areas to be stimulated, (2) have spasticity, and (3) present additional difficulties in determining good or appropriate control methods. Current research addressing these issues, which are different across each population, will enable individuals who have sustained these disabilities to benefit from the advances of neuroprostheses.

2.6 SUMMARY

Neuroprostheses are playing an ever increasing role in the restoration of function of individuals with spinal cord injury. Basic knowledge has been developed that provides the basis for safe and reliable activation of neural tissue and generation of stable and controllable muscle contractions. First generation technology has been developed which provides stimulation of multiple muscle groups using an implanted multichannel, externally powered and controlled receiver-stimulator. This technology has been utilized for restoration of hand grasp in persons with mid-cervical level injuries and has been reliable and well received by those individuals who have utilized it. The system has received FDA pre-market approval and the CE mark and is available worldwide. The implantable technology is also being employed in a system for standing and transfer for individuals with incomplete tetraplegia and paraplegia. A second generation technology has been developed and is in clinical testing. This technology provides control of additional channels of stimulation for activation of more muscle groups and is controlled by an implanted sensor. Both of these systems have been demonstrated by their users to have significantly increased their functional capabilities and to be reliable and functional over extended periods. Additional research is ongoing in areas of technology development (e.g., new electrodes, new implantable devices), new physiological techniques (e.g., spasticity suppression, selective activation), and new clinical applications, in extending indications both within the spinal cord injury population and to other disability groups. Neuroprostheses are providing an increasingly important method for restoring function in individuals with central nervous system injuries.

ACKNOWLEDGMENT

The authors wish to acknowledge the Department of Veterans Affairs Rehabilitation Research and Development Service and the National Institutes of Health, National Institute of Neurological Diseases and Stroke (Neural Prosthesis Program) for the primary support of the research reported in this chapter. Support was also provided by the National Institutes of Health Clinical Research Center at MetroHealth Medical Center Grant No. 5M01RR 00080 for conduct of the clinical studies.

REFERENCES

1. Mortimer, J. T., Motor prostheses, in *Handbook of Physiology — The Nervous System II,* Brookhart, J. M. and Mountcastle, V. B., Eds., American Physiological Society, Bethesda, MD, 1981, 155.
2. Beard, J. E., Long, C., Follow-up study on usage of externally-powered orthoses, *Orthotics and Prosthetics,* 35, 85, 1970.
3. Allen, V. R., Follow-up study of wrist-driven flexor-hinge-splint use, *American Journal of Occupational Therapy,* 25, 420, 1971.
4. Hentz, V.R., Ladd, A. L., Functional restoration of the upper extremity in tetraplegia, in *Diagnosis and Management of Disorders of the Spinal Cord,* Young, R. R. and Woolsey, R. M., Eds., W. B. Saunders Co., Philadelphia, 1995.
5. Moberg, E., Surgical treatment for absent single-hand grip and elbow extension in quadriplegia. Principles and preliminary experience, *Journal of Bone and Joint Surgery,* 57A, 196, 1975.
6. Triolo, R., Nathan, R., Handa, Y., Keith, M., Betz, R., Carroll, S., Kantor C., Challenges to clinical deployment of upper limb neuroprostheses, *Journal of Rehabilitation Research and Development,* 33, 111, 1996.
7. Peckham, P. H., Marsolais, E. B., Mortimer, J. T., Restoration of key grip and release in the C6 tetraplegic patient through functional electrical stimulation, *Journal of Hand Surgery,* 5(5), 462, 1980.
8. Peckham, P. H., Mortimer, J. T., Marsolais, E. B., Controlled prehension and release in the C5 quadriplegic elicited by functional electrical stimulation of the paralyzed forearm musculature, *Annals of Biomedical Engineering,* 8, 369, 1980.
9. Keith, M. W., Peckham, P. H., Thrope, G. B., Stroh, K. C., Smith, B., Buckett, J. R., Kilgore, K. L., Jatich, J. W., Implantable functional neuromuscular stimulation in the tetraplegic hand, *Journal of Hand Surgery,* 14A, 524-530, 1989.
10. Peckham, P. H., Keith, M. W., Motor prostheses for restoration of upper extremity function, in *Neural Prostheses: Replacing Motor Function After Disease or Disability,* Stein, R. B., Peckham, P. H., and Popovic, D. B., Eds., Oxford University Press, New York, 1992.
11. Kilgore, K. L., Peckham, P. H., Keith, M. W., Thrope, G. B., Wuolle, K. S., Bryden, A. M., Hart R. L., An implanted upper extremity neuroprosthesis: A five patient follow-up, *Journal of Bone and Joint Surgery,* 79A, 533, 1997.
12. Marsolais, E. B., Kobetic, R., Functional walking in paralyzed patients by means of electrical stimulation, *Clinical Orthopedics,* 175, 30, 1983.
13. Johnson, M. W., Peckham, P. H., Evaluation of shoulder movement as a command control source, *IEEE Transactions on Biomedical Engineering,* 37, 876, 1990.
14. Kralj, A., Bajd, T., Turk, R., Krajnik. J., Benko, H., Gait restoration in paraplegic patients: a feasibility demonstration using multichannel surface electrode FES, *Journal of Rehabilitation Research and Development,* 20, 3, 1983.
15. Bajd, T., Kralj, A., Turk, R., Benko, H., Sega, J., Use of functional electrical stimulation in the rehabilitation of patients with incomplete spinal cord injuries, *Journal of Biomedical Engineering,* 11, 96, 1989.

16. Graupe, D., EMG pattern analysis for patient-responsive control of FES in paraplegics for walker-supported walking, *IEEE Transactions on Biomedical Engineering*, 36, 7, 1989.

17. Chizeck, H. J., Crago, P.E., Kofman, L.S., Robust closed-loop control of isometric muscle force using pulsewidth modulation, *IEEE Transactions on Biomedical Engineering*, 35, 510, 1988.

18. Crago, P. E., Chizek, H. J., Neuman, M. R., Hambrecht, F. T., Sensors for use with functional neuromuscular stimulation, *IEEE Transactions on Biomedical Engineering*, 33, 256, 1986

19. Handa, Y., Handa, T., Ichie, M., Murakami, H., Hoshimiya, N., Ishikawa, S., Ohkubo, K., Functional electrical stimulation (FES) systems for restoration of motor function of paralyzed muscles — versatile systems and a portable system, *Frontiers of Medical and Biological Engineering*, 4, 214, 1992.

20. Nathan, R. H., Control strategies in FNS systems for the upper extremities, *Critical Reviews in Biomedical Engineering*, 21, 485, 1993.

21. Peckham, P. H., Mortimer, J. T., Restoration of hand function in the quadriplegic through electrical stimulation, in *Functional Electrical Stimulation: Applications in Neural Prosthesis*, Reswick, J. B. and Hambrecht, F.T., Eds., Marcel Dekker, 1977, 83.

22. Perkins, T. A., Brindley, G.S., Donaldson, N. N., Polkey, C. E., Rushton, D. N., Implant provision of key, pinch and power grips in a C6 tetraplegic, *Medical and Biological Engineering and Computing*, 32, 367, 1994.

23. Saxena, A., Nikolic, S., Popovic, D., An EMG-controlled grasping system for tetraplegics, *Journal of Rehabilitation Research and Development*, 32(1), 17, 1995.

24. Wieler, M., Kenwell, Z., Gauthier, M., Isaacson, G., Prochazka, A., Electronic glove augments tenodesis grip and hand opening in people with quadriplegia. *Physiotherapy Canada*, 46, 94, 1994.

25. Keith, M. W., Restoration of tetraplegic hand function using an FES neuroprosthesis, in *Tendon and Nerve Surgery in the Hand — A Third Decade*, Hunter, J. M., Schneider, L. H., Mackin, E. J., Eds., Mosby-Year Book, Inc., St. Louis, 1997, 226.

26. Wijman, C. A., Stroh, K. C., Van Doren, C. L., Thrope, G. B., Peckham, P. H., Keith, M. W., Functional evaluation of quadriplegic patients using a hand neuroprosthesis, *Archives of Physical Medicine and Rehabilitation*, 71, 1053, 1990.

27. Smith, B., Buckett, J. R., Peckham, P. H., Keith, M. W., Roscoe, D. D., An externally powered, multichannel, implantable stimulator for versatile control of paralyzed muscle, *IEEE Transactions on Biomedical Engineering*, 34, 499, 1987.

28. Buckett, J. R., Peckham, P. H., Thrope, G.B., Braswell, S.D., Keith, M. W., A flexible, portable system for neuromuscular stimulation in the paralyzed upper extremity, *IEEE Transactions on Biomedical Engineering* 35, 897, 1988.

29. Betz, R. R., Mulcahey, M. J., Smith, B. T., Triolo, R. J., Weiss, A. A., Moynahan, M., Keith, M. W., Peckham, P. H., Bipolar latissimus dorsi transposition and functional neuromuscular stimulation to restore elbow flexion in an individual with C4 quadriplegia and C5 denervation, *Journal of American Paraplegia Society*, 15, 220. 1992.

30. Handa, Y., Kameyama, J., Hoshimiya, N., FES-control of shoulder motion in hemiplegic and quadriplegic patients, in *Proceedings of the Vienna International Workshop on Functional Electrostimulation*, Department of Biomedical Engineering and Physics, University of Vienna, 1992, 127.

31. Zancolli, E., *Structural and Dynamic Basis of Hand Surgery*, 2nd ed., JB Lippincott, Philadelphia, 1979, 17.

32. Keith, M. W., Kilgore, K. L., Peckham, P. H., Wuolle, K. S., Creasey, G., Lemay, M., Tendon transfers and functional electrical stimulation for restoration of hand function in spinal cord injury, *Journal of. Hand Surgery*, 21A, 89, 1996.

33. Letechipia, J. E., Peckham, P. H., Gazdik, M., Smith, B., In-line lead connector for use with implanted neuroprostheses, *IEEE Transactions on Biomedical Engineering*, 38, 707, 1991.

34. Kilgore, K. L., Peckham, P. H., Thrope, G. B., Keith, M. W., Stone, K. A., Synthesis of hand movement using functional neuromuscular stimulation, *IEEE Transactions in Biomedical Engineering*, 36, 761, 1989.

35. Lemay, M. A., Crago, P. E., Restoration of pronosupination control by FNS in tetraplegia: experimental and biomechanical evaluation of feasibility, *Journal of Biomechanics,* 29, 435, 1996.

36. Hart, R. L., Kilgore, K. L., Peckham, P. H., A comparison between control methods for implanted FES hand grasp systems, *IEEE Transactions on Rehabilitation Engineering*, 6, 1, 1998.

37. Mohammed, K. D., Rothwell, A. G., Sinclair, S. W., Willems, S. M., Bean, A. R., Upper-limb surgery for tetraplegia, *Journal of. Bone and Joint Surgery*, 74B, 73, 1992.

38. Brandt, E. N. Jr., Pope, A. M., Eds., *Enabling America — Assessing the Role of Rehabilitation Science and Engineering,* Committee on Assessing Rehabilitation Science and Engineering, Institute of Medicine, National Academy Press, Washington, D.C., 1997.

39. Wuolle, K. S., Van Doren, C. L., Thrope, G. B., Keith, M. W., Peckham, P. H., Development of a quantitative hand grasp and release test for patients with tetraplegia using a hand neuroprosthesis, *Journal of Hand Surgery*, 19A, 209, 1994.

40. Mulcahey, M. J., Smith, B. T., Betz, R. R., Triolo, R. J., Peckham, P. H., Functional neuromuscular stimulation: outcomes in young people with tetraplegia, *Journal of the American Paraplegia Society,* 17, 20, 1994.

41. Smith, B., Tang, Z., Johnson, M., Pourmehdi, S., Gazdik, M., Buckett, J., Peckham, P., An externally powered, multichannel, implantable stimulator-telemeter for control of paralyzed muscle, *IEEE Transactions on Biomedical Engineering,* 45, 463, 1998.

42. Lauer, R. T., Kilgore, K. L, Peckham, P. H., Bhadra, N., Keith, M. W., The function of the finger intrinsic muscles in response to electrical stimulation, *IEEE Transactions on Rehabilitation Engineering,* 7, 19, 1999.

43. Lemay, M. A., Crago, P. E., Restoration of pronosupination control by FNS in tetraplegia: experimental and biomechanical evaluation of feasibility, *Journal of Biomechanics*, 29, 435, 1996.

44. Grill, J. H., Peckham, P. H., Functional neuromuscular stimulation for combined control of elbow extension and hand grasp in C5 and C6 quadriplegics, *IEEE Transactions on Rehabilitation Engineering,* 6, 190, 1998.

45. Crago, P. E., Memberg, W. D., Usey, M. K., Keith, M. W., Kirsch, R. F., Chapman, G. J., Katorgi, M. A., Perreault, E. J., An elbow extension neuroprosthesis for individuals with tetraplegia, *IEEE Transactions on Rehabilitation Engineering,* 6, 1–6, 1998.

46. Johnson, M. W., Peckham, P. H., Bhadra, N., Kilgore, K. L., Gazdik, M. M., Keith, M. W., Strojnik, P., An implantable transducer for two-degree-of-freedom joint angle sensing, *IEEE Transactions on Biomedical Engineering,* 7, 349, 1999.

47. Tang, Z., Smith, B., Schild, J. H., Peckham, P. H., Data transmission from an implantable biotelemeter by load-shift keying using circuit configuration modulator, *IEEE Transactions on Biomedical Engineering*, 42, 525, 1995.

48. Memberg, W. D., Peckham, P. H., Keith, M.W., A surgically implanted intramuscular electrode for use with an implantable neuromuscular stimulation system, *IEEE Transactions on Rehabilitation Engineering,* 2, 80, 1994.
49. Scott, T. R. D., Peckham, P. H., Kilgore, K. L., Tri-state myoelectric control of bilateral upper extremity neuroprostheses for tetraplegic individuals, *IEEE Transactions on Rehabilitation Engineering* 4(4), 251, 1996.
50. Pourmehdi, S., Strojnik, P., Peckham, H., Buckett, J., Smith, B., A custom designed chip to control an implantable stimulator and telemetry system for control of paralyzed muscles, *Artificial Organs,* 23, 396, 1999.
51. Wolpaw, J., McFarland, D., Neal, G., Forneris, C., An EEG-based brain computer interface for cursor control, *Electroencephalography and Clinical Neurophysiology,* 78, 252, 1991.
52. Lauer, R. T., Peckham, P.H., Kilgore K. L., EEG-based control of a hand grasp neuroprosthesis, *Neuroreport,* 10, 1767, 1999.
53. Lauer, R., Peckham, P., Kilgore, K., Heetderks, W., Applications of cortical signals to neuroprosthetic operation: a critical review, *IEEE Transactions on Rehabilitation Engineering,* 8, 205–207, 2000.
54. Crago, P. E., Nakai, R. J., Chizeck, H. J., Feedback regulation of hand grasp opening and contact force during stimulation of paralyzed muscle, *IEEE Transactions on Biomedical Engineering,* 38, 17, 1991.
55. Haugland, M. K., Hoffer, J. A., Sinkjaer, T., Skin contact force information in sensory nerve signals recorded by implanted cuff electrodes, *IEEE Transactions on Rehabilitation Engineering,* 2, 18, 1994.
56. Haugland, M., Lickel, A., Haase, J., Sinkjaer, T., Control of FES thumb force using slip information obtained from the cutaneous electroneurogram in quadriplegic man, *IEEE Transactions on Rehabilitation Engineering,* 7, 215, 1999.
57. Van Doren, C. L., Menia, L. L., Representing the surface texture of grooved plates using single-channel electrocutaneous stimulation, in *Sensory Research Multimodal Perspectives,* Verrillo, R. T., Ed., Lawrence Erlbaum Associates, Hillsdale, NJ, 1993, 177.
58. Mortimer, J. T., Agnew, W. F., Horch, K., Citron, P., Creasey, G., Kantor, C., Perspectives on new electrode technology for stimulating peripheral nerves with implantable motor prostheses, *IEEE Transactions on Rehabilitation Engineering.* 3, 145, 1995.
59. Kim, J. H., Manuelidis, E. E., Glenn, W. W. L., Kaneyuki, T., Histopathological changes in phrenic nerve following long-term electrical stimulation, *Journal of Thoracic Cardiovascular Surgery,* 72, 602, 1976.
60. Brindley, G. S., An implant to empty the bladder or close the urethra, *Journal of Neurology, Neurosurgery, and Psychiatry,* 40, 369, 1977.
61. Grill, W. M, Mortimer, J. T., Quantification of recruitment properties of multiple contact cuff electrodes, *IEEE Transactions on Rehabilitation Engineering,* 4, 49, 1996.
62. Yoshida, K., Horch, K., Selective stimulation of peripheral nerve fibers using dual intrafascicular electrodes, *IEEE Transactions on Biomedical Engineering,* 40, 494, 1993.
63. Tyler, D. J., Durand, D. M., Slowly penetrating interfascicular electrode for electrical stimulation of nerves, *IEEE Transactions on Rehabilitation Engineering,* 5, 51, 1997.
64. Sweeney, J. D., Mortimer, J. T., An asymmetric two electrode cuff for generation of unidirectionally propagated action potentials, *IEEE Transactions on Biomedical Engineering,* 33, 541, 1986.

65. Loeb, G. E, Zamin, C. J., Schulman, J. H, Troyk, P. R., Injectible microstimulator for functional electrical stimulation, *Medical and Biological Engineering and Computing,* 29, NS13, 1991.

66. Hines, A. E, Crago, P. E., Billian, C., Hand opening by electrical stimulation in patients with spastic hemiplegia, *IEEE Transactions on Rehabilitation Engineering,* 3, 193, 1995.

3 BION™ Implants for Therapeutic and Functional Electrical Stimulation

Gerald E. Loeb and Frances J.R. Richmond

CONTENTS

3.1 OVERVIEW

Since the experiments of Luigi Galvani two centuries ago, it has been known that electrical currents can be used to stimulate muscle contraction. Achieving function-ally useful movements of a multiarticular limb has been more elusive. It is only within the past 30 years that robotics engineers and neurophysiologists have fully recognized and started to address the demands that such movements place on sensors, actuators, and control systems. It now appears feasible to graft robotic and electro-physiological instrumentation onto a biological system to repair it, but this will require many channels of information transmission in each direction. These channels must be installed and function safely and reliably in one of the most challenging environments conceivable — the human body.

For this discussion, we shall consider three classes of treatment, based on the nature of the interaction between the patient and the technology:

- TES (Therapeutic Electrical Stimulation) — electrically induced exercise in which the beneficial effect occurs primarily offline as a result of trophic effects on muscles and perhaps the central nervous system (CNS);
- NMS (Neuromodulatory Stimulation) — preprogrammed stimulation that directly triggers or modulates a function without ongoing control or feedback;
- FES (Functional Electrical Stimulation) — precisely controlled muscle contractions that produce specific movements required by the patient to perform a task.

This chapter describes the philosophy, strategies, and technologies of a multi-disciplinary and multi-institutional project that has been underway since 1988 to build bidirectional interfaces between biological limbs and electronic controllers. Three classes of BION implants have been identified and are in various stages of development.

- BION1 provides stimulation only and operates only when powered and controlled by an external transmission coil in the vicinity of the implants. These implants have completed preclinical testing and have been approved as investigational devices for a specific clinical application; three patients have been implanted to date.
- BION2 provides stimulation and sensing via bidirectional telemetry but only when powered and controlled by an external transmission coil. The feasibility of this technology has been demonstrated and fully functional implants are now in design.
- BION3 is programmed and charged by an external coil but can generate stimulation programs autonomously while drawing energy from an internal, rechargeable battery. The enabling technology is now under development.

Collaborating organizations and key personnel have included:

- Queen's University, Kingston, Ontario; G.E. Loeb and F.J.R. Richmond (former address)
- Illinois Institute of Technology, Chicago, IL; P.R. Troyk
- A.E. Mann Foundation for Scientific Research, Santa Clarita, CA; J.H. Schulman
- Advanced Bionics Corp., Sylmar, CA
- Aztech Associates, Inc., Kingston, Ontario

Principal sources of funding have included:

- National Institutes of Health (NIH) Neural Prosthesis Program
- Medical Research Council of Canada
- Canadian Neuroscience Network of Centres of Excellence
- Ontario Rehabilitation Technology Consortium
- A.E. Mann Foundation for Scientific Research

Most of the technical information contained in this report has been described in greater detail in the various Quarterly Progress Reports of NIH Contracts #N01-NS-9-2327, #N01-NS-2-2322, and #N01-NS-5-2325 to the A.E. Mann Foundation during the period 1989–1999.

3.2   OBJECTIVES OF THE PROJECT

3.2.1   Ease of Deployment and Use

There have been many attempts to use noninvasive interfaces with nerves and muscles based on skin-surface electrodes (discussed elsewhere in this volume) and, more recently, magnetic fields.[1-3] While such approaches avoid the expense and morbidity of surgery, they impose significant daily burdens on the patient, who must carefully don and maintain the interface and manage various external leads and connections in order to achieve the desired function. This burden is particularly onerous for patients with the very sensorimotor disabilities that the interface is supposed to address. Such interfaces often entail a degree of visible intrusiveness that patients find unacceptable cosmetically. At best, these interfaces tend to be poorly selective, generating unwanted sensory and motor effects even when they are applied and functioning optimally.

Surgically implanted electrodes and stimulators can overcome the technical limitations and daily use problems of external stimulation, but the costs are high and may not be justified if clinical utility is limited or uncertain. A large part of the costs associated with implantable medical devices arises from the surgical implantation itself. For many rehabilitative applications of electronic interfaces with muscle, the prescribing and treating physician is not a surgeon. The involvement of a referring surgeon, the scheduling of an operating room and anesthesia support, and the preoperative and postoperative care attendant upon surgical procedures all pose enormous logistical and financial obstacles to the widespread deployment of such rehabilitative measures. Thus, one key objective has been the development of a technology that would combine the reliability and simplicity of use of an implanted interface with the low cost and low morbidity of a nonsurgical approach.

3.2.2   Modular Design

The costs of developing and testing a Class III (long-term implant) medical device have risen enormously over the past 25 years. Although some of the expense may be due to the increasing sophistication of the embedded technologies, much of it reflects the burden of regulatory approval. At the same time, "third party" payers of medical expenses are increasingly reluctant to approve or reimburse novel treatments until they have been proven to produce a clear cost-benefit enhancement in a particular disorder. In order to recoup increasing costs in such an environment, medical device manufacturers concentrate on large, proven applications with a high probability of success and low regulatory, liability, and fiscal risks. Unfortunately, many of the most promising clinical applications for TES or FES have no such precedents.

One of the goals of the present project has been to produce a generic and modular implant technology that can be configured and applied to a wide range of anatomical sites and physiological functions. Generic technologies are possible because all neural signaling uses the same transmission scheme. The biophysics of action potentials are the same whether they occur in muscle fibers, motor neurons, muscle spindles, touch receptors, or special sense organs such as the eye and ear.

A related goal that is particularly important in FES is to be able to assemble and build increasingly complex systems for a single patient over time. Most of the neurological disorders that underlie neuromuscular disabilities tend to change over time as a result of progression of the underlying pathology or adaptation to prior damage. Furthermore, the lifestyles and clinical needs of patients tend to evolve as they come to grips with their disabilities and learn to take advantage of the available rehabilitative technology. Thus, it is highly desirable to be able to add or replace interface channels without resorting to major surgery or endangering the function of the previously implanted components.

3.2.3 LONG-TERM RELIABILITY

The roots of the implantable medical device industry lie in orthopedic and cardiovascular applications in which the majority of recipients are elderly, seriously ill, or both. Perhaps the most notable widespread application of such technology to very young and otherwise healthy individuals is the cochlear implant, now the treatment of choice for children with profound or severe congenital or neonatal hearing loss[4,5] (see Chapter 1). This experience has heightened awareness of the substantial difficulty of designing and building sophisticated medical devices that must survive without direct maintenance for many decades in the hostile environment of the body. The problem is likely to be particularly severe in the peripheral musculoskeletal system, where implanted devices are subject to continuous motion and direct mechanical stress as a consequence of their intended motor function.

3.3 DESIGNING FOR SAFETY AND EFFICACY

3.3.1 LEADLESS INJECTABLE PACKAGE

In most electronic systems, the largest assembly costs and most likely sources of failure tend to be in the physical connections between components. This is particularly so when those connections must function in a pool of warm salt water that is constantly moving.[6] These considerations led us to design a system based on individual modules with the following key properties:

- Small enough to be located directly at the site where stimulation and/or sensing is required, with self-contained electrodes;
- Form factor suitable for injection through a hypodermic needle;
- Powered by inductive coupling of energy from an externally generated magnetic field;
- Capable of receiving and transmitting data by modulated radiofrequency (RF) telemetry;
- Digitally encoded device address and stimulus parameters.

The color figure shows the package for the BION1 stimulation module, which consists of a cylindrical glass capsule with two rigidly mounted electrodes on its ends. The electronic subassembly in the capsule is dominated physically by the antenna coil, which consists of about 200 turns of 1 mil insulated copper wire wound over a cylindrical ferrite form to maximize the capture of energy from the magnetic field. It is wound so as to be self-resonant at the 2 MHz carrier frequency generated by the external primary coil, which uses Class E amplification and synchronous modulation to achieve high magnetic field strength with high electrical efficiency.[7] The cylindrical ferrite is actually a sandwich of two hemicylindrical ferrites glued to a custom integrated circuit (IC) mounted on an alumina ceramic printed circuit board (PCB), as shown in the center portion of Figure 3.1. This arrangement maximizes the surface area available for the IC and provides a stable platform for its wire bonds plus a discrete diode chip and soldered termination of the coil windings.

The PCBs are built in wafers of 16, using two-sided, gold-plated through-hole metallization on an alumina wafer. Before metallization, the wafer is laser-drilled in a pattern that becomes the edge-conductive detents at the ends of the PCBs after the wafer is diced. One end of the wafer makes electromechanical contact with a platinum-iridium washer welded to the tantalum stem of the tantalum electrode. The other end makes contact to a gold-plated elgiloy spring that provides an electromechanical connection to the iridium electrode at the other end, via a hollow tantalum feedthrough (Figure 3.1, right). The electronics are sealed into the capsule by sliding them into an open-ended capsule, which is formed onto the Ta electrode (Figure 3.1, left), and welding the tubular feedthrough (Figure 3.1, right) to the glass capillary walls of the capsule.

3.3.2 Achieving and Demonstrating Hermeticity

The most important requirement for the package is to protect the electronic circuitry from the deleterious effects of water. Sophisticated electronic circuitry such as the tuned RF power and data receiver and the digitally controlled stimulus pulse generator are particularly vulnerable to condensed moisture. Failure mechanisms include detuning of the self-resonant receiver coil, poisoning of semiconductor junctions from solubilized contaminants, and shorting of circuitry as a result of corrosion and dendrite formation between conductors with voltage differences.[6] At body temperature, water vapor will condense when it reaches 6% concentration (dew point) in the free space of the capsule, which is about 50% of the total inside volume of the glass capsule (0.01 cc).

BION™ Glass Package Assembly Tree:

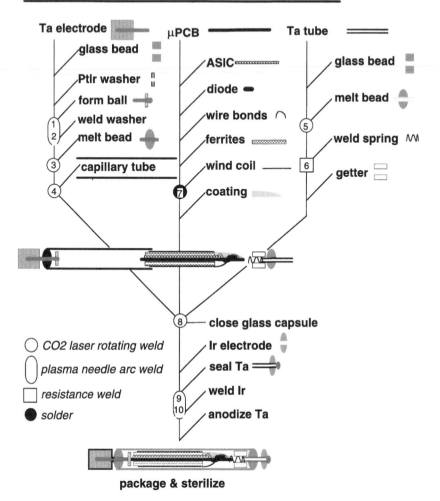

FIGURE 3.1 The fabrication sequence of the BION1 implant consists of three major subassemblies (top, left to right): capsule subassembly built onto the Ta electrode, electronic subassembly stacked onto the ceramic micro-printed circuit board (μPCB), and feedthrough subassembly built onto a Ta tube. After the subassemblies are brought together, the final fabrication process includes hermeticity testing and vacuum bake-out through the still-open Ta tube followed by final closure of the Ta tube and attachment of the Ir electrode. The last process is anodization of the Ta electrode, which is accomplished by probing the Ir electrode and passing current backward through the output circuit with the Ta electrode dipped in 1% phosphoric acid. Numbers refer to ten critical joints accomplished with four different technologies as shown in the key at the lower left. The glass-to-metal (3, 5) and glass-to-glass (4, 8) seals are made by melting the glass in an infrared laser beam that provides well-controlled and very localized heating. The metal-to-metal joints (1–2, 9–10) are made in an argon plasma needle arc, which provides the intense but localized heat needed to melt Ta (2700°C) and Ir (2400°C). A resistance weld is used to attach the spring to the Ta tube without loss of temper (6) and the fine copper coil windings are terminated by microsoldering to the μPCB (7).

There are well-developed sealing techniques and test methods for implantable electronic devices such as pacemakers and cochlear implants, but these devices are much larger than BIONs (see nomogram in Figure 3.2). The various test methods differ in their sensitivity, which is expressed as a gas leakage rate in cc/s for one atmosphere of pressure gradient. The most sensitive method is the "helium sniffer" in which a high vacuum is applied to one side of the seal to be tested and helium gas is squirted over the other side of the seal. Trace helium gas leaking through the seals is detected by a specially tuned mass spectrometer in the vacuum line. The practical limit of this test in the laboratory is about 2×10^{-11} cc atm/s; the sensitivity level for production testing is usually derated to 1×10^{-9} cc atm/s. The equivalent leak rate for water vapor would be about $\frac{1}{2}$ that of helium (it is related to the square root of the molecular weight of the gas). Thus it is possible to compute the length of time it would take for water vapor to reach the dew point in a capsule of a given volume if there were a leak in one of the seals at a rate just below the detection limit of the test method. For the tiny BION, hermeticity testing at even the laboratory limit could not guarantee a functional life of more than one year.

The problem of guaranteeing sufficient moisture resistance for each manufactured BION has been addressed by incorporating a "getter" that absorbs water vapor. It consists of anhydrous magnesium sulfate powder molded into a small cylinder of silicone elastomer that fits over the spring inside the BION capsule (see Chapter 3, Color Figure 1* and Figure 3.1). Magnesium sulfate is one of a class of salts that can be used as desiccants because its crystal structure tends to bind water molecules in its interstices. Fully hydrated magnesium sulfate holds up to 8 moles of water per mole of salt. Even the small amount of magnesium sulfate in the BION getter (10% of the package volume containing 2.3 µg = 2×10^{-5} moles) can absorb a very large volume of water vapor (>1 cc). In effect, the internal volume of the package for water vapor is about 100 times larger than the volume of the BION package, resulting in a projected 30-yr survival time even if the package is leaking at just below the sensitivity limit for production testing (10^{-9} cc atm/s).

Extensive qualification testing of the completed BIONs and of individual components of the seals indicates that the seals are, in fact, hermetic to a degree far beyond the sensitivity of laboratory leak-testing. BIONs built without a water getter have been soaked in saline at 160 atm pressure for over four months without evidence of detectable condensation when cooled. Other BIONs without getters have been operated for over a year with continuous output pulsing in saline while cycling between 37°C (3 h) and 77°C (9 h). In another test, BION capsules were fitted with a bare PCB in place of the usual electronic subassembly, resulting in an open circuit between the adjacent traces of metal on the PCB connected to the output electrodes. High impedance spectrograms at room temperature were compared at various intervals during one year of soaking in saline at 85°C, with no detectable change (this method is sensitive to a layer of condensed water vapor only 10 molecules thick).

The hermeticity of the glass-to-metal seals in the BION probably depends on chemical bonding between the borosilicate glass (Kimbel N51A) and the native oxide on the tantalum metal of the Ta electrode stem and the Ta tubular feedthrough.

---

* Chapter 3, Color Figure 1 follows page 112.

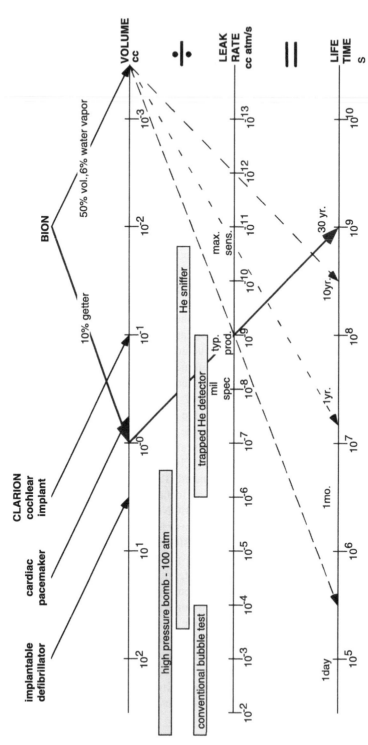

FIGURE 3.2 Nomogram to determine the minimal guaranteed life time (s) of an implanted device from its volume (cc) divided by the minimal leak rate (cc atm/s) that can be detected during testing of its hermetic seals, as shown by log scales. Volume: Product names of various commercial implants are positioned over scale to indicate total package volume; black solid arrows indicate the volume of water vapor required to reach dew point at 37°C (6%) assuming 50% free volume. Leak Rate: Assumes 1 atm pressure head for test and use conditions. The equivalent sensitivity ranges of various conventional test methods are shown in the boxes above and parallel to the leak rate scale. Life Time: Straight lines from a given volume through a leak-rate detection limit intersect the life time scale to show the minimal reliable working life of the device; log time scale calibrated in calendar time (above) and seconds (below).

In order to prevent excess oxidation, the glass-to-metal seals are made by melting the glass using a CO_2 infrared laser under an argon gas curtain (Model F48-2-28W, Synrad). The hermeticity of these seals can be lost if excess residual oxygen or longitudinal grooves are left in the Ta metal from the drawing process. Earlier versions of the BION package used a tubular feedthrough of 90%Pt–10%Ir, which also produced seals that tested hermetic but tended to fail catastrophically during prolonged soaking and temperature cycling in saline because of differences in the coefficient of thermal expansion between the glass ($5.5 \times 10^{-6}/°C$) and PtIr ($8.7 \times 10^{-6}/°C$). Excess residual stress in the walls of the sealed glass capsules was measured using the photoelastic effect on the rotation of polarized light (Model 33 Polarimeter, Polarmetrics, Inc., Hillsborough, NH). By contrast, the coefficient of Ta is $6.5 \times 10^{-6}/°C$. Even low amounts of residual stress, however, can result in catastrophic package failure if there are "stress-risers" in the package, such as scratches or irregularly melted regions of the glass capillary. Fortunately, these failures can be greatly accelerated by soaking in saline pressurized to 160 atm. At this pressure, the dipolar water molecules insinuate themselves rapidly into the glass defect, resulting in a propagating crack that leads to catastrophic failure of the package almost immediately.

Each seal in the BION package is tested for hermeticity and robustness at several points in the fabrication process. Each open capsule subassembly is mounted in an O-ring fixture to perform a He sniffer test of the bead-to-stem and capillary-to-bead seals. Each closed capsule is tested for hermeticity of both glass-to-glass and glass-to-Ta seals by fixturing on the open Ta tube in the He sniffer (Model ASM 110 Turbo, Alcatel). Any residual moisture in the capsule is removed by vacuum bake-out and back-filling with an inert gas mix containing 25% He. The final seal is made by melting the Ta tube closed in a plasma needle arc welder (Ultima 150, Thermal Dynamics, West Lebanon, NH) (Figure 3.1). Slow leaks in the final seal are screened by trying to detect this captured He escaping from the Ir electrode end of the package. Finally, all finished BIONs are bombed in saline at 160 atm pressure for 24 h and examined for visible cracks or condensed moisture before being cleaned and sterilized.

The mechanical integrity of the BION package and seals has also been tested in several destructive qualification tests. The glass capsule is most susceptible to breakage by three-point bending over its long axis; it will fail catastrophically if forces greater than 2 kg are applied at rigid fixation points. The maximal stresses that the BION can experience *in situ* are limited, however, by the structural strength of the soft muscle tissue in which it is intended to be implanted (about 30 N/cm^2 even in tetanically contracting fast-twitch muscle[8]). Using a combination of static and dynamic load testing and instrumentation of the capsules with strain gauges, we have demonstrated that the capsule has a safety margin of at least 4:1 once it is surrounded by at least 1 cm thick muscle. We have also been unable to induce failures of the hermetic seals when loaded axially to 1.3 kg force, which is over four times the maximal insertion force. BIONs have also been subjected to multiple episodes of random severe handling such as might occur prior to implantation, including dropping bare devices onto a steel instrument tray from a height of 20 cm and dropping packaged devices onto a tile floor from a height of 1 m, all without failure. The same devices have then survived five cycles of autoclaving and freezing without loss of function or hermeticity.

3.3.3 ELECTRODE-TISSUE INTERFACE

Electrical currents in metallic conductors such as the BION circuits and electrodes are carried by free electrons. Electrical currents in biological tissues are carried by the motion of ions. There are two basic mechanisms for passing current from the one medium to the other: capacitive "double-layer" charging and electrochemical reactions. Electrochemical reactions occur whenever the voltage across the metal-electrolyte junction reaches the free energy barrier for the reaction in question. For "noble" metals such as platinum, this tends to occur at about ±0.8 VDC, resulting in electrolysis of water, denaturation of proteins, and gradual dissolution of the electrode into heavy metal ions.[6] All of these reactions are irreversible electrochemically, and they tend to generate toxic products that are severely deleterious to the adjacent tissue. Capacitive charging permits alternating currents to be passed indefinitely across the metal–electrolyte junction as long as the charge density (coulombs per square centimeter of electrode surface area) in any single phase of the waveform does not polarize the interface to this ±0.8 VDC threshold. In conventional clinical stimulators, this is usually accomplished by using charge-regulated AC waveforms with symmetrical alternating phases and by incorporating a capacitor between the stimulus generator and the electrodes to block any residual DC leakage. Such circuitry is not practical for the BION.

In order to maximize the power efficiency and safety of the BION within the constraints of its unusually small package, we developed a novel electrode–tissue interface system based on the special properties of tantalum[9] and iridium.[10] The instantaneous power required during a single stimulus pulse at maximal intensity (30 mA through 0.5 kilohms = 0.45 W) is far higher than can be transmitted over the relatively weak inductive link between the external RF coil and the tiny implant coil. The BION takes advantage of the low duty cycle required for muscle stimulation (pulse width <0.5 ms times pulse rate <20 pps yields <1% duty cycle). During interpulse intervals, it stores energy from the RF field on an electrolytic capacitor. However, the BION package has little room available for such a discrete component. Furthermore, a component failure resulting in DC leakage would result in severe electrolysis if the system were powered. Instead, the BION employs a fail-safe electrolytic capacitor consisting of its output electrodes and the body fluids (Chapter 3, Color Figure 2*).

The Ta electrode is made from a sintered powder, resulting in a porous structure with a very high surface area. It is anodized to +70 VDC (four times the maximal operating voltage of +17 VDC) in the last step of BION fabrication, resulting in an electrolytic capacitor with about 4 μF capacitance and less than 1 μA leakage current. The resulting tantalum pentoxide surface is biocompatible and capable of self-healing by reanodization in saline if damaged.[9] The counter-electrode is iridium, which develops a porous, electrically conductive oxide layer in its "activated" state.[10] Iridium exhibits a range of positive valence states with essentially no polarization, allowing each layer of the oxide coating to absorb and release hydroxyl ions from solution while simultaneously shifting electrons out of and into the metal, respectively. Thus,

---

* Chapter 3, Color Figure 2 follows page 112.

it acts like a metal-electrolyte capacitance with an extremely large surface area and with a tendency to dissipate any net polarization by shifting the distribution of valence states among the constituent iridium atoms in the oxide layers.[11]

The normal operating mode of the BION is to charge continuously the output electrodes to the regulated high voltage supply powered by the RF field (+17 VDC Ta vs. Ir) and to generate stimulation pulses by discharging this capacitor through a current-regulated circuit for the desired number of clock cycles. As long as the mean rate of charge dissipation (stimulus current times pulse duration times pulse rate) is lower than the recharging capability, the electrodes stay charged to the full 17 V compliance voltage. This compliance voltage limits the maximal stimulus current that can be generated through a given load, which tends to be dominated by the resistance of the tissue in which the implant is located. This is generally in the range of 500–1000 ohms for a long term implant in muscle.[12,13]

3.3.4 STIMULUS OUTPUT AND RECRUITMENT PROPERTIES

The present BION1 devices receive a 36-bit Manchester-encoded command that contains three 8-bit bytes of data embedded in various formatting and parity bits to avoid erroneous responses.[11] The first byte specifies one of 256 possible addresses. If the address matches the 8-bit value that is specified by the metallization pattern of the IC chip in a particular implant, then the remaining two bytes are decoded to specify a single output pulse emitted by that device at the end of a valid command. The second byte specifies the pulse duration of 2–512 μs in 2 μs steps, and the third byte specifies the pulse current of 0-30 mA in two ranges of 16 each (steps of 0.2 and 2.0 mA, respectively). Two bits of the last byte are also used to specify the constant current (0, 10, 100, 500 μA) that recharges the capacitor electrodes between output pulses, as described above.

Intramuscular stimulation tends to produce a relatively gradual increase in the twitch contractile force of the muscle as the stimulus intensity is increased (Figure 3.3A). This is because stimulus current from the electrodes spreads throughout the volume-conductive muscle. At threshold, the first recruited motor axons are those lying nearest the Ta electrode (the effective cathode) because the current density is highest there. As stimulus intensity is increased, recruitment spreads to more distant axons. There is a general tendency for the largest axons, innervating the fast, fatigable muscle fibers, to be recruited at lower stimulation intensities than the smaller axons, innervating slower, fatigue-resistant muscle fibers. However, recruitment in muscle tends to be dominated by the relative distance between the axons and the stimulation electrodes, resulting in a more gradual recruitment of a mixture of muscle fiber types (Singh et al., in press). By contrast, when stimulation pulses are applied directly to the nerves that innervate the muscles, there is only a narrow range of stimulation strength from threshold for the first recruited largest axon to the last recruited small axon.

The actual recruitment curves for the BION in feline biceps femoris are shown in Figure 3.3B.[14] In this plot and in all of the software used by the clinician, stimulus strength is normalized to a single threshold intensity that is calculated as the product of pulse duration and current, which is stimulus charge. Over the range of about

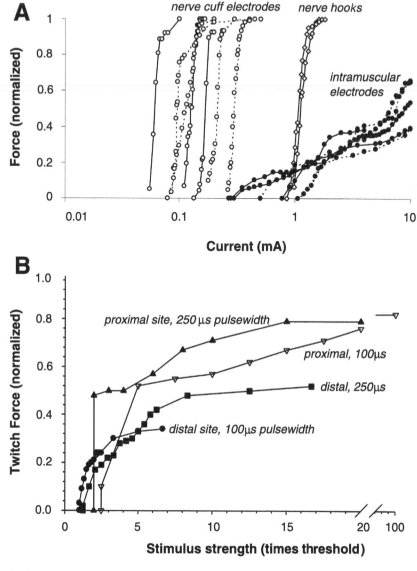

FIGURE 3.3 Recruitment curves for single twitch contractions at various stimulus intensity levels, measured as force normalized to the maximal twitch for 100% recruitment of the muscle. A. Comparison of recruitment of cat medial gastrocnemius by nerve cuff electrodes (open circles) and nerve hook electrodes (open diamonds) on the common sciatic nerve, as opposed to intramuscular electrodes configured as in the BION (solid circles) for acutely implanted (solid lines) and chronically implanted (dashed lines) electrodes (adapted from Singh et al., in press). B. Recruitment of cat posterior biceps muscle by a BION1 implanted acutely in a proximal or distal location in the muscle belly, with stimulus intensity (abscissa) normalized to the threshold for a just palpable twitch at each of two pulse widths (adapted from Cameron et al., 1997). The plateau effects at less than 100% recruitment represent stimulus current levels at which compliance voltage limits were reached for an earlier version of the BION1 that had 8 VDC rather than 17 VDC compliance.

30–300 µs, pulse duration and current are interchangeable as a way to modulate stimulus strength. Thus the ranges of these output variables available in the BION can be used to produce very fine control of recruitment. The details of recruitment depend on the location of the device in the muscle. A stimulator located near the entry point of the nerve into the muscle is able to recruit more of the axons before they fan out in their course toward more distal parts of the muscle.[14]

3.4   PRECLINICAL BIOCOMPATIBILITY TESTING

Before any new medical device can be tested in humans, it must pass a series of tests to demonstrate that the complete device and its constituent materials are unlikely to cause adverse reactions when implanted in the body. The specific tests required depend on the nature and duration of contact with the patient. The BION is a Class III device according to the U.S. Food and Drug Administration (Risk Category III in the Canadian Therapeutic Products Program), which applies to a long-term implant that delivers energy to body tissues but is not in direct contact with circulating blood. The required regulatory testing varies somewhat between jurisdictions, but conforms generally to international guidelines published in ISO standards #10993 and #14971. The more routine testing is available through various commercial contract laboratories.

3.4.1   *In Vitro* Testing

Assessments of device biocompatibility generally start with tests to examine the capacity of the device to elicit cytotoxic and genotoxic effects. To evalute cytotoxicity, we exposed L-929 mouse fibroblast cells to extracts made by soaking devices in minimal essential medium for 1–3 days. These tests showed no evidence of cell lysis or toxicity compared to negative controls. Three types of genotoxicity testing were conducted to evaluate the capacity of leachable extracts to cause damage to DNA, chromosomes, or genes. These tests included *Salmonella typhimurium* reverse mutation tests[15] and the chromosomal aberration tests and sister chromatid exchange assay.[16] No evidence of mutagenicity was seen.

3.4.2   Short-Term Tissue Exposure

To test the biological reactivity of BIONs, tests of acute toxicity, sensitization, and irritation were carried out in experimental animals. Acute toxicity was evaluated by injecting saline or cottonseed oil extracts from soaked devices into mice, then observing them over a 1–3 day period for evidence of toxicity compared to animals injected with control materials. Potential to produce irritation was evaluated by intracutaneous injection of saline and cottonseed oil extracts into rabbits. Injection sites of experimental and control extracts were examined daily for erythema and edema for 3 days. Sensitization was studied by injecting similar experimental and control extracts intradermally in guinea pigs. All device extracts produced reactions similar to negative controls.

3.4.3   Long-Term Passive Implantation

A particularly salient test of biological reactivity depends on replicating the conditions under which devices will perform over a longer term in an appropriate experimental animal model. Thus, in a series of nine cats, devices were implanted in paraspinal muscles around lumbar vertebrae and in tibialis anterior muscles of the hindlimb for survival periods of one, six, and thirteen months. Different sites in the paraspinal muscles of the same animals were also implanted with negative control articles, including USP-recommended polyethylene rods, silicone rods custom-fabricated to reproduce the form of the tested devices, and BIONs whose glass package was sheathed in thin-walled silicone tubing (0.1 mm wall thickness). Each test and control article was inserted into a small pocket made in the muscle with the tips of fine scissors. The insertion sites of the devices were closed and marked with 6-0 multistranded polybutylate-coated polyester nonresorbable suture (Ethicon Ethibond). This suture also provided a test article against which the bioreactivity of the tested devices could be compared (Figures 3.4 and 3.5).

After all tested survival times, implant sites were evaluated histologically in fixed sections stained using haematoxylin and eosin, and in frozen sections reacted for adenosine triphosphatase activity (Figure 3.4). None of the implanted articles were found to change the enzyme reactivity of surrounding muscle fibers. No obvious difference could be found in the degree of inflammatory reaction around the devices and the negative control articles when the implant sites were evaluated according to their contents of inflammatory cells of different types and the extent of fibrotic and necrotic change using scales similar to those described elsewhere.[17] In addition, the thicknesses of capsules around the implants were measured and found to be similar around devices and control articles (Figure 3.5). In the first few months, implants were surrounded by a loosely structured connective tissue capsule about 100–500 microns thick that contained many inflammatory cells. In longer-term implants of 6–13 months, capsules were much thinner and were composed of flattened connective tissues with few associated inflammatory cells. No obvious difference was observed between reactions around the devices at six and thirteen months, suggesting that the use of a six-month survival period may be adequate to establish a stable interface between the device and the biological tissue. Interestingly, the reactions around suture material appeared to be greater than those around devices and control articles; capsules were somewhat thicker and less organized, and inflammatory cells could be identified in close approximation to strands of suture even at the longest survival times. The somewhat heightened reaction might have resulted because the braided material of the suture was more difficult to encapsulate cleanly by a muscle scar.

FIGURE 3.4 Histological appearance of the connective tissue capsules around long-term implants removed from cat muscles. A. Encapsulation around silicone rod (6 month survival). Note the dense inner capsule and loose connective tissue around the implant site. B. Fibrosis in interstices between muscle fibers close to implant site (13 month survival). C. Capsule around non-coated BION (6 month survival). D. Capsule around non-coated BION (13 month survival); tissue stained with ATPase to show reactivities typical of normal muscle fibers.

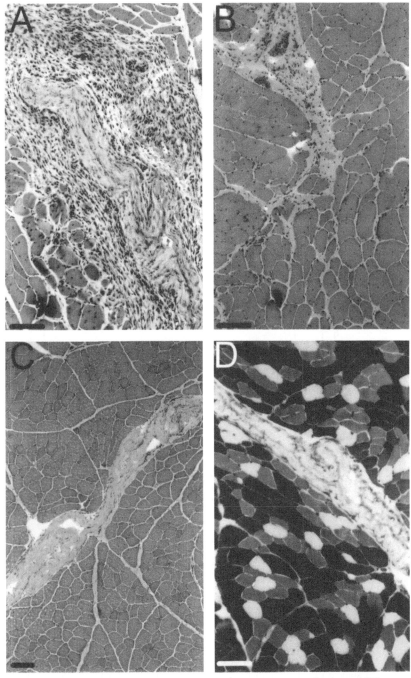

bar = 100 μm

FIGURE 3.4

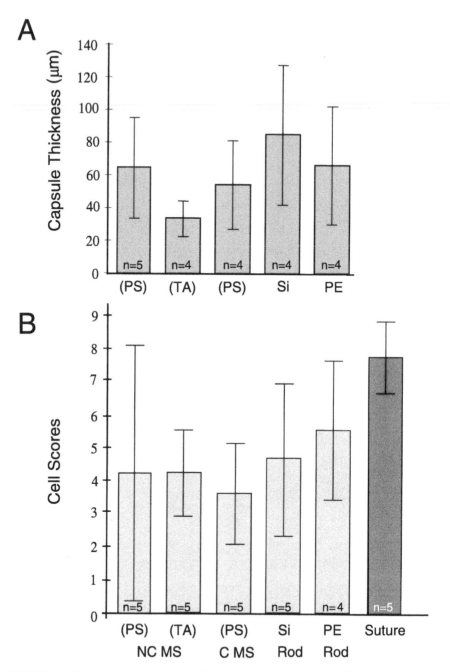

FIGURE 3.5 Quantitative comparison of foreign body reactions around long-term implants (means with standard deviations in n cats) showing capsule thickness in A (corrected for collapse after removal of the implants) and inflammatory cell scores in B using a method modified from Black.[17] Implant code: NC MS = non-coated BION1 implant, C MS = BION1

Previous literature has suggested that the form factor of devices may contribute significantly to the nature and thickness of the tissue response around implanted materials.[18,19] In this study, however, the invaginated profiles of the devices did not seem to stimulate the production of thicker capsules than those around the smoothly contoured silicone rods. With time, the devices with invaginated surfaces around the electrodes appeared to integrate quite securely into the muscle tissue. At the histological level, streams of thickened perimysial tissue appeared to emanate from the ends of the capsules around the devices into the interstices of surrounding muscle fascicles (Figure 3.4B). We hypothesize that this web of connective tissue helps to reduce the mechanical shear between the rigid device and the surrounding muscle during muscle contractions and serves to improve the tolerance of the implanted foreign body by the muscle. In three instances, silicone or polyethylene rods were found to have migrated from their sites of implant. The thin, rather sharp ends of the polyethylene rods may have encouraged the penetration of the rod through the muscle tissue. The propensity of the glass capsules to migrate was evaluated in a separate study in cats implanted with BION-sized, smooth glass capsules. The initial implantation sites were permanently stained with a fluorescent marker that remained colocalized with the connective tissue capsules surrounding the glass capsules at the end of three months of normal physical activity.[20]

3.4.4   Chronic Active Testing

The aforementioned study was carried out using passive devices. However, when BIONs are used for their intended therapeutic purpose, they emit electrical pulses to stimulate motor axons and produce muscle contraction. To test whether the active devices would have greater foreign-body responses than passive implants, an additional series of four cats was implanted with active BIONs in four muscles of one hindlimb and passive BIONs in the other.[12] The active devices were used to stimulate muscles for two hours each day, five days per week for three months. The biological reactions to the implanted devices were similar for active and passive implants, according to the quantitative outcome measures used in the passive trials described above. Stimulation was carried out in fully alert, behaving cats so that their reactions to the stimulation could be observed. Typically, the cats tolerated the stimulation well. They generally ignored evoked movements of their limb and slept through much of the stimulation period. In only one animal was a negative reaction observed when stimulus currents were increased. This device had low stimulus thresholds and was retrieved from a site very close to the parent nerve, where small movements of the electrode with respect to the nerve might have great impact on the threshold of the stimulation and the recruitment of sensory fibers in the parent nerve. Such a

---

FIGURE 3.5 (continued) with a 75 μm thick sheath of silicone rubber tubing over the glass capsule, Si Rod = medical silicone in the size and shape of the BION1, PE Rod = standard USP negative control article, suture = Ethibond™ 6-0 multistranded polybutylate-coated polyester. Site codes: (PS) = paraspinal muscle, (TA) = tibialis anterior muscle.

reaction was not seen in other animals in which devices presumably were placed at a greater distance from the parent nerve.

In two cats active devices were also placed in the muscles sartorius and semi-tendinosus, which are known to be composed of muscle fibers that are shorter than the muscle fascicles and are arranged in series within the fascicles. After a few weeks of stimulation, one animal began to lick the skin overlying the semitendinosus muscle in the periods between stimulation trials. This animal was sacrificed within a few days of the onset of the licking. Histopathological analysis of semitendinosus suggested a modest degree of damage to a small region of muscle tissue extending for several mm beyond the end of the device. The circumscribed region was marked by a modest increase in fibrosis of the connective tissues. Some muscle fibers appeared shrunken and had centrally placed nuclei.[12] We speculate that this small region of muscle damage may be similar to a common type of exercise-related injury (sometimes called a "hamstring pull injury") that has been reported in long, strap muscles of exercising humans.[21]

3.5 CLINICAL APPLICATIONS AND THEIR TECHNOLOGICAL CHALLENGES

The three classes of treatment described at the beginning of this chapter (TES, NMS, and FES) each include many specific clinical applications for which BION technology in its present or future form might be suitable. The selection of a particular application requires a major commitment of resources to the development of external hardware and software, the design and conduct of a clinical trial, and the regulatory process for investigational device approval and eventual commercialization of a product if the trial is successful. We examined twenty different applications according to a cost-benefit analysis based on such factors as the amount of research required to demonstrate efficacy, the clinical value of the treatment if successful, the size of the market, and the likelihood of reimbursement.[22] The following is a technical analysis of two of the more promising applications to emerge from the cost-benefit analysis.

3.5.1 TES for Shoulder Subluxation in Stroke

The presently available BION1 implants provide accurate and reproducible muscle stimulation, but they do not provide detection of volitional command signals or sensory feedback to permit fine control (see below). They also require a transmitting coil to be in close proximity to the implants during stimulation. The present version of the RF power and control system consumes approximately 2 W of power and is designed to be powered by an AC/DC converter plugged into a conventional receptacle rather than a portable battery. Thus the BION1 system is most suitable for TES applications in which a sedentary patient uses the system to exercise and strengthen muscles for relatively brief periods each day.

For the initial clinical testing of the new BION technology, we selected an application in which it is relatively easy to implant the devices, assess their functionality, and achieve a clear and quantitatively measurable clinical outcome in a reasonably short period of time. This is the prevention of shoulder subluxation in stroke patients by TES-induced muscle exercise. The most common residual defect of a stroke is paralysis of the contralateral arm, with atrophy of the affected muscles. The majority of these patients go on to develop chronic pain in the affected shoulder as the weight of the arm gradually stretches the flaccid muscles and pulls the humeral head out of its socket (the glenoid fossa).[23] Conventional treatments with slings and painkillers is generally ineffective and tends to interfere with any rehabilitation effort. Electrical stimulation of the affected muscles via skin surface electrodes (so-called deep transcutaneous electrical neuromuscular stimulation) has been reported to prevent and reverse this problem,[24] but each treatment session requires a trip to the clinic to apply the electrodes and adjust the stimulation parameters.

Figure 3.6 shows the clinical system for use of BION implants to prevent shoulder subluxation. One or two BIONs are implanted into each of the two key muscles: the middle head of the deltoid and the supraspinatus. The location can be determined accurately before each BION is implanted by delivering stimulation pulses from a conventional stimulator through the modified trochar of the insertion tool. The insertion tool is based on a conventional 12 gauge Angiocath™ needle, consisting of a plastic sheath about 10 cm long over a sharp, removable trochar for penetrating skin and fascia. By adjusting the stimulation intensity as the target muscle is palpated, it is possible to identify that the tip of the insertion tool is within the target muscle and close enough to its innervation zone to permit reasonably strong recruitment. The trochar is then removed while holding the sheath in place and the BION is slipped down the sheath and extruded from the tip by pushing a blunt plunger while withdrawing the sheath.

After allowing the BIONs to settle *in situ* for a few days, the clinician determines the threshold for muscle activation at each site and develops an exercise program based on a sequence of ramped and overlapping trains of stimulation at the various sites. A software package has been written in Visual Basic 5 to provide an intuitive graphical user interface for the clinician. It automatically tracks all of the clinically relevant data in a time-stamped relational database, including the serial numbers, addresses, and locations of each implant; the electrical thresholds over time; the various exercise programs devised for the patient's use; and the compliance of the patient in using the system at home. The exercise programs are downloaded from the clinician's personal computer into a microcontroller-based portable device (called a Personal Trainer™) that the patient operates simply by hitting start and stop buttons. The elapsed usage is uploaded into the database when the patient returns to the clinic with the Personal Trainer for testing and reprogramming as needed.

Several outcome measures have been validated in preclinical testing or adopted from the clinical literature on shoulder subluxation. The most direct demonstration of efficacy is the prevention of the loss of muscle mass that normally accompanies disuse atrophy. Imaging techniques such as magnetic resonance (MR) and x-ray computed tomography (CT) have been used to measure the cross-sectional area of individual muscles,[25] but they are lengthy and expensive procedures to perform repeatedly. Measurement of diameter from well-positioned ultrasound scans has

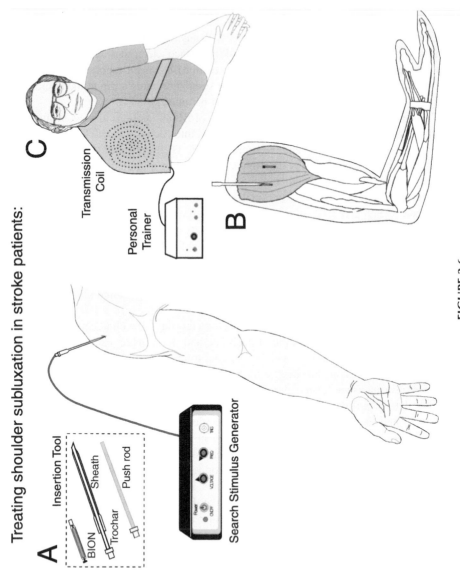

Treating shoulder subluxation in stroke patients:

A Insertion Tool

BION
Trochar
Sheath
Push rod

Search Stimulus Generator

B

C

Transmission Coil

Personal Trainer

FIGURE 3.6

been validated by comparison with MR in the same normal subjects.[29] Such scans will be used to measure atrophy by comparison with the unaffected arm. Other important outcome measures include distance of subluxation as measured by palpation and oblique radiographs, and range of passive motion without pain, which declines rapidly as subluxation develops.

At the time of this writing, two stroke patients have been implanted with two BIONs and are receiving regular TES to build up atrophic deltoid and supraspinatus muscles. The study will eventually include 30 subjects in a randomized, prospective, controlled, cross-over paradigm.

3.5.2 FES FOR ASSISTED GRASP IN STROKE AND SPINAL CORD INJURY

BION2 technology will provide outgoing telemetry of data from various modalities of sensing in order to achieve FES. The sensing modalities that appear immediately feasible include bioelectrical signal recording from the existing electrodes, range-finding between implants (based on implants quantifying the strength of the outgoing RF transmission of another implant), and acceleration (based on incorporating a microelectromachined silicon [MEMS] sensor in the implant package).

The BION2 incorporates a novel power and data transmission scheme called "suspended carrier" that reduces the power requirement about 5-fold from the BION1 amplitude-modulated 2-MHz carrier. The power oscillator can be switched completely off and on again within two carrier cycles and with minimal power loss by electronically opening the tank circuit at the instant when all of the energy in the circuit is in the form of charge stored on its tuning capacitor (Troyk, Heetderks and Loeb, U.S. Patent #5,649,970, July 22, 1997). All of the outgoing telemetry and some of the low-level sensing will occur during periods of carrier suspension.

Figure 3.7 illustrates a typical scheme for using multiple BION2 implants to achieve the relatively simple clinical FES function of assisting a patient with weak voluntary grasp. This application has the advantage that all of the muscles that must be monitored and stimulated are located in the forearm, where BIONs can be easily powered and controlled by a transmission coil embedded in the sleeve of a garment. A somewhat simpler scheme has been tested clinically, using transcutaneous stimulation of digit flexors triggered by mechanical monitoring of voluntary wrist flexion.[26]

Functional grasp of objects includes several distinct postural strategies, each requiring the coordinated recruitment of many muscles.[27] The example chosen is a relatively crude palmar prehension task in which an object is captured by curling

---

FIGURE 3.6 A. Insertion of BION1 implants into supraspinatus (not illustrated) and middle deltoid muscle using an insertion tool that permits application of search stimuli to identify the correct site. B. Extrusion of BION1 implant into desired site in middle deltoid. C. Daily self-administered TES session requires donning of transmission coil configured similar to a heating pad and push-button operation of Personal Trainer™ into which clinician has downloaded one or more customized exercise programs.

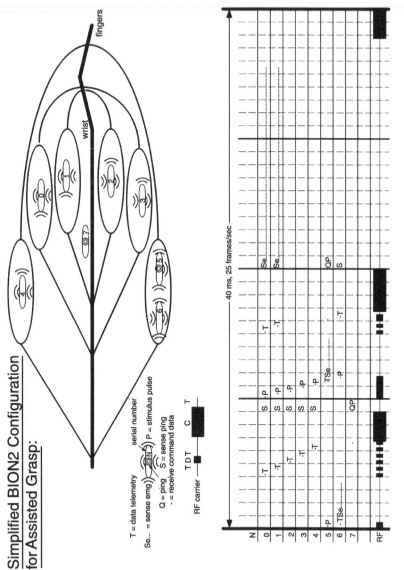

FIGURE 3.7

the fingers around the object and forcing it into the palm. Many patients with strokes and cervical spinal cord injury can initiate the first phase of such a grasp, which is to stabilize the wrist in extension so that the subsequent contraction of the long finger flexors is not dissipated by their flexion action at the wrist. In Figure 3.7, BIONS #0 and #1 in the two wrist extensors detect the onset of this voluntary activity by monitoring their electromyographical (EMG) signals. The external controller then coordinates the electrical stimulation of the other wrist and digit muscles, monitoring the resulting hand motion by measuring the distances between the various implants as the muscles change length. Changes in the shape of a contracting muscle may cause changes in the recruitment properties of an intramuscular stimulator and in the force-generating capabilities of stimulated motor units. In the example, BIONS #5 and #6 in the critical long digit flexor muscle each monitor the M-wave produced by the stimulation from the other implant as well as the distance between the two implants. The external controller can then compensate for such effects in order to maintain the desired level of grip force.

Many other FES applications can be imagined using different combinations of stimulation and sensing (see Chapters 2 and 5 for additional examples). All are likely to require some form of sensory feedback to replace the reflexive control of the spinal circuitry for normal voluntary behaviors and some form of command signal detection based on mechanical or electrical sensing of a part of the neuromuscular apparatus that remains under voluntary control. As the peripheral interfaces available through BION technology become more sophisticated, the limiting factor in performance will shift first to the control algorithms for using the sensory feedback and then to the integration of the sensory and motor components with the cognitive and volitional functions of the brain itself. These challenges are covered in other chapters in this volume.

FIGURE 3.7 BION2 implants will be capable of various sensing modalities as well as generating stimulation pulses. As illustrated in the icon key, a given implant can be commanded to sense bioelectrical signals such as EMG for a programmable sensing interval and programmable gain. While the power RF carrier is turned off (state T), the implant telemeters out a digitized value corresponding to the area of the EMG waveform. For position sensing, a given implant can emit a standardized RF burst (ping) while other implants record and digitize the local field strength of the ping for later outgoing telemetry. The RF carrier can exist in one of two on-states: D for continuous modulation to encode command data addressed to individual implants and C for continuous unmodulated carrier, which maximizes field strength to precharge the power storage capacitors in the implants to function during subsequent carrier-off periods. Data sent to each implant include synchronization codes to permit various implants to begin sensing and sending data back upon detection of particular carrier states. The state diagram at the bottom shows the sequence of functions in each implant for one typical frame of control, with frames repeated at 25 frames/sec. Each implant can be sent a command in 160 to 270 µs, depending on the number of new parameters required for its next function. Back-telemetry of one 8-bit data byte (plus formatting bits) from one implant requires 256 µs.

REFERENCES

1. Galloway, N. T. M., El-Galley, R. E. S., Sand, P. K., Appell, R. A., Russell, H. W., and Carlan, S. J., Extracorporeal magnetic innervation therapy for stress urinary incontinence, *Urology,* 53, 1108, 1999.
2. Ishikawa, N., Suda, S., Sasaki, T., Yamanishi, T., Hosaka, H., Yasuda, K., and Ito, H., Development of a non-invasive treatment system for urinary incontinence using a functional continuous magnetic stimulator (FCMS), *Medical and Biological Engineering and Computing,* 36, 704, 1998.
3. Lin, V. W., Hsieh, C., Hsiao, I. N., and Canfield, J., Functional magnetic stimulation of expiratory muscles: a noninvasive and new method for restoring cough, *Journal of Applied Physiology,* 84, 1144, 1998.
4. Osberger, M. J., Current issues in cochlear implants in children, *The Hearing Review,* 4, 29, 1997.
5. Severens, J. L., Brokx, J. P. L., and van den Broek, P., Cost analysis of cochlear implants in deaf children in the Netherlands, *American Journal of Otology,* 18, 714, 1997.
6. Loeb, G. E., McHardy, J., Kelliher, E. M., Neural prosthesis, in *Biocompatibility in Clinical Practice,* Vol. II, CRC Press, Boca Raton, FL, 1982, 123.
7. Troyk, P. R. and Schwan, M. A., Closed-loop class E transcutaneous power and data link for microimplants, *IEEE Transactions on Biomedical Engineering,* 39, 589, 1992.
8. Brown, I. E., Satoda, T., Richmond, F. J. R., and Loeb, G. E., Feline caudofemoralis muscle: muscle fiber properties, architecture and motor innervation, *Experimental Brain Research,* 121, 76, 1998.
9. Guyton, D. L. and Hambrecht, F. T., Capacitor electrode stimulates nerve or muscle without oxidation-reduction reactions, *Science,* 181, 74, 1973.
10. Robblee, L. S., Lefko, J. L., and Brummer, S. B., Activated Ir: an electrode suitable for reversible charge injection in saline solution, *Journal Electrochemical Society,* 130, 731, 1983.
11. Loeb, G. E., Zamin, C. J., Schulman, J. H., and Troyk, P. R., Injectable microstimulator for functional electrical stimulation, *Medical and Biological Engineering and Computing,* 29, NS13, 1991.
12. Cameron, T., Liinamaa, T. L., Loeb, G. E., and Richmond, F. J. R., Long-term biocompatibility of a miniature stimulator implanted in feline hind limb muscles, *IEEE Transactions on Biomedical Engineering,* 45, 1024, 1998.
13. Cameron, T., Loeb, G. E., Peck, R. A., Schulman, J. H., Strojnik, P., and Troyk, P. R., Micromodular implants to provide electrical stimulation of paralyzed muscles and limbs, *IEEE Transactions on Biomedical Engineering,* 44, 781, 1997.
14. Cameron, T., Richmond, F. J. R., and Loeb, G. E., Effects of regional stimulation using a miniature stimulator implanted in feline posterior biceps femoris, *IEEE Transactions on Biomedical Engineering,* 45, 1036, 1998.
15. Ames, B. N., McCann, J., and Yamasaki, E., Methods for detecting carcinogens and mutagens with salmonella/mammalian-microsome mutagenicity test, *Mutation Research,* 31, 347, 1975.
16. Galloway, S. M., Bloom, A. D., Resnick, M., Margolin, B. H., Nakamura, F., Archer, P., and Zeiger, E., Development of a standard protocol for *in vitro* cytogenetic testing with Chinese hamster ovary cells, *Environmental Mutagens,* 7, 1, 1985.
17. Black, J., *Biological Performance of Materials,* 2nd ed., Markel Dekker, Inc., New York, 1992.

18. Matlaga, B. F., Yasenchak, L. P., and Salthouse, T. N., Tissue response to implanted polymers: the significance of sample shape, *Journal of Biomedical Materials and Research (New York)*, 10, 391, 1976.
19. Salthouse, T. N. and Matlaga, B. F., Some cellular effects related to implant shape and surface, in *Biomaterials in Reconstructive Surgery*, Rubin, L. R., Ed., Mosby, London, 1983, 40.
20. Fitzpatrick, T. L., Liinamaa, T. L., Brown, I. E., Cameron, T., and Richmond, F. J. R., A novel method to identify migration of small implantable devices, *Journal of Long-Term Effects of Medical Implants*, 6, 157, 1996.
21. Oakes, B. W., Hamstring muscle injuries, *Australian Family Physician*, 13, 587, 1984.
22. Loeb, G. E. and Richmond, F. J. R., FES or TES: how to start an industry?, *Proceedings of the 4th Annual Conference of the International Functional Electrical Stimulation Society*, 169, 1999.
23. Caillet, R., *Shoulder Pain*, 3rd ed., F.A. Davis, Philadelphia, 1991.
24. Faghri, P. D., Rodger, M. M., Glaser, R. M., Bors, J. G., Ho, C., and Akuthota, P., The effects of functional electrical stimulation on shoulder subluxation, arm function recovery, and shoulder pain in hemiplegic stroke patients, *Archives of Physical Medicine and Rehabilitation*, 75, 73, 1994.
25. Engstrom, C. M., Loeb, G. E., Reid, J. G., Forrest, W. J., and Avruch, L., Morphometry of the human thigh muscles. A comparison between anatomical sections and computer tomographic and magnetic resonance images, *Journal of Anatomy*, 176, 139, 1991.
26. Prochazka, A., Gauthier, M., Wieler, M., and Kenwell, Z., The bionic glove: an electrical stimulator garment that provides controlled grasp and hand opening in quadriplegia, *Archives of Physical Medicine and Rehabilitation*, 78, 608, 1997.
27. Kilgore, K. L. and Peckham, P. H., Grasp synthesis for upper-extremity FNS. Part 1. Automated method of synthesising the stimulus map, *Medical and Biological Engineering and Computing*, 31, 607, 1993.
28. Singh, K., Richmond, F. J. R., and Loeb, G. E., Recruitment properties of intramuscular and nerve-trunk stimulating electrodes, *IEEE Transactions on Rehabilitation Engineering* (in press).
29. Dupont, A. C., Sauerbrei, E. E., Fenton, P. V., Shragge, P. C., Loeb, G. E., and Richmond, F. J. R., Real-time ultrasound imaging to estimate muscle thickness: A comparison with MRI and CT, *Journal of Clinical Ultrasound* (in press).

4 Intraspinal Microstimulation: Techniques, Perspectives and Prospects for FES

*Simon F. Giszter, Warren M. Grill,
Michel A. Lemay, Vivian K. Mushahwar,
and Arthur Prochazka*

CONTENTS

4.1   INTRODUCTION

4.1.1   GOALS

The goals of this article are to bring together advances made in several laboratories, describe the perspectives of these laboratories and assess the prospects for using intraspinal functional electrical stimulation (FES) as a rehabilitative technique (see Chapters 2 and 3). Three lines of investigation are represented: a more basic research perspective drawing on work in frog and rat and two now long-standing efforts aimed at clinical applications, both using the cat. While we fully acknowledge the clinical and quality-of-life gains for the patient that are likely to arise from intraspinal FES applications to micturition, ventilation or autonomic control, our article focuses on control of limb motions, which we believe is probably the most taxing and difficult FES endeavor.

4.1.2   BACKGROUND

Sherrington[1] demonstrated that the neuronal circuitry of the spinal cord is capable of autonomously generating coordinated muscle activity. Since then, an abundance of evidence has shown that the spinal cord can mediate a number of motor behaviors including standing,[1] locomotion,[1-4] scratching,[5] micturition,[6-8] defecation,[9] and reaching-like limb movements.[10] Additionally, recent evidence indicates that locomotor control circuits are present in the human spinal cord.[11-15] These results suggest that electrical activation of spinal neural networks and artificial control of the behaviors they subserve may simplify the generation of complex motor acts using neural prosthetic devices.

Neural motor prostheses use electrical activation or recording of the nervous system to restore lost functions to individuals with neurological impairment.[16-19] All motor system neural prostheses to date electrically excite the "lower" or segmental motor neuron. They thus access the nervous system at the lowest possible level and act by stimulating the motor axons directly. An unfortunate effect of stimulating motor axons directly is that the normal size-ordered recruitment of motor units is reversed, and large fast-fatiguing twitch units are recruited first, followed by smaller axons.[20,21] However, if recruitment order problems can be overcome, interfacing with the nervous system at the motor neuron level provides the most flexible access to the motor apparatus. It allows arbitrary activation of muscles in arbitrary combinations. However, the cost of this enormous flexibility is that each and every detail of the control must be solved by the FES designer and then implemented by the FES controller. These include the solution of numerous ill-posed problems using the motor apparatus. These problems represent fundamental problems for control of complex behaviors involving the interaction of multiple muscles. However, these same sets of control problems have been solved by evolution and tuned during ontogeny by the individual organism. Viable solutions to the control and selection problems posed by the limbs and trunk are at least in part embedded in the spinal motor apparatus itself. To restore complex motor functions it may be advantageous to access the nervous system at a higher level and use the intact neural circuitry to control the individual elements of the motor system.[22] However, it is also important to acknowledge that the spinal embedded motor functions are most often coarse representations of behaviors such as locomotion or grooming and reaching. Thus while it is conceivable that reaching behaviors might be restored to the upper limb by using FES to access intraspinal "reaching circuits," it is also not unlikely that those finely adjusted details of reach and grasp behaviors which we associate with skilled execution require fractionated control of muscles in the cervical enlargement. Ideally each of these capacities should be restored, together with broader controls of the spinal motor apparatus, such as neuromodulatory state, that normally occur in parallel with voluntary actions in intact individuals.

4.1.3 Perspectives

This chapter combines descriptions of several perspectives and experimental approaches bearing on intraspinal stimulation as an FES technique. In Section 4.2 Giszter examines the recent findings in frog and rat that support existence of a collection of modular primitives in the vertebrate spinal cord that may be described as force-fields. These primitives can be elicited by microstimulation but also underlie natural behaviors. Current understanding of the combination and control of such elements in natural behavior is described. Some new results for microstimulation and for construction of reflex behaviors that impact on FES applications and designs are also described. In Section 4.3 Grill and Lemay describe their detailed analyses of acute microstimulation in cats which show that results in lower vertebrates and rats are applicable to an FES application in larger animals. They explore force-field approaches in this preparation. In Section 4.4 Mushawhar and Prochazka describe the use of implanted microwires applied for use in chronic microstimulation in cats. Issues of pain, stability and spinal cord state in the awake behaving condition are addressed together with the kinds of whole limb movements evoked by intraspinal FES. The techniques that may be needed to gain fractionated control of motor pools are discussed. Section 4.5 discusses prospects and future issues for development of FES and explores some of the more speculative aspects of an intraspinal FES program.

4.2 SPINAL RECRUITMENT OF PRIMITIVES IN FROG AND RAT: BASIC OBSERVATIONS AND IMPLICATIONS FOR INTRASPINAL FES*

4.2.1 BACKGROUND

Several lines of work have suggested that the spinal cord of lower vertebrates and mammals may be organized into different kinds of movement-control modules.[23-29] In 1991, based on microstimulation of frog spinal cord, Bizzi, Giszter and Mussa-Ivaldi suggested that spinal circuits are in part organized into modules that support and control specific force patterns in the limb.[10,30] We proposed that the spinal cord organized a small set of force patterns or "force-field primitives." Related ideas and analyses of modularity of reflexes were proposed separately by Nichols, which led to a different but related line of research.[31] Eight years later, the idea of the force-field primitive in frog and rat has survived several sets of critical tests and corroborating experiments. However, it is still not well understood how these spinal primitives are recruited and combined and how they interact with one another in the construction of reflex and voluntary limb movements. For any intraspinal FES design that aims to control spinal primitives in a manner that emulates the biological control, this is a critical issue. There are good reasons to use intrinsic spinal circuits and modularity. Modularity may provide ease of design, ease of control, and, in the context of motor learning, ease of use and rapid skill development with a neural prosthesis. In this section we first review the framework that was developed based on force-field primitives. We then describe some recent data from our laboratory bearing on how force-field primitives are put to use in real behaviors. Finally we describe some new data more directly related to FES use of primitives involving bilateral simulations.

Spinal microstimulation provided the initial evidence that "premotor" regions in the frog spinal cord may be organized into modules that generate force-field primitives.[10,30,32] A force-field is defined here as a function mapping limb position to endpoint force. Force-fields were constructed in our experiments by measuring isometric forces at the ankle with the limb held in a range of positions. While the stimulating electrode remained fixed in its location in the spinal cord, the same stimulus was delivered at several limb positions. This data collection process allowed data to be represented as a force-field and showed that the intraspinal stimulation resulted in production of a force-field of conserved structure and stability. The force-field could be recruited by a range of stimulus amplitudes by varying the stimulating current, but its structure was preserved at each stimulus amplitude both over the period of activation and on repeated trials.

Force-fields with these properties were found to be consistently recruited by microstimulation of the intermediate grey matter from frog to frog. While the results of microstimulation in the spinal cord were consistent among frogs, it was still not clear what neural processes were being recruited by the stimulation. Since frog motonueuron dendrites are extensive, it was important for us to determine the targets of the intraspinal stimulation to completely understand which neural processes are

---

* By Simon F. Giszter

recruited and the full extent of that recruitment along the length of the spinal cord. Sulforhodamine uptake confirmed that the focal microstimulation eliciting a force-field acted in part through interneurons. Stimulation was calculated to activate cells in a 50–100 micron radius (stimulation parameters were at maximum 10 µA 0.5-ms pulses in a 40-Hz train). However, sulforhodamine uptake indicated that this focal stimulation led to distributed motor pool and muscle activations involving several spinal segments. This could only have occurred by interneuronal relays of excitation. Interneuronal recruitment of motor pools is expected to occur in the more natural size order since it is mediated by synaptic processes, not axonal activation.

Given an ability to recruit muscle assemblies in size order with microstimulation, we were interested in how the CNS or experimenter might use the biomechanical properties organized in this way. We sought to characterize primitives in more detail. Primitives were defined as active force-fields that exhibited invariant force vector directions and magnitude balances over time, i.e., primitives had conserved structures through time.[10] We established the invariant structure using several comparison methods, ultimately using an inner product measure.[32,33]

According to the definition used in Kargo and Giszter,[33] a force-field primitive is observed as a structurally invariant force-field over time. In its most general form, a force-field primitive would thus be a function of both position and velocity expressed in the form

$$F(r, \dot{r}, t) = a_i(t)\phi_i(r, \dot{r}) \tag{4.1}$$

$F(r, \dot{r}, t)$ is the observed field derived from a primitive which can be expressed as a scaling through time, $a_i(t)$, of a fixed field structure $\phi_i(r, \dot{r})$, a function only of position and velocity. The scaling through time represents the waxing and waning recruitment of the primitives during activation by microstimulation or natural reflex processes.

Under isometric conditions, at a single limb position, the force vectors generated by a primitive will increase and decrease in magnitude with the activation-deactivation dynamics of the primitive. At the same time, because there is no movement, these descriptions can be written without velocity terms. If a primitive, as defined here, is recruited by intraspinal microstimulation, the force magnitude among the sampled limb positions remains constant (i.e., forces at a pair of positions with force magnitudes in the ratio 2:1 maintain this ratio through time), and a primitive's time evolution in isometric conditions can be expressed in the form $a_i(t) \cdot \phi_i(r)$.

To test the consistency of isometric data from microstimulation with these equations we examined the direction of the resultant force vectors $F_R(r,t)$ produced at a single limb position and the organization of an entire force. At single locations we measured the variance of the resultant force directions for each frog tested. Microstimulation recruited a synchronously activated group of muscles. Our data showed that the ensemble forces generated by these muscles at the ankle during stimulation exhibit 12° or less variance (see Table 4.1 in Giszter et al.[10]). We took force direction variance measures of less than 12° for an individual frog to support the hypothesis that the resultant forces (i.e., correction response forces) have a fixed direction over time.

For force-field testing and comparison, an inner product measure was calculated between two force-fields (see Mussa-Ivaldi[32]). For example, to compare fields through time at two times, t and $t+m$, a field inner product was obtained as:

$$\langle F_t, F_{t+m} \rangle = \sum_i (F_t(x_i) * F_{t+m}(x_i)) \tag{4.2}$$

where * represents the inner product of two vectors. F_t and F_{t+m} denote the two resultant force vectors that are compared at times t and $t+m$ at N locations $x_1, x_2, \ldots, x_N$.

The cosine of the angle between the two sampled fields is calculated using the following:

$$\cos(F_t, F_{t+m}) = \frac{\langle F_t, F_{t+m} \rangle}{|F_t| \cdot |F_{t+m}|} \tag{4.3}$$

where

$$|F_t| = \langle F_t, F_t \rangle^{1/2} \tag{4.4}$$

represents the norm of a sampled force-field. A value of 1 indicates that the two fields tested are simply scaled versions of one another. In actual application of the test, if the value of the cosine measure from Equation 3 was greater than 0.9 for two test fields, these were considered to have a similar force-field structure, as described in Mussa-Ivaldi et al.[32] A consistent value of greater than 0.9 over time thus indicates similar force-field structure over time. The stability of field structure over time and the possibility of amplitude scaling make the spinally organized biomechanical function represented by the force-field primitive an ideal target for FES. In an FES application a tightly conserved and controlled set of elastic properties can be evoked and scaled by an intraspinal stimulating electrode, which can be used to determine limb position or interaction with the environment.

In addition to observing stable field structures through time, we found that only a few types of primitives were present in any individual frog. Individual types of primitives from this limited set converged to different locations in the limb's work-space. This implied that a set of postures and patterns of stable limb mechanics were represented in the frog's spinal cord and were accessible through microstimulation. Further, the types and the locations of primitives were similar among frogs.[10,30] To better examine this force-field clustering we tested complete force-fields that were recruited by microstimulation and also tested individual limb configurations' force directions at higher sampling densities. Both types of data clustered into a few types. K-means, nearest neighbor and other cluster analyses all confirmed a small (4–6) set of field types. Recently, we have resolved force patterns into component joint torques at hip and knee. We found that the force clusters represent pure hip extension, pure hip flexion, combined hip extension and knee extension, combined knee flexion and hip flexion, and a similar hip/knee flexor combination with a hip elevator component.[34] The total absence of pure knee joint torques was in keeping with the importance of the various roles of double joint hip and knee musculature in limb properties and energy exchanges. These data suggest that to generate isolated knee torques in the frog spinal system using intraspinal stimulation it would be necessary to either (1) directly activate single joint muscles' motor pools or (2) recruit combinations of multi-joint primitives.

4.2.2 SUMMATION OF PRIMITIVES IN MICROSTIMULATION

Either the motions represented and embedded in the spinal cord of frogs as primitives were extremely limited, or the force-field patterns were extensible by some mechanism such as combinations or interactions of primitives. We explored the rules for combining the force-field primitives. We found that the effect of coactivating primitives could be simply described as directly proportional to the linear sum of the individually activated force-fields in 80% of trials.[32] In the remaining 20% of microstimulation trials we observed a winner take all effect, in which only one force-field of the pair was expressed. Nonlinear interactions were notably absent.

Theoretical studies have shown that summation and magnitude scaling of a few force-field types could in principle be used to generate a large range of force-fields. In particular, this set of results and theoretical considerations motivated Mussa-Ivaldi to explore methods of synthesis of arbitrary smooth control fields.[35] He showed that arbitrary smooth fields could be generated by appropriate combinations of nonlinear conservative and rotational basis fields. Conservative fields can do no external work. A closed path through the field conserves energy. In contrast, rotational fields can be used to do external work by choosing appropriate paths through the field. Arbitrary smooth fields could be built as follows:

$$F(r) = \Sigma_i A_i \phi_i(r) + \Sigma_i B_i \psi_i(r) \tag{4.5}$$

where the field $F(r)$ is constructed from conservative primitives $\Phi_i(t)$ and rotational or circulating primitives $\Psi_i(r)$ modulate by the individual scaling parameters A_i and B_I, respectively.

This theoretical framework was particularly interesting for two reasons. First, these studies raised the interesting possibility that descending and segmental systems might produce the range of dynamic force patterns needed for real limb behaviors by recruiting, scaling, and combining the force-field primitives revealed by microstimulation.[35-40] Second, the theory showed that to completely span the range of smooth fields it was essential to include nonconservative fields in the basis set (second term in Equation 5 $\Psi_i(r)$).

The primitives embedded in spinal cord all appeared to be solely conservative. Nonconservative fields were not observed in microstimulation. Thus the repertoire of force patterns that could be constructed from the primitives embedded in spinal cord appeared more limited. Colgate and Hogan had showed previously that the class of conservative and passive systems could be coupled to other passive systems without danger of instability. The set of fields that could be constructed by combinations of frogs' primitives belong to this class. The spinal cord fields observed may represent a compromise between a completely flexible force pattern construction system and the need for stability. These stability properties would also be highly desirable in an FES-based system and its controller.

The prevalence of combination by vector superposition in spinal microstimulation suggested that a simple principled FES design based around force-field primitives recruited by intraspinal FES was feasible and might offer specific stability guarantees.

4.2.3 PRIMITIVES IN SPINAL BEHAVIORS

In an FES design it can be argued that a central goal is the most "natural" replication of behaviors. Any scheme based around recruitment of primitives should therefore be concerned with how they are used in natural behaviors by both segmental and descending mechanisms. These issues are important in order to understand how the intraspinal FES is likely to interact with local reflex controls and with spared descending pathways in an injury.

Initially it was observed that force-fields produced during wiping and withdrawal reflex behaviors in the frog were similar to microstimulation-generated fields.[10] Recent data indicate that reflex behaviors (and sequential phases of these reflex behaviors) do not correspond in a simple one-to-one fashion with a force-field primitive, but instead correspond to a combination of such elements.[33,41,42,43] Tresch et al. applied stimuli to a large number of skin areas. They showed that the range of distinct withdrawal responses that were elicited could be described by different weighted combinations of a common set of muscle synergies. These were similar to the muscle synergies evoked by spinal microstimulation. Similarly, although muscle patterns were not examined, combination of fields by descending controls is suggested by the data of d'Avella and Bizzi.[44]

Our laboratory[33,41,42] has also examined flexion withdrawal but has focused much more on the precisely targeted multiphase wiping reflex. We have now shown that both the formation of the aimed trajectory and the rapid on-line corrections of the trajectory that circumvent obstacles can be described concisely as the summation of primitives. This summation could be shown at both force-field and electromyographic (EMG) levels. Perhaps more remarkable was the observation that the temporal dynamics of activation of muscle groups and primitives that we observed in the correction seemed to be a property that was conserved across the motor program. This property, if confirmed in mammals and in microstimulation-based recruitment of primitives in mammals, might considerably simplify an FES control design. Our data from the frog currently support the idea that all primitives in a spinalized frog can be described as having similar temporal dynamics of force-field amplitude scaling, i.e., a common function $a(t)$.[33] The data suggest any primitive in isometric conditions can be expressed in the form:

$$\theta(r, t) = A_i \cdot a(t)\phi_i(r) \qquad (4.6)$$

where $a(t)$ is similar in all primitives and A_i is an amplitude scaling. A method of construction of a time varying force-field $F(r,t)$ for generation of a behavior then becomes selection of the scalings A_i and the phasing τ_i of the component primitives through the motor pattern.

In this formulation a time varying field $F(r,t)$ is constructed as

$$F(r, t) = \sum_i A_i a(t + \tau i)\phi_i(r) \tag{4.7}$$

Either feedback or feedforward control algorithms could be used to determine the scaling and phasing parameters for constructing the actual force pattern $F(r,t)$. Feedback organization and control of a natural behavior could modulate the strength, balance and sequence of primitives via proprioceptive triggering and gain scheduling or fuzzy logic chaining (see below). Alternatively, feedforward construction of natural behaviors could be achieved through a pattern generator recruiting primitives centrally. These types of algorithms would bear some similarity to the reflex chaining and fuzzy control schemes of Prochazka,[45] or to pulse step models and the work of Mussa-Ivaldi,[37] respectively.

In summary, recent experiments support a decomposition of spinal behaviors into functional modules that encompass (in our estimation) a group of circuits that combine muscle recruitment and feedback systems acting as a unit. These circuits act to stabilize a force pattern and its time evolution. To us, issues of key concern in an FES system based on recruiting and combining such elements are the following:

1. How stable and systematic is the recruitment of these elements within and between frogs (rats, cats, humans)?
2. Is a particular force-field pattern always generated by the same component muscles at all sites in a target spinal cord (1:1 relationship of field to muscles)?
3. How stable are summations of these elements in different conditions?
4. Are winner take all and summation the only observed effects of combinations of sites?

What is the likelihood of stable and unique force/muscle patterns? Many studies suggest that the pattern generating and motor circuitry of the spinal cord may be highly distributed, may be reconfigured "on-line," and may represent use of population coding strategies.[46] However, despite this, it is clear that topographic maps of afferents in frog spinal cord exist[43] and that motor pools have clear topography. Conceivably, microstimulation of different sites in intermediate regions of spinal cord may recruit motor primitives in quite different ways. For example, we might expect differing degrees of recruitment of supporting circuitry and sensory feedback circuits under different conditions. In addition, recruitment of motor primitives may vary in state-dependent ways or the relationships of force patterns to muscle patterns contained in the spinal cord may vary and may not be unique. Although not the primary focus of our laboratory, we have conducted microstimulation experiments that bear on these issues in the frog. Similar issues to the ones we will describe are likely to arise in mammalian spinal cord, as discussed in subsequent sections.

4.2.4   Muscle Maps in Frog: Force/Muscle Relationships

Maps of both forces and muscle recruitment by microstimulation have been collected at fine spatial in unanaesthetized spinal frogs. Force maps are presented in Tresch et al.[43] and Giszter et al.[34] In Giszter et al.[34] we focused on measurement of force samples at a fixed limb configuration and mapped the entire lumbar enlargement of the frog. We chose a limb configuration at which several two-dimensional force-fields characterized in earlier work could be distinguished using a single measurement. This restriction of data collection to single force vectors rather than complete fields was necessary to make the extensive mapping experiments feasible. Full force-field collection at the density of sampling and current variations used here would not have been possible. This is due to the data explosion of at least an order of magnitude that is required for two-dimensional field description. Stimulation sites were arrayed in a three-dimensional grid of 200-micron mediolateral separation, 200-micron depth variation and 1mm rostrocaudal separation throughout the spinal grey. Locations were later visualized using iron deposition stained using Prussian blue. The grid completely spanned the three segments comprising the majority of the lumbar spinal cord in the frog. Each site was stimulated at 1, 2, 4 and 8 microAmps using 0.5-ms pulses in 300-ms trains of 40 Hz.

Forces and EMG patterns recruited across the grid of spinal cord we sampled were examined by generating regular lattices of scalar or vector data using Delaunay tesselation of the space and piecewise linear interpolation. Figure 4.1 shows areas of high force production in one such map. Muscle patterns underlying such field responses have not been presented in detail for these data and are briefly discussed here.

Mechanically antagonistic muscles are often coactivated in the frog reflex behaviors. This coactivation was also observed in most microstimulation. Regions of low force activation that were observed by microstimulation do not therefore always imply a corresponding low muscle activation. We therefore first analyzed EMG magnitudes regardless of force magnitudes.

Chapter 4, Color Figure 1* shows the variation of activation of individual muscles in a single frog, expressed as contours at the parasagittal surface 400 microns from the midline. We examined contours in parasagittal planes at 200, 400 and 600 microns lateral to the midline, but only display the contours for the 400-micron surface. While these planes differed in detail, the 400-micron plane characterizes all three planes for the purpose of these analyses. Peaks of activation of individual muscles appeared to be arranged in repeating or segmentally localized regions. Patterns of activation in dorsal and ventral spinal cord differed. The differences we observed are exemplified by the activation of biceps or iliofibularis muscles (BI). The rostrocaudal pattern of muscle recruitment obtained from the stimulation map altered with depth. Ventral horn regions of cord are expected to represent more direct motoneuron recruitment by stimulation. In some regions of ventral cord, and continuing up to the lower intermediate zone (900–1000 microns), muscles were activated strongly over broad regions probably representing motor pools. In dorsal horn regions of cord we observed that activation could be staggered or more rapidly

---

* Chapter 4, Color Figure 1 follows page 112.

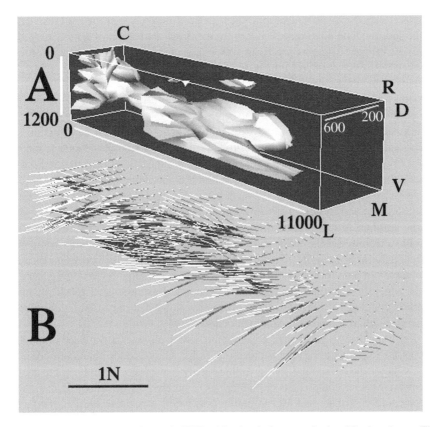

FIGURE 4.1 A. High force sites (>0.67 N) with stimulation magnitude of 8 microAmps. The force magnitude surface drawn contains all stimulation sites >0.67 N. The surface was calculated in the Explorer software package (Numerical Algorithms Group). B. Peak forces at 500 ms after stimulation onset, in lateral view. Force vectors are shown in the figure as grey and white lines in 3D. Each force is drawn radiating from a base located at the point in the spinal cord at which it was recruited by the stimulation. The vector tips are white and the bases grey. Forces in several directions can be seen. The Z axis is upwards and the Y axis is directed along the rostroaudal axis of the spinal cord's coordinate system. (Data re-plotted from Giszter et al. 2000.)

varying in relation to the variations in the motor pools in ventral horns. We believe this variation in upper cord areas represented the repeating segmental organization of the lumbar cord and the dorsal root entry zones. The upper intermediate zone from 400–600 microns frequently showed higher thresholds and lower responses and exhibited a different pattern from corresponding depths in dorsal or ventral horns. This can be seen in the two separate rostrocaudal regions of biceps recruitment by stimulation at this depth. Prior force-field collection was from 600–800 microns depth, usually 300 microns from the midline. This region usually showed more frequent EMG variation rostrocaudally compared to ventral horn or upper intermediate zone but with good muscle activation. It also differed from the dorsal horn. To us, these variations in upper dorsal and intermediate zone and the higher threshold or lower response region separating them make it unlikely that the patterns elicited in the intermediate zone represent simply the direct activation of motoneuron dendrites, or primary afferents processes. While muscle recruitment shows clear topography, points of highest recruitment of individual muscles do not correspond to sites of maximum force. Maps in Chapter 4, Color Figure 2* show the distributions in the stimulated spinal cord volumes of two primitive force directions extracted from the ensemble map data by statistical means using K-means clustering. Thus each region represents recruitment of forces with directions close to the direction of a statistically identified force-field primitive. From comparison of the muscle maps in Color Figure 1 and the maps in Color Figure 2, it appears that the relation between force pattern and muscle activity may be complex and redundant.

We have used factor analysis,[47] K-means,[48] and independent component analysis[49,50] of these force and muscle maps to examine how muscle patterns and forces relate to one another. In this presentation we focus on the results of factor analysis. We sought to discover if the maps show mixtures of statistically separable component muscle patterns at each site, and whether there were sites at which components could be obtained in pure form. Finally we sought to discover if there were sites generating patterning following microstimulation. N-methyl-D-asportic acid (NMDA) iontophoresis coupled with microstimulation studies suggested that stimulation of some sites might elicit temporally patterned activity.[51]

Factor analysis finds a small set of factors which explain the variance of the data. In contrast to principal component analysis, factor analysis does not require the component factors be orthogonal. In the Varimax analysis we used, we chose not to attempt to constrain factor components or weightings to purely positive values. Because each factor could contribute to EMG in both a positive or a negative fashion, the factors extracted could not be considered to represent muscle synergies per se. However the factors could be thought of as pre-motoneuronal excitatory and inhibitory drives.

In this analysis we found that more than 80% of the variance of the EMG patterns throughout the cord could be explained by either five or six factors in each frog. Using the factor weightings we could predict assignment to a force direction cluster. The factor weightings successfully predicted force/torque class in 68% of the force samples collected at 8 microAmp currents in our largest map. This was equivalent to the success of the full rank EMG. Further, we found that force direction and magnitude could be directly predicted by the extracted EMG factors using median least squares linear regression (r-squared of 0.62).

Our results from this analysis suggested that most sites represented mixtures of the factors we obtained. This is in keeping with many stimulation sites in the spinal cord representing activation of a mixture of distributed circuitry associated with one or more primitives. Most sites represented combinations of factors in the intermediate layers in the frog. It remains possible that appropriate sets of constraints on the analysis or rotations of the extracted factors might allow a purely positive set of factors to be used to capture the data, thereby representing "synergies."[43] The EMG activity itself can clearly only be positive (i.e., excitatory to muscle).

---

* Chapter 4, Color Figure 2 follows page 112.

What was remarkable was that despite the mixture of component muscle activities and factor weights at most sites, the maps of force-fields or force directions showed clear segregation into a few types seen in Color Figure 2. Loeb et al.[52] showed in Monte Carlo simulations that some muscle groups generate very robust fields. Thus the force stability in the face of muscle mixtures may be due to the robustness of force-fields generated by some muscle groups. However, it must be acknowledged that Saltiel's work[51] suggests some pure sites may exist.

Obtaining pure recruitment of the primitives (i.e., the biological functional muscle groups and controls) organized naturally by spinal cord is clearly ideal for an FES system. How many such pure sites are accessible remains to be seen. Microstimulation in the frog activates cells near the electrode which are both excitatory and inhibitory but also excites axons and dendrites of remote cells which are passing close by the stimulation site. It therefore may not be surprising that mixtures of effects are observed when the data is examined at the muscle level. However it is also worth noting that mixtures of primitives may represent the spinal cord's normal output when interpreting sensory input and generating flexion withdrawal movements[43] or aimed trajectories for wiping.[33,42]

Possibly the more restricted dendritic arbors of motoneurons and the more clearly laminated organization of mammalian spinal cord will allow for purer recruitment of individual primitives by microstimulation.

4.2.5 SUMMATION VIOLATIONS IN THE FROG SPINAL MICROSTIMULATION OBTAINED WITH BILATERAL INTERACTIONS

Perhaps the most important observation for FES applications that arose from the frog work described above was force-field summation. This is the basis of the ideas presented above regarding motion synthesis, and vector summation of force-field primitives may in the future play an important role in the design of controllers. In all summation tests described thus far the co-stimulation involved unilateral combinations of stimuli. One to four electrodes were placed on the same side of the spinal cord and combinations among them were tested. Many sites when stimulated have bilateral force responses. In the experiments summarized in this section we (Giszter, William Kargo and Michelle Davies) used force sensors on both hindlimbs held in the same standard configuration as in the maps in Section 4.2.4.

In four frogs we tested the effects of bilateral stimulation. We first placed a mapping electrode at each position in a high-density grid on one side of the spinal cord (~600 sites). We stimulated through this one electrode at each location of the grid. We then separately stimulated a single fixed test electrode placed contralaterally. Finally we tested the force elicited when the selected location was stimulated in combination with the single fixed test electrode on the contralateral side of the cord. It should be noted that we did not record force-fields. However, this was unnecessary in order to test force superposition when using the full translational force exerted at the ankle (3 degrees of freedom). In a weighted vector summation the combination of the two vectors must lie in the plane defined by the two individual stimulations' force vectors.

If site A generates vector **a** and site B generates vector **b** then if simultaneous stimulation of A+B causes some arbitrary weighted combination of **a** and **b**, **c**, where:

$$\mathbf{c} = A'\mathbf{a} + B'\mathbf{b} \tag{4.8}$$

then the inner product of $\mathbf{c}$ with the vector product of $\mathbf{a}$ and $\mathbf{b}$ must always be zero. ($\mathbf{c}$ must lie in the plane of $\mathbf{a}$ and $\mathbf{b}$.) Our test therefore examined the value of an inner product measure which indicated the fraction of out-of-plane force components in the dual stimulation condition:

$$R = \frac{c \cdot a^\wedge b}{|c||a^\wedge b|} \tag{4.9}$$

Using the full translation force (3 degrees of freedom) at both ankles, we could examine whether the combined stimulus effects at either ankle had any significant out-of-plane components. We chose as a criterion a value of R representing a 30% or greater out-of-plane component. Values greater than 30% were taken to represent a clear failure of either summation or winner-take-all mechanisms to apply to the co-stimulation. In our experiments with bilateral maps and combinations we found that there were definite areas of the spinal cord in which combined bilateral stimuli violated simple superposition or winner-take-all effects. Examples of these data are presented in Figure 4.2 as a volume map similar to Figures 4.1 and Color Figure 2. It can be seen that the areas with nonlinear results of co-stimulation represent substantial volumes, though linear summation is possible in many regions. Sites of violations of summation also generated significant forces. These violations were not at high threshold/low force sites. We have still to determine whether the sites at which violations of summation occurred represent sites where state changes based on the contralateral stimuli reveal new primitives and muscle groups or whether the new vectors simply represent replacement or combined addition of one more of the original set primitives. Bilateral interactions are likely to play a major role in using intraspinal FES effectively.

FIGURE 4.2 Regions of spinal cord in a frog in which we examined how the maps shown in the preceding figures were modulated or summed with forces induced by contralateral stimulation. In this figure the volumes of spinal cord shown represent cord regions where the joint bilateral stimulation altered behavior. Stimulation (8-microAmps magnitude at each site) of sites shown and a contralateral site generated a change in force direction that could be explained by neither winner-take-all nor simple summation. The volumes were created by thresholding the magnitude of the scalar projection of the force vector elicited by joint stimulation onto a perpendicular to the plane defined by the two individual stimulation force vectors. Threshold used was 30% of normalized magnitude or greater. This map is for the alterations of the ipsilateral leg. A similar type of map of summation and non-summation sites exists for the contralateral leg. Peak forces at 500 ms after stimulation onset are overlaid with regions of change. Force vectors are shown in the figure as grey and white lines in 3D. Each force is drawn radiating from a base located at the point in the spinal cord at which it was recruited by the stimulation. The vector tips are white and the bases grey. Forces in several directions can be seen. The Z axis is upwards and the Y axis is directed along the rostroaudal axis of the spinal cords coordinate system.

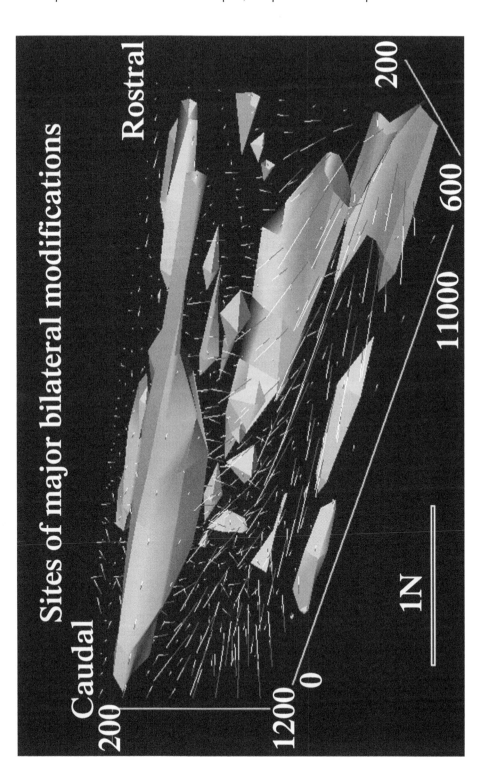

4.2.6 PERSPECTIVE AND CONCLUSIONS

"Primitives" may form a useful description of intraspinal microstimulation results. Designing and controlling behaviors with primitives may provide advantageous in FES. To control the spinal cord precisely through FES is likely to require a concerted effort to fully understand the combination rules and state changes in primitives and their circuits that occur during interaction among sites of stimulation. In addition, in an FES design there may be alterations of behavior as a result of natural reflex behaviors, which may coexist and co-occur during the neural prostheses' imposed controls. The rules and state changes that determine these interactions are likely to involve both ipsilateral and bilateral effects, as well as interactions among cervical and lumbar enlargements.

4.3 HINDLIMB MOTOR RESPONSES EVOKED BY INTRASPINAL MICROSTIMULATION IN THE ANAESTHETIZED AND ACUTELY PREPARED CAT*

4.3.1 BACKGROUND

The long-term goal of our work is to assess the feasibility of motor system neural prostheses based on electrical activation of spinal neural networks. In this section we report on two series of experiments designed to characterize the hindlimb motor responses evoked by intraspinal microstimulation of the lumbar spinal cord in the cat. In the first series of experiments, high-resolution mapping was used to identify the areas of the spinal cord that produce motor responses when stimulated. The second series of experiments was designed to characterize further the hindlimb motor response evoked at the identified locations. For this purpose, measurements were made of the end-point force evoked at the paw at a number of hindlimb positions in the workspace pattern. The results indicate that microstimulation at some regions in the lumbar spinal cord can generate organized, multiple-joint, multiple-muscle motor responses, which are not obtained by direct stimulation of muscles or by intraspinal stimulation of motor neurons.

4.3.2 HIGH RESOLUTION SPATIAL MAPS OF KNEE TORQUE GENERATION

Experiments were conducted to map the hindlimb motor responses evoked by intraspinal microstimulation of the lumbar segments (L5–L7) in neurologically intact, chloralose-anaesthetized adult cats. The isometric torque generated about the knee joint and intramuscular EMGs from knee flexors and extensors were recorded in response to intraspinal stimuli (1s, 20Hz, 5–100 µA, 100 µs) applied with metal microelectrodes. Stimuli were applied in ipsilateral and contralateral segments with a 250 µm mediolateral resolution and a 200 µm dorsoventral resolution to create fine grained maps at half-segment intervals (i.e., L6, L6/L7 boundary, L7, etc.).

---

* By Warren M. Grill and Michel A. Lemay.

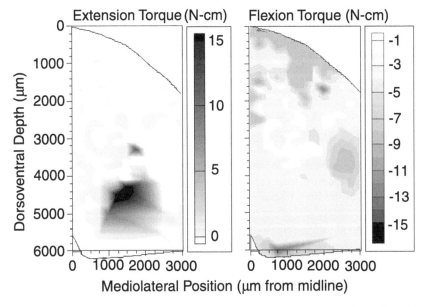

FIGURE 4.3 Isometric knee extension and knee flexion torques evoked by intraspinal micro-stimulation of the ipsilateral L6 spinal cord in a chloralose-anaesthetized cat (50 µA, 100 µs, 20 Hz, 1 s). The ordinate corresponds to the midline of the spinal cord; the bold lines indicate the dorsal surface of the spinal cord and the ventral extent of the mapped region. Torques were mapped with a dorsoventral resolution of 200 µm and a mediolateral resolution of 250 µm, and are plotted as continuous grey-scale values through triangular interpolation.

Torque maps were repeatable across experiments with a strong congruence between the spatial locations where torques were evoked and similar relative magnitudes of flexion and extension torques. Maps of the torques evoked by intraspinal microstimulation revealed a rostrocaudal, mediolateral, and dorsoventral organization of regions of the spinal cord that evoked motor responses about the knee joint (Figure 4.3). At L6 and L7, stimulation in the ipsilateral dorsal aspect of the cord produced strong flexion torques that accommodated rapidly during the 1 s stimulus. Weaker flexion torques were also produced by microstimulation at the lateral aspect of the intermediate region/medial aspect of the white matter. Stimulation in the contralateral dorsal horn produced extension torques that accommodated rapidly. The spatial maps of torques evoked by microstimulation in the ipsilateral ventral horn varied at different rostrocaudal levels. The spatial representation of flexors in the ventral horn increased at more caudal levels as compared to more rostral levels, and there were differences in absolute torque magnitudes. For example, stronger flexion torques were produced by microstimulation over a larger area in L7 than in L6. In the lateral ventral horn, microstimulation produced strong extension torques, while microstimulation at more ventral and medial locations produced flexion torques. Thus, extension torques were produced at locations that were lateral to locations producing flexion torques. EMG recordings indicated that microstimulation in the ipsilateral dorsal, intermediate, and medial ventral locations activated knee flexors selectively, while microstimulation in lateral ventral locations activated knee extensors selectively. However, microstimulation at 50–100 µA did not provide selective activation of individual synergists. From L5 though L7 there was a large area in the intermediate region of the cord which produced no torques (Figure 4.3).

4.3.3 CHARACTERIZATION OF LIMB MOTOR RESPONSE

Fine mapping of the knee joint response evoked by intraspinal microstimulation identified regions of the spinal cord that produce hindlimb motor responses and established surgical and stimulation methods in the cat model, but these results provided only a limited picture of the limb motor response. As demonstrated by Giszter et al.[10] in the frog, measurements of endpoint forces provide a more complete representation of the limb responses to stimulation, and with careful measurements of the limb geometry, still allow recovery of the individual joint torques.

The end-point forces elicited by intraspinal microstimulation (0.5-s train of 40-Hz 100-µA 100-µs biphasic current pulses) of the L5–L7 spinal cord were measured in adult cats either anaesthetized with chloralose or decerebrated. End-point force vectors in the sagittal plane were measured at 9–12 positions of the hindlimb that spanned the range of knee and ankle angles encountered during locomotion. Bifilar wire electrodes were inserted into four hindlimb muscles (knee flexor, knee extensor, ankle flexor and extensor) to record the EMG activity: *biceps femoris, vastus medialis* or *lateralis, tibialis anterior,* and *medial gastrocnemius,* with electrode location verified via post-mortem dissection. The raw EMG signal was amplified, filtered (10–1000 Hz), and sampled at 2500 Hz. The animal's pelvis and femur were held with bone pins, and the paw was attached to a gimbal mounted on a movable six-axis force transducer. The kinematic linkage was thus knee and ankle as opposed to hip and knee as in the previous experiments in the frog. Force-fields were constructed by dividing the workspace into triangles and estimating the force vectors within a triangle by a linear interpolation based on the vectors measured at the vertices (Figure 4.4). We limited our analysis to the sagittal plane since this is the primary plane of limb motion during locomotion. Forces were divided into a passive component (force measured before the onset of activation) and an active component (total forces measured minus the passive portion).

The number of response types elicited via intraspinal microstimulation was limited, similar to what has been found in the frog[10] and rat.[53] With ipsilateral stimulation, force fields (FFs) were of four types (Figure 4.5): flexion withdrawal, rostral flexion, caudal extension, and rostral extension. Flexion withdrawal was the most common field type encountered (33 of 54 fields), especially at superficial depths of stimulation (800 µm), and is probably the traditional flexion withdrawal reflex.[1] Extensor fields were most commonly encountered at ventral depths (5000 µm), and appeared to be the result of direct activation of motoneurons. FFs found at intermediate depths (1500 µm) presented the largest variety.

The characteristics of the FF differed between ventral vs. dorsal and intermediate locations (Figure 4.6). Endpoint forces at dorsal or intermediate locations varied both in direction and magnitude with changes in the limb's configuration. For some stimulation sites, the end-point forces varied so as to produce a FF which converged

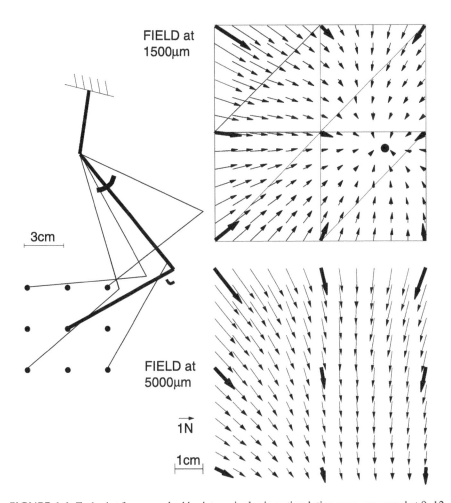

FIGURE 4.4 Endpoint forces evoked by intraspinal microstimulation were measured at 9–12 different positions of the hindlimb shown in the left panel. With the femur held fixed via pins, the knee and ankle were moved through a large range of motion (joint range of motion during walking shown by the thick arcs). Force-fields (FFs) were constructed from measured force vectors (dark arrows) by interpolation (light arrows). Forces represented are the total forces (active and passive) minus the component due to gravity. The FF evoked by stimulation at 1500 μm exhibited a point of convergence in the workspace, where the net endpoint forces were zero, while the FF evoked at 5000 μm did not.

to a point in the workspace where the net active end-point force was zero (Figure 4.4). Electromyographic activity of four hindlimb muscles (*tibialis ant.*, *med. gastroc*, *bicep fem.*, *vastus med.* in the example in Figure 4.4) indicated that the convergent FFs were produced by coactivation of multiple muscles. In contrast, while the magnitude of end-point forces evoked by ventral microstimulation varied with limb position, their directions were largely invariant, and the resulting FFs were parallel or divergent (Figure 4.4). Similar parallel or divergent patterns were evoked by intramuscular stimulation of individual muscles.

ENDPOINT FORCE FIELD TYPES

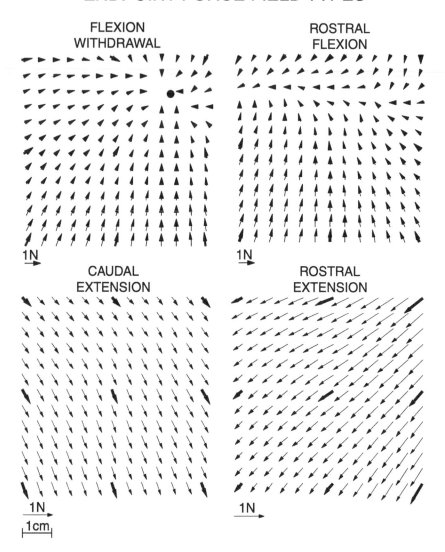

FIGURE 4.5 Endpoint force-field types produced by intraspinal microstimulation. All active responses were of one of the types shown: flexion withdrawal, rostral flexion, caudal extension, and rostral extension. EMG activity indicated that the fields were produced by coactivation of muscles (knee flexor, knee extensor, ankle flexor, and ankle extensor), rather than by activation of a single muscle.

It is well established that the spinal cord is organized with many crossed reflexes and connections. Therefore, we also mapped the spinal cord contralateral to the instrumented limb. The responses evoked by contralateral stimulation were primarily extension responses: caudal (5 of 16 fields) or rostral (7 of 16 fields), and were more

FIELD TYPE DISTRIBUTION BY DEPTH

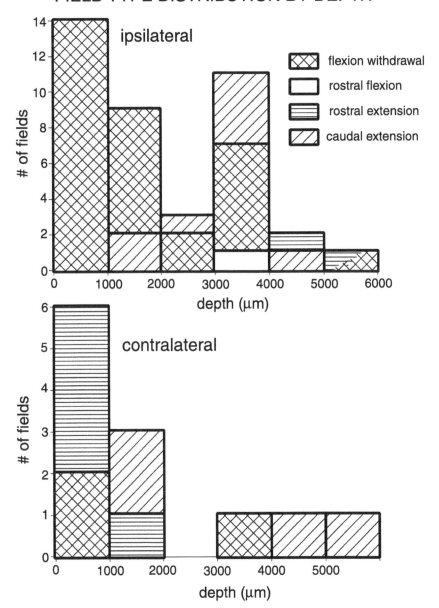

FIGURE 4.6 Field type distribution by depth for intraspinal microstimulation applied to the side of the cord ipsilateral to the limb (top graph), and contralateral to the limb (bottom graph). The height of the bar represents the number of a particular field type found in the depth interval indicated on the horizontal axis. Flexion withdrawal was the most frequent response observed for ipsilateral stimulation, while extensor responses were more frequent for contralateral stimulation.

readily evoked by stimulation in the dorsal aspects of the cord as shown in Figure 4.6, which illustrates the response distribution by depth of penetration. No new types of responses were identified, nor were any contralateral responses classified as rostral flexion (1 of 54 on the ipsilateral side). Interestingly, the contralateral extensor responses exhibited a point of zero active force within the measured workspace. A number of contralateral rostral extension responses showed a point of convergence in the work-space, while none of the flexion responses did. Since the extension responses were evoked when stimulating in the more dorsal aspects of the cord, it is likely that they are related to the crossed-extension reflex and may be due to afferent activation. How-ever, responses evoked in the dorsal aspects of the cord (both ipsi- and contra-lateral) may also be due to activation of interneurons, as Tresch and Bizzi[53] demonstrated that responses evoked by dorsal intraspinal microstimulation are preserved in deafferented rats. The stimulation may be activating the afferent terminals, interneurons, or most likely a combination of both. Based on the results of Tresch and Bizzi,[53] it seems likely that interneurons are part of the circuitry responsible for the force patterns.

4.3.4 PERSPECTIVE AND CONCLUSIONS

The results of the high-resolution mapping studies identified regions of the lumbar spinal cord that generate hindlimb motor responses when electrically stimulated. The maps demonstrated spatial segregation of responses in the ventral horn that correlated well with previous anatomical tracing studies.[54,55] The endpoint force measurements revealed that intraspinal microstimulation can generate organized, multiple-joint, multiple-muscle motor responses, which are not obtained by direct stimulation of muscles or by intraspinal stimulation of motor neurons. Stimulation in the dorsal aspect of the cord generated ipsilateral flexion and contralateral exten-sion. These results suggest that electrical microstimulation can generate two classical spinal reflexes: flexion withdrawal and crossed extension.

Organized motor behaviors evoked by microstimulation support the hypothesis that intraspinal microstimulation and activation of spinal interneuronal circuitry may simplify neural prosthetic control of complex motor behaviors.[22] Control of end-point behavior via activation of individual muscles is possible, but the complexity of the hardware and software necessary to perform the task increases significantly, and systems must be devised to coordinate the action of the controllers around each joint. The spinal circuitry coordinates the activation of multiple limb muscles, and this coactivation gives the end-point an inherent stability that is not affected by either controller delay loops or muscle activation time.

Since this work was conducted in spinal intact animals, further research is necessary to establish the effects of cellular and network plasticity on the evoked motor responses. The results of Tresch et al. in rats[53] indicate that responses are evoked at lower stimulation thresholds in chronically spinalized animals, but the patterns of responses are remarkably similar to those we measured in the cat. Possible differences in the structure of the fields between intact and spinalized animals have not been investigated, but it seems likely that the release of the supraspinal influences would affect the threshold, time course, and relative distribution of the fields, rather than their structures. The development of neural prostheses based on the technique of intraspinal microstimulation awaits answers to these questions, as well as further research into the dynamic and control properties of the evoked motor responses.

4.4 HINDLIMB MOTOR RESPONSES INDUCED IN AWAKE BEHAVING CATS*

4.4.1 BACKGROUND

Electrical stimulation of muscles or nerves has long been used for the purpose of restoring motor function following spinal cord injury, head trauma or stroke.[56] Neuroprostheses based on this approach have been reasonably successful especially in the areas of diaphragm pacing[57] and foot-drop correction.[58] But restoring locomotion using this approach has proved to be rather difficult.[18,59] We proposed the use of electrical microstimulation of the lumbosacral spinal cord for improving motor function following injury for several reasons.[60,61] The spinal cord is distant from contracting muscles, so electrodes implanted therein would be subjected to significantly less stress and strain than epimysial or nerve cuff electrodes. The region of the spinal cord containing the cell bodies of motor neurons innervating the lower extremity muscles is relatively small (~5 cm in humans). Therefore, by implanting electrodes in a compact and protected region, most of the muscles in the lower extremities could be accessed. Finally, tapping into the movement control centers in the spinal cord would allow for the activation of intact subsets of locomotor networks. However, for various reasons, it is still somewhat debatable whether spinal cord microstimulation (SCμStim) would generate functional and controllable movements in people with spinal cord injury.

4.4.2 ACUTE MAPPING OF THE LUMBOSACRAL SPINAL CORD

In 1992 the University of Utah Neuroprosthetics Laboratory initiated the investigation of the feasibility of SCμStim for restoring functional movements following spinal cord injury.[60] The long-term objective of the study was to develop and evaluate an electrode array and a stimulator suitable for chronic implantation and stimulation of motor pathways in the mammalian lumbosacral spinal cord. Prior to this, spinal cord segmental and intersegmental networks involved in producing stereotyped and programmed reflexive movements had been extensively studied using methods ranging from intracellular recordings to epidural spinal cord stimulation and systemic administration of locomotor-inducing drugs. Yet, the ability to generate fractionated, controllable movements by electrically stimulating the spinal cord had not been demonstrated. We posited that the generation of fractionated movements is essential for restoring functional mobility in paraplegia since such movements allow for the performance of simple daily acts like rising from a wheelchair, getting in and out of bed and immediately correcting for unanticipated obstacles. Generation of fractionated movements requires that individual muscles, or synergistic muscle groups, are activated in isolation and that the produced forces are graded through a large

---

* By Vivian K. Mushahwar and Arthur Prochazka.

range of stimulus strengths. Furthermore, functional movements require the maintenance of muscle force for relatively long periods without excessive muscle fatigue.

To determine whether selective activation of muscles can be obtained using SCμStim, the lumbosacral portion of the spinal cord was mapped in pentobarbital-anaesthetized adult cats, and locations generating activation within the main knee and ankle flexor and extensor muscles (i.e., hamstrings, quadriceps, tibialis anterior, triceps surae/plantaris) were noted. The maps demonstrated that there were regions within the spinal cord from which single muscles (i.e., tibialis anterior) or synergistic muscle groups (i.e., quadriceps) can be activated in isolation.[60-63] We referred to these regions as "activation pools." Within these pools, muscles are selectively activated with low stimulus strengths (≤40 μA) and the selectivity is maintained with further increases in stimulus amplitude (≥2× threshold). The results therefore affirmed that the overlap between various motoneuronal pools within the lumbar cord is minimal and functional muscle groups could be selectively activated using SCμStim. A recent comprehensive horseradish peroxidase map of motoneuronal pools in the lumbosacral spinal cord further confirmed these findings.[55]

The characteristics of force recruitment using SCμStim were studied by stimulating numerous locations within the activation pool of each of the muscles of interest with varying stimulus strengths and measuring the resulting force. The constructed force recruitment curves demonstrated that smooth and graded control of muscle force could be obtained using SCμStim. Correlations between stimulus strength and duration of force twitches suggested that SCμStim could also produce a near-normal physiological order of motor unit recruitment.[63] This comes in contrast to the mixed or reversed motor unit recruitment order observed in peripheral functional neuromuscular stimulation systems using epimysial or motor point electrodes. Finally, when two electrodes implanted in a given activation pool were stimulated with interleaved pulse trains, muscle fatigue was significantly reduced.[64] Interleaved stimulation mimics the natural asynchronous activation order of motor units.[59,65,66] Muscle fatigue is significantly reduced because fused contractions of whole muscles are produced without having to sustain tetanic contractions in single motor units by stimulating them at high rates.

The findings of the acute studies affirmed that functional limb movements could be produced through SCμStim. They demonstrated that individual muscles or synergistic muscle groups can be selectively activated; that the generated muscle force can be smoothly graded by varying stimulus strength; and that the rate of muscle fatigue can be reduced using simple interleaved stimulation. Therefore, it was concluded that SCμStim may be feasible for restoring functional mobility in paraplegia, and specifications for a high-density electrode array to be implanted in the spinal cord were proposed based on the dimensions of activation pools and the amount of effective stimulus spread.[64] Yet, whether the acute results could be reproduced in awake, chronic preparations remained unclear.

4.4.3 CHRONIC FIXATION TECHNIQUE

Given the delicate nature of the spinal cord and its ability to move within the vertebral column, special attention was given to electrode design and fixation to the vertebrae and dura mater. Floating microwires were used and the techniques of their implantation and stabilization were based on those developed for chronic recordings from single neurons in spinal dorsal roots.[67] Healthy adult cats were anaesthetized with sodium pentobarbital and the hindlimbs and back were shaved. A skin incision was made from the L4 to L6 vertebral spinous processes and the dorsal surface of vertebra L5 was removed to expose spinal cord segments L5 to S1. Six to twelve microwire electrodes contained within a silastic tube and attached to a connector on one end were passed subcutaneously from the back incision to the head. The wires were 30-μm diameter stainless steel or platinum-iridium (80%–20%) with 30–70 μm tip exposure and 10–30 kΩ impedance. They were pre-cut and bent to an appropriate angle and length. The head connector was embedded in dental acrylic secured to the skull by screws and the silastic tube was anchored to the L4 spinous process.[67] The microwires emerging from the tube were tethered to the dura mater with 8/0 ophthalmic sutures and further bonded with cyanoacrylate glue. A 5-cm length of bared multistrand stainless steel wire was used as the indifferent electrode and was placed in the back muscles close to the vertebral column. The microwires were inserted through the dura mater on both sides of the spinal cord, 2 mm from the midline. Their placement was based on maps established in the acute experiments,[62,63] and stimulation through each electrode was used to guide final positioning. Once inserted in the spinal cord (to a depth of 3.5 to 4.5 mm), the epidural portions of the wires lay flat on the dura mater (Figure 4.7). A small piece of plastic thin film, serving as a barrier to the growth of connective tissue in and around the electrodes, was spot-glued over the dura mater and microwires with cyanoacrylate. The back wound was sutured closed in layers and the cats recovered in an intensive care unit for 1 to 2 days. Animals received doses at 12-hour intervals of a strong opioid analgesic (buprenorphine) during the recovery period. Animals were maintained from 2 to 24 weeks following surgery. Twice a week throughout the duration of implantation, the cats were stimulated through individual spinal cord microwires. Visual inspection, palpation and intramuscular EMG recordings were used to detect muscle activation and the generated movements were noted.

4.4.4 TYPES OF MOVEMENTS EVOKED BY SPINAL CORD MICROSTIMULATION

In the chronic experiments, we were faced with two challenges. First, very few electrodes (six per side) were used to target activation pools spanning a 30-mm length of lumbosacral spinal cord. On average, individual motoneuronal pools are <1mm in diameter and ~10 mm in length.[61,62,64] The acute experiments indicated that several electrodes implanted in a single pool and stimulated in a patterned manner produce optimal recruitment of whole muscles while maintaining selectivity of activation, smoothness of force generation and minimal fatigue. Second, in awake, behaving animals, the spinal cord is more excitable than in pentobarbital-anaesthetized preparations. Given the large number of interneurons in the cord ventral horn, it was questionable whether SCμStim could generate coordinated and useful limb movements. Yet, functional movements *were* generated when spinal cord microwires were stimulated in the awake animals (nine implanted to date). Taken collectively, 60% of the implanted electrodes predominantly generated movements across a single

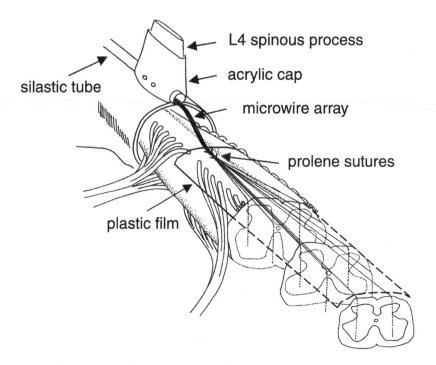

FIGURE 4.7 *Microwire Implantation Technique.* The microwires targeted motoneuron pools located in the ventral horn of the spinal cord. Their epidural portions lay flat on the spinal cord and were tethered to the dura mater with a fine ophthalmic suture. The silastic tube containing the microwires was anchored to the L4 spinous process.

joint throughout the duration of the experiments. Stimulation through 30% of the electrodes generated whole limb synergies (involving hip, ankle and knee) with torques large enough to lift the animals' hindquarters. The final 10% of the implanted electrodes elicited cocontraction of multiple muscles resulting in stiffening of one or two joints without producing any net movement. Figure 4.8 gives examples of the single-joint and whole limb synergistic movements generated with SCμStim in awake animals. In (A) and (B), knee extension and ankle dorsiflexion were generated primarily due to the activation of quadriceps and tibialis anterior, respectively. In some cats, quadriceps contractions generated knee torques up to 1 Nm (an estimated torque of about 0.6 Nm is required to support the hindquarters in stance). In (C), stimulation through a single electrode produced a powerful hip-knee-ankle extensor synergy capable of bearing the weight of the animal's hindquarters. These whole limb synergies may represent the force-field primitives that arise from focal stimulation of premotoneuronal areas in the spinal cord.[10] Several electrodes implanted on both sides of the spinal cord generated synergies similar to the one shown in Figure 4.8C. Therefore, patterned stimulation through as few as two or three electrodes on each side of the spinal cord could, in principle, restore simple functional activities such as maintaining a standing posture in human spinal cord injured subjects.

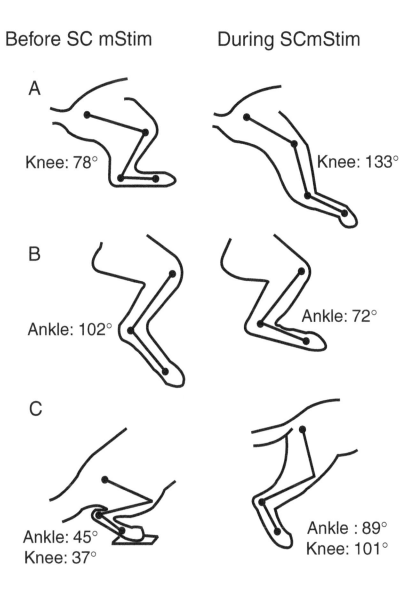

FIGURE 4.8 *Movements Evoked by SCμStim in Awake Animals.* A. An example of knee extension obtained by stimulating through a single electrode in an awake animal one month after implantation (maximum of 144 μA, 300 μs, 50°/s train). B. An example of ankle dorsiflexion elicited by stimulating through a single electrode in an awake animal one month after implantation (maximum of 200 μA, 300 μs, 50°/s train). C. An example of a whole-limb extension synergy generated when stimulating through one electrode in an awake, weight-bearing animal two months after implantation (maximum of 240 μA, 300 μs, 50°/s train). The generated torques were large enough to lift the animal's hindquarters. The examples in A, B and C are from three different animals.

With some electrodes, the characteristics of the generated movements substantially changed as the stimulus pulse amplitude was increased. Starting with single joint movements at low stimulus strengths, more intense stimulation through these electrodes (up to 240 µA) often resulted in a whole-limb extensor synergy, similar to that shown in Figure 4.8C. Though the rostral-caudal location of these electrodes varied along the lumbosacral spinal cord, the extensor synergy seemed to be the "default" synergy elicited with higher stimulus strengths. Such default synergy may reflect the existence of a dominant extensor interneuronal network within the lumbosacral cord, the presence of which provides an advantageous extensor bias needed for postural and antigravity movement control. The default extensor synergy may also be an artifact of electrode location in the ventral portion of the cord. Given that flexor motoneuronal pools tend to have a more dorsal location in the cord than extensor pools, it is possible that these electrodes were positioned deep in the cord and close to ventral rootlets in regions activating extensor muscles.[53,55,61,62] Therefore, the default synergy may be the result of stimulus spread to these rootlets. Closer examination of the relationship between the location of the implanted microwires and the resulting movement synergy is needed.

The high proportion of electrodes producing selective muscle activation in the awake cat reaffirms previous results in acute studies[54,55,61,62] and provides the means for designing flexible control strategies which allow for the generation of novel and functional movements. Combined with the whole limb movements seen when stimulating through one third of the electrodes, a large repertoire of stereotypical and novel movements could be generated using SCµStim.

Across all animals (n = 9), at least two thirds of the implanted electrodes (range: 67 to 100%, mean ± SD: 85.7 ± 13.2%) consistently elicited contractions in the same muscle(s) throughout the period tested. This indicates that, with suitable fixation techniques, microwires implanted in the spinal cord remain securely in place for long periods of time. For all electrodes, stimulus threshold doubled during the first ten days following implantation. This was presumably due to electrode encapsulation, a product of the physiological reaction to implanted foreign objects. Following this initial increase, stimulus thresholds remained, on average, constant throughout the duration of implantation.

SCµStim did not cause any apparent discomfort in the intact animal even when powerful muscle contractions were evoked. This finding is very important for any future clinical application of the work since it implies that pain pathways are not activated with SCµStim, thus removing a potential barrier to clinical implementation and acceptance by patients.

4.4.5   SENSORY INTERACTION WITH SPINAL CORD MICROSTIMULATION

Though, on average, stimulus threshold remained constant after the first ten days of microwire implantation, day-to-day variations in threshold of ± 15 µA were detected.

The oscillations in threshold were found to be dependent on the postural state and activity level of the animal. We observed that in many electrodes, thresholds for activating muscles were lowered by palpating the muscles and their synergists, by imposing joint rotation that stretched the muscles or, in some cases, by lightly touching the skin over and around them. This reduction in threshold is presumably due to the depolarization of motor and premotor neurons by sensory inputs.

The effect of afferent input on stimulus threshold was quantified. In awake animals, implanted microwires were independently activated at stimulus levels producing threshold EMG responses in one of the muscles of interest (e.g., quadriceps), and the activated muscle was stretched and shortened by manually imposed cyclical joint rotation. A gyroscope was used to measure the angular velocity of the imposed movements. At least 50 consecutive 2/s pulses were delivered through each electrode during the imposed movements, and the elicited EMG responses from all electrodes were pooled. EMG records were rectified and activity within a 20-ms window starting 10 ms following a stimulus pulse was integrated and plotted against the instantaneous angular velocity measured 2 ms prior to the stimulus pulse. The mean and standard error of integrated EMG activity within 50°/s bins were calculated and the data were curve-fitted using a power-law relationship. Figure 4.9 shows quadriceps EMG responses obtained during imposed knee flexion in an awake animal. Imposed knee flexion (lengthening of muscle) increased quadriceps EMG responses in a power-law relationship that is consistent with motoneuronal depolarization resulting largely from spindle Ia afferent input.[68]

4.4.6 WIRE CORROSION TESTING

Once the stability of implanted wires and evoked responses was established, it was necessary to find a suitable microwire for long-term implantation and SCμStim in human subjects. A suitable wire was defined as one that would maintain stable impedance for at least 10 years of use.

Accelerated *in vitro* corrosion testing was performed on stainless steel (SS) and platinum-iridium (Pt-Ir) wires with varying diameters and stiffness. The testing was performed on five 30–40 μm 304 SS wires, three 25 μm 90%–10% Pt-Ir wires and four 30 μm 80%–20% Pt-Ir wires. The microwires (insulated except for 30–80 μm at the tip) were placed in separate saline (0.9% NaCl) baths and a multistrand 316 SS wire in each bath served as the reference electrode. Constant-current, biphasic pulses (300-μA amplitude, 100-μs duration) were delivered at a rate of 125–500/s through each of the wires 24 hrs/day and the impedance of the wires was periodically measured.

Figure 4.10 tracks the changes in microwire impedance over time. The *in vitro* stimulation time is expressed in terms of anticipated human use time. This conversion was based on an anticipated human use of 4 hrs/day with stimuli delivered at a rate of 50°/s. The figure shows that four of the five SS wires corroded within the first four days of testing (i.e., within the first three months of human use). The impedance of the fifth SS and all Pt-Ir wires remained constant throughout the three months' duration of testing. These results demonstrate that the Pt-Ir microwires can survive more than the equivalent of 50 years of human stimulation. We therefore conclude that Pt-Ir may be a suitable wire for long-term SCμStim in human subjects.

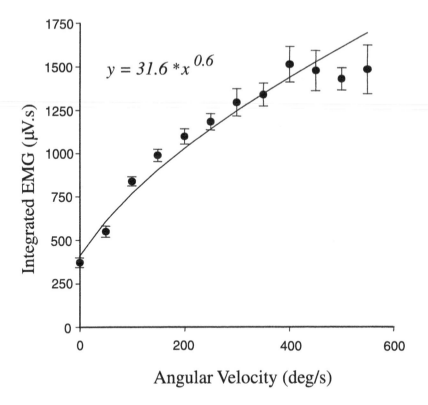

FIGURE 4.9 *Effect of Afferent Input on SCµStim Threshold.* Mean and standard error of integrated quadriceps EMG responses are plotted against angular velocity of imposed knee flexion (divided into 50°/s bins). Responses were obtained from an awake animal in which six spinal cord microwires activating quadriceps were independently stimulated at levels producing threshold EMG responses in the stationary limb. At least 50 consecutive 2/s pulses were delivered through each electrode. Phases of knee flexion (positive angular velocities) were associated with increased quadriceps EMG responses. The fitted curve indicates a power-law relationship with an exponent of 0.6 ($r = 0.98$, $p < 0.0001$).

4.4.7 TISSUE REACTION TO IMPLANTED WIRES AND LONG-TERM STIMULATION

Implanted animals were maintained for up to six months. Throughout this time, no deficits were noticed in activities of daily life. This suggests that the implanted microwires were not physically damaging the spinal cord, and the overall tissue displacement by the electrodes was minimal. Even though the animals actively played and jumped around the laboratory and housing area, the stability of evoked contractions and gross histology suggest that the electrodes remained fixed in place and caused no significant tissue damage. Therefore, it is anticipated that complications due to the fixation of electrodes in the spinal cord of paralyzed individuals who are much less physically active would be minimal. Gross examination of the spinal cord upon termination of the experiment showed mild thickening of the dura

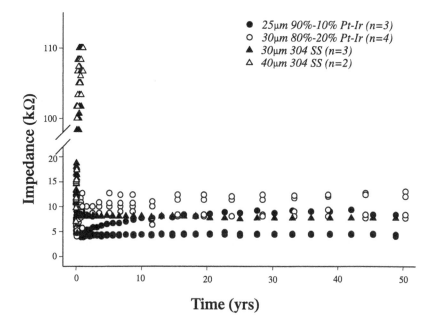

FIGURE 4.10 *Accelerated In Vitro Wire Corrosion Testing.* Microwire impedance is plotted against anticipated duration of human use. Four of the five SS wires corroded within the first three months of human use. The remaining SS wire and all Pt-Ir wires survived more than 50 years of anticipated human use.

mater on the dorsal surface of the cord immediately above the implanted region in two animals. In the remaining seven animals, the spinal cord looked normal. Previous studies of chronically implanted microwires performed by the Huntington Medical Research Institute in Pasadena, CA, indicated that the cellular damage in the spinal cord was limited and localized.[69]

4.4.8 PERSPECTIVE AND CONCLUSION

Our chronic experiments tested the feasibility of SCµStim under the most demanding conditions: intact animals, free to perform normal bodily movements. We were particularly interested in the quality of the movements that could be elicited from within the active, non-anaesthetized spinal cord, in assessing the stability of the evoked movements over time and in determining whether the stimulation caused discomfort.

Our results demonstrated that single-joint and coordinated multi-joint movements can be obtained in the awake or deeply anaesthetized animal by stimulating the spinal cord through implanted microwires. The majority of electrodes generating extensor movements produced torques adequate for weight-bearing. The selective activation of individual muscles (such as tibialis anterior) combined with the whole-limb synergistic movements provide a range of functional movements that could be useful in restoring some mobility in paraplegia. The long-term stability of implanted wires, evoked muscle responses and stimulus thresholds affirm that the intraspinal wires remain securely in place and cause no damage to the spinal cord. The lack of any apparent discomfort experienced by the animals during stimulation of the ventral cord, even when powerful contractions were elicited, indicates that pain pathways are not activated with SCµStim in this region. The results therefore support the idea that SCµStim could be a feasible clinical tool for restoring motor function following spinal cord injury, head trauma or stroke. The suitability of Pt-Ir for long-term microstimulation suggests that human implants may be implemented sooner than originally anticipated.

4.5 DISCUSSION AND CONCLUSIONS

4.5.1 PROSPECTS

Prospects for successful FES systems based on intraspinal implanted electrodes appear reasonable. In experiments ranging across several types of tetrapod vertebrates it seems that similar organizational and functional principles may apply in spinal cord and limb control across species, albeit with variations in detail and complexity. This chapter has described methods for gaining control of the spinal circuits that support movement and also gaining direct control of motor pools. It seems clear from the data presented that recruiting ensemble circuits is possible, obtaining "for free" some of the controls organized in the spinal cord. These ensemble circuits may be matched to the musculoskeletal complex and organized into modules that make sense in the context of tasks that frequently recur. Evolution may have constructed sets of such modules that are pre-adapted for commonly occurring tasks (locomotion, grooming, reaching). An FES control scheme that recruits natural circuit modules in the spinal cord may be better suited to motor learning by a user of FES than a system based around individual muscles alone. Issues of stability, muscle redundancy and order of recruitment might be solved at the outset by the spinal cord rather than posing problems for the user. Users of such intraspinal neuroprostheses might control stimulation patterns of recruitment in spinal cord through a more conventional interface or else via a second implant.[70]

On the basis of the work reviewed and presented here it seems clear that to build a successful and optimum FES intraspinal stimulating system we will need a detailed understanding of the collection of spinal reflex patterns, pattern generators and their integration. Spinal reflex systems interact with one another and will interact with FES recruited circuits. Similarly, in a partial lesion, any spinal elements which FES recruited would have to be integrated with spared descending controls and with fine fractionated movements. This type of data must be an underpinning of intraspinal stimulation research programs.

4.5.2 FUTURE ISSUES AND DIRECTIONS

Several things have been omitted from the presentation in this chapter which bear directly on use of intraspinal stimulation and neural prostheses. Among these are in-depth discussion of integration of the intraspinal neural prostheses with other devices such as cortical implants, neuromodulating pumps or other delivery systems; integration of an FES implant with rehabilitative therapies; and the integration of FES implants with regeneration promoting transplants or gene therapies.

4.5.2.1   Integration With Other Prostheses: Cortical Recording

It seems clear to some that the ideal system of intraspinal stimulation would draw its control signals from a cortical implant (see Chapter 6), which was recording intentional and motor control activity at the "upper motoneuron" or support circuits[70] (see Chapter 8). In effect this would represent a kind of neural bypass. The meshing of cortical and spinal motor representations and enabling appropriate learning at the interface will be critical to such an effort. This must depend in part on how movement and interaction with the environment are represented and controlled in these brain areas. "Primitives," "synergies" or "reflex assemblies" that are embedded in the spinal cord may embody principles of muscle recruitment that take into account the special biomechanical roles of particular muscle architectures.[71] Restoring or reeducating the integration of voluntary, cortical and spinal elements and representations after regeneration or an implanted artificial stimulating/recording bridge will not be a trivial task. In part this may depend on the developmental stage. For example, although young animals develop cortical representations and learn strategies to cope with very severe disabilities,[72,73] it currently appears that adult animals do not. Clearly, the more normal are the properties and controls that are implemented by an FES system, the more facile will be subsequent control and learning by the user of a cortical implant. Intraspinal FES offers the promise of a more intuitive interface for a user who is learning operation of an "FES bypass" type of system.

4.5.2.2   Cord State: Neuromodulators, Inhibition and Specificity of Stimulation

In complete spinal injuries it is very likely that intraspinal FES can be greatly augmented by appropriately restored levels of neuromodulators. Modulation by descending systems is likely to be an important aspect of normal control and motor function.[2,28,46,74,75] Loss of some neuromodulation following injury is one of many factors responsible for the spinal cord's worsened initial state. The effects of neuromodulator and transmitter restoration have been shown in several studies.[28,76] Baclofen is commonly used to reduce spinal spasticity and can be expected to form a useful adjunct to FES in the clinic. Which combinations of neuromodulation may be best to support an FES system, and more specifically an intraspinal implant, are relatively unexplored.

4.5.3   NEW DIRECTIONS: FIBER OPTIC METHODS IN THE FUTURE?

Recently Giszter and colleagues have begun to consider the feasibility of a fiber optic system for delivering stimulation, inhibitory stimuli and neuromodulators in an intraspinal implant arrangement. In the past several years caged compounds have become routinely available. In caged compounds the bioactivity of a "caged" agent is masked by a chemical "cage" that can be removed by various means, thus allowing the agent to act in the target tissue. Light excitation using specific wavelengths including dual photon techniques (depending on the cage agent/structure) is one means of uncaging. Using light for uncaging may allow a novel type of FES device to be constructed. In an FES application we propose using fiber optic delivery of light pulses for uncaging. We believe that fiber optic release of chemically caged transmitters, modulators, and receptor agonists will provide a method of rapid multisite and precisely targeted release of excitatory, inhibitory and modulatory agents. This scheme is particularly attractive because it is likely that when the intact spinal cord is controlled by the brain there are continuously varying focal changes. Neuromodulation, excitation and inhibition are varied at different sites and in different ways depending on the motor task.[77] Rapid switching of light-pulses and delivered wavelength on an implanted fiber, and rapid switching of sources among several implanted fibers in different targets may ultimately allow different patterns of uncaging of multiple compounds. Currently, as a result of communications and other applications, fiber switching technologies and fiber-optic sensors are undergoing rapid and continuous improvement. It is possible that feedback control of local neurotransmitter levels may be feasible. Even in non-FES applications (such as control of spasticity) it is conceivable that these technologies could represent a considerable advance over single intrathecal pumps for baclofen, just as the pump is an advance over oral ingestion. In an FES application of these technologies the limiting issues are likely to be (1) possible caging agent byproduct side effects, (2) excitation wavelength side effects (heat, mutagenesis, protein degradation), (3) the engineering of appropriate cages of appropriate wavelength selectivity, (4) the avoidance of spurious uncaging and release, and finally (5) the long term tissue compatibility and materials issues associated with implanted fibers. However, despite these concerns it is clear that fiber optic uncaging is becoming an established stimulation method[78] and (at least in frogs) focal transmitter delivery at intraspinal sites is effective in recruiting patterned motor responses and "primitives."[51]

4.5.4 CONCLUSIONS

The prognosis for applications of intraspinal FES in the clinic at some future time seems good. Various neuroprosthetic designs are being tested, and a concerted and coherent effort is being applied to this approach to the FES problem by several independent groups.

ACKNOWLEDGMENTS

Simon Giszter acknowledges grants NIH NS34640 and NIH NS09343 for supporting different aspects of the work described in Section 4.2. He would like to thank William Kargo and Michelle Davies for assistance with collection and analysis of bilateral microstimulation data.

Warren Grill and Michel Lemay acknowledge NIH neural prostheses program NIH NS-5-2331, NS-8-2300 for work described in Section 4.3.

Arthur Prochazka and Vivian Mushahwar would like to thank Michel Gauthier and Allan Denington for their design and implementation of the multichannel microstimulator used in the project described in Section 4.4. Their work was funded by the Alberta Paraplegic Foundation and the Alberta Heritage Foundation for Medical Research.

REFERENCES

1. Sherrington, C.S. (1910) Flexion-reflex of the limb, crossed extension reflex and reflex stepping and standing. *J. Physiology.* 40: 28-121.
2. Barbeau, H., S. Rossignol (1987) Recovery of locomotion after chronic spinalization in the adult cat. *Brain Res.* 412:84-95.
3. Lovely, R.G., R.J. Gregor, R.R. Roy (1990) Weight-bearing hindlimb stepping in treadmill exercised adult chronic spinal cats. *Brain Res.* 514:206-218.
4. Gossard, J.-P., H. Hultborn (1991) "The organization of the spinal rhythm generator in locomotion" in *Plasticity of Motoneuronal Connections*, A. Wernig, Ed., Elsevier Science Publishers, pp. 385-404.
5. Berkinblit, M.B., T.G. Deliagina, A.G. Feldman, I.M. Gelfand, G.N. Orlovsky (1978) Generation of scratching I. Activity of spinal interneurons during scratching. *J. Neurophysiol.* 41:1040-1057.
6. Rampal, G., P. Mignard (1975) Behavior of the urethral striated sphincter and of the bladder in the chronic spinal cat. *Pflügers Arch.* 353:33-42.
7. de Groat, W.C., I. Nadelhaft, R.J. Milne, A.M. Booth, C. Morgan, K. Thor (1981) Organization of the sacral parasympathetic reflex pathways to the urinary bladder and large intestine. *J. Autonomic Nervous System* 3:135-160.
8. Shefchyk, S.J., R.R. Buss (1999) Urethral afferent-evoked bladder and sphincter reflexes in decerebrate and acute spinal cats. *Neurosci. Lett.* 20:137-140.
9. de Groat, W.C., J. Krier (1978) The sacral parasympathetic pathway regulating colonic motility and defecation in the cat. *J. Physiol.* 276:481-500.
10. Giszter, S.F., F.A. Mussa-Ivaldi, and E. Bizzi (1993) Convergent force fields organized in the frog's spinal cord. *J. Neuroscience* 13:467-491.
11. Calancie, B., B. Needham-Shropshire, P. Jacobs, K. Willer, G. Zych, B.A. Green (1994) Involuntary stepping after chronic spinal injury. *Brain* 117:1143-1159.
12. Dietz, V., G. Colombo, D.M. Jensen, L. Baumgartner (1995) Locomotor capacity of spinal cord in paraplegic patients. *Ann. Neurol.* 37:574-582.
13. Dimitrijevic, M.R., Y. Gerasimenko, M.M. Pinter (1998) Evidence for a spinal central pattern generator in humans. *Ann. N.Y. Acd. Sci.* 860:360-376.
14. Nicol, D.J., H.G. Malcolm, R.H. Baxendale, and S.J.M. Tuson (1995) Evidence for a human spinal stepping generator. *Brain Res.* 684:230-232.
15. Bussel, B., A. Roby-Brami, O.R. Neris, A. Yakovleff (1996) Evidence for a spinal stepping generator in man. Electrophysiological study. *Acta Neurobiologiae Experimentalis.* 56(1):465-8.
16. Hambrecht, F.T. (1979) Neural prostheses. *Ann. Rev. Biophys. Bioeng.* 8:239-267.
17. Loeb, G.E. (1989) Neural prosthetic interfaces with the nervous system. *Trends in Neuroscience* 12:195-201.
18. Stein, R.B., P.H. Peckham, D.P. Popovic (1992) *Neural Prostheses Replacing Motor Function After Disease or Disability.* Oxford University Press, New York.
19. Grill, W.M., R.F. Kirsch (1999) "Neural Prostheses", in *Encyclopedia of Electrical and Electronics Engineering,* Vol. 14, J.G. Webster, Ed., Wiley & Sons, New York, pp. 339-350.
20. Henneman, E. (1957) Relation between size of neurons and their susceptibility to discharge. *Science* 126:1345-1346.
21. Cope, T.C. and M.J. Pinter (1995) The size principle: Still working after all these years? *News in Physiological Sciences* 10:280-286.

22. Barbeau, H., D.A. McCrea, M.J. O'Donovan, S. Rossignol, W.M. Grill, M.A. Lemay (1999) Tapping into spinal circuits to restore motor function. *Brain Research Reviews* 30:27-51.
23. Alstermark, B. and S. Sasaki (1986) Integration in descending motor pathways controlling the forelimb in the cat 15. Comparison of the projection from excitatory C3-C4 propriospinal neurones to different species of forelimb motoneurones. *Exp. Brain Res.* 63: 543-566.
24. Alstermark, B., A. Lundberg, M. Pinter, and S. Sasaki (1987) Long C3-C5 propriospinal neurones in the cat. *Brain Res.* 404:382-388.
25. Arshavsky, Y.I., G.N. Orlovsky, G.A. Pavlova, and L.B. Popova (1986) Activity of C3-C4 propriospinal neurons during fictitious forelimb locomotion in the cat. *Brain Res.* 263:354-357.
26. Grillner, S. and P. Wallen (1985) Central pattern generators for locomotion, with special reference to vertebrates. *Ann. Rev. Neurosci.* 8: 233-261.
27. Loeb, G.E. (1985) Motoneurone task groups: coping with kinematic heterogeneity. *J. Exp. Biol.* 115:137-46.
28. Rossignol, S. (1996) Neural control of stereotypic limb movements. In *Handbook of Physiology,* sec. 12, L. Rowell, J. Shepherd, Eds., Oxford University Press, NY.
29. Stein, P.S.G. (1983) The vertebrate scratch reflex. *Symp. Soc. Exp. Biol.* 37: 398-403.
30. Bizzi, E., F.A. Mussa-Ivaldi, and S.F. Giszter (1991) Computations underlying the execution of movement:a novel biological perspective. *Science* 253: 287-291.
31. Nichols, T.R. (1994) A biomechanical perspective on spinal mechanisms of coordinated muscular action: an architecture principle. *Acta Anat.* 151(1):1-13.
32. Mussa-Ivaldi, F.A., S.F. Giszter, and E. Bizzi (1995) Linear superposition of primitives in motor control. *Proceedings of National Academy of Sciences.* 91:7534-7538.
33. Kargo, W.J. and S.F. Giszter (2000) Rapid correction of aimed movements by summation of force-field primitives. *J. Neurosci.* 20(1):409-426.
34. Giszter, S.F., E. Loeb, F.A. Mussa-Ivaldi, and E. Bizzi (2000) Repeatable spatial maps of a few force and joint torque patterns elicited by microstimulation applied throughout the lumbar spinal cord of the spinal frog. *Human Movement Science* (in press).
35. Mussa-Ivaldi, F.A. (1992) From basis functions to basis fields: vector field approximation from sparse data. *Biol. Cybern.* 67(6):479-89.
36. Mussa-Ivaldi, F.A. and S.F. Giszter (1992) Vector field approximation: a computational paradigm for motor control and learning. *Biol. Cybern.* 67(6): 491-500.
37. Mussa-Ivaldi, F.A. (1997) Nonlinear force fields: a distributed system of control primitives for representing and learning movements. *Proc. IEEE Int. Symp on Computational Intelligence in Robotics and Automation.* 84-90.
38. Cannon, M. and J.J.E. Slotine (1995) Space frequency localized basis networks for nonlinear estimation and control. *Neurocomputing* 9(3).
39. Massaquoi, S.G. and J.E. Slotine (1996) The intermediate cerebellum may function as a wave-variable processor. *Neuro Letts* 215: 60-4.
40. Mataric, M.J., M.M. Williamson, J. Demiris, and A. Molan (1998) Behavior-based primitives for articulated control. *Proceedings SAB 98,* Zurich, Switzerland.
41. Kargo, W.J. and S.F. Giszter (2000) Afferent roles in hindlimb wipe reflex trajectories: free-limb kinematics and motor patterns *J. Neurophysiol.* (In press.)
42. Giszter, S.F. and W.J. Kargo (2000) Conserved temporal dynamics and vector superposition of primitives in frog wiping reflexes during spontaneous extensor deletions. *Neurocomputing* (In press).
43. Tresch, M.C., P. Saltiel, and E. Bizzi (1999) The construction of movement by the spinal cord. *Nat. Neurosci.* 2(2):162-7.

44. d'Avella, A. and E. Bizzi (1998) Low dimensionality of supraspinally induced force fields. *PNAS* 95(13):7711-4.

45. Prochazka, A. and D. Gillard (1997) Sensory control of locomotion. *Proc. American Control Conf.* 2846-2850.

46. Kiehn, O., J. Hounsgaard, and K. Sillar (1997) Basic building blocks of vertebrate spinal central pattern generators. In *Neurons, Networks, and Motor Behavior,* P. Stein, S. Grillner, A. Selverston, D. Stuart, Eds., MIT Press, Boston, pp. 47-60.

47. Harman, H.H. (1976) *Modern Factor Analysis,* 3rd Ed. University of Chicago Press, Chicago.

48. Hartigan, J.A. and M.A. Wong (1979) A k-means clustering algorithm. *Applied Statistics* 28:100-108.

49. Bell, A.J. and T.J. Sejnowski (1995) An information maximization approach to blind separation and blind deconvolution. *Neural Computation* 7:1004-1034

50. Bell, A.J. and T.J. Sejnowski (1996) Edges are the "independent components" of natural scenes. In *Advances in Neural Information Processing Systems,* 9, MIT Press, Boston.

51. Saltiel, P., M.C. Tresch, E. Bizzi (1998) Spinal cord modular organization and rhythm generation: an NMDA iontophoretic study in the frog. *J. Neurophysiol.* 80(5):2323-39.

52. Loeb, E.P., S.F. Giszter, P. Saltiel, F.A. Mussa-Ivaldi, and E. Bizzi (1999) Output units of motor behavior: an experimental and modeling study. *J Cognitive Neuroscience* 12:1-20.

53. Tresch, M.C. and E. Bizzi (1999) Responses from spinal microstimulation in the chronically spinalized rat and their relationship to spinal systems activated by low threshold cutaneous stimulation. *Experimental Brain Res.* 129:401-416.

54. Romanes, G.J. (1951) The motor cell columns of the lumbo-sacral spinal cord of the cat. *J. Comp. Neurol.* 94:313-358.

55. vanderHorst, V.G. and J.M. Holstege (1997) Organization of lumbosacral motoneuronal cell groups innervating hindlimb, pelvic floor, and axial muscles in the cat. *J. Comp. Neurol.* 382:46-76.

56. Campbell, J.M. (1999) Technology transfer in FES: The clinician's perspective, presented at 4th Annual Conference of the International Functional Electrical Stimulation Society, Sendai, Japan.

57. Baer, G.A., P.P. Talonen, V. Hakkinen, G. Exner, and H. Yrjola (1990) Phrenic nerve stimulation in tetraplegia, *Scand. J. Rehabil. Med.,* 22:107-111.

58. Waters, R.L., D.R. McNeal, W. Faloon, and B. Clifford (1985) Functional electrical stimulation of the peroneal nerve for hemiplegia. Long-term clinical follow-up, *Journal of Bone and Joint Surgery, America,* 67, 792-793.

59. Prochazka, A. (1993) Comparison of natural and artificial control of movement, *IEEE Trans. Rehabil. Eng.,* 1, 7-16.

60. Mushahwar, V.K. and K.W. Horch (1993) Selective activation of functional muscle groups through stimulation of spinal motor pools, presented at Proc. 15th Ann. Intl. Conf. IEEE Eng. Med. Biol. Soc., San Diego, CA.

61. Mushahwar, V.K. and K.W. Horch (2000) Selective activation of muscles in the feline hindlimb through electrical microstimulation of the ventral lumbo-sacral spinal cord, *IEEE Trans. Rehabil. Eng.,* vol. in press.

62. Mushahwar, V.K. and K.W. Horch (1998) Selective activation and graded recruitment of functional muscle groups through spinal cord stimulation, *Ann. NY Acad. Sci.,* 860, 531-5.

63. Mushahwar, V.K. and K. W. Horch (2000) Muscle recruitment through electrical stimulation of the lumbo-sacral spinal cord, *IEEE Trans. Rehabil. Eng.,* vol. in press.

64. Mushahwar, V.K. and K. W. Horch (1997) Proposed specifications for a lumbar spinal cord electrode array for control of lower extremities in paraplegia, *IEEE Trans. Rehabil. Eng.*, 15, 237-243.

65. Rack, P.M.H. and D.R. Westbury (1969) The effects of length and stimulus rate on tension in the isometric cat soleus muscle, *J. Physiol.*, 204, 443-460.

66. Yoshida, K. and K. Horch (1993) Reduced fatigue in electrically stimulated muscle using dual channel intrafascicular electrodes with interleaved stimulation, *Annals of Biomed. Eng.*, 21, 709-714.

67. Prochazka, A. (1984) Chronic techniques for studying neurophysiology of movement in cats, in *Methods for Neuronal Recording in Conscious Animals (IBRO Handbook Series: Methods in the Neurosciences, Vol. 4)*, R. Lemon, Ed., Wiley, New York, pp. 113-128.

68. Prochazka, A. and M. Gorassini (1998) Models of ensemble firing of muscle spindle afferents recorded during normal locomotion in cats, *J. Physiol.* (London), 507, 277-91.

69. Woodford, B.J., R.R. Carter, D. McCreery, L.A. Bullara, and W.F. Agnew (1996) Histopathologic and physiologic effects of chronic implantation of microelectrodes in sacral spinal cord of the cat, *J. Neuropathol. Exp. Neurol.*, 559, 82-91.

70. Chapin, J.K., K.A. Moxon, R.S. Marlowitz, and M.A.L. Nicolelis (1999) Realtime control of a robot arm using simultaneously recorded neurons in the motor cortex. *Nature Neurosci.*, 2:7.

71. Gielen, S. (1993) Muscle activation patterns and joint angle coordination in multijoint movements. In: *Multisensory Control of Movement*, C. Gielen, V. Henn, K.P. Hoffmann, M. Imbert, F. Lacquaniti, A. Rocoux, P. Viviani, and J. Van Gisbergen, Eds., Oxford, pp 293-313.

72. Miya, D., S. Giszter, F. Mori, V. Adipudi, A. Tessler, and M. Murray (1997) Fetal transplants alter the development of function after spinal cord transection in newborn rats. *J. Neurosci.* 17(12):4856-4872.

73. Giszter, S.F., W.J. Kargo, M.R. Davies, and M. Shibayama (1998) Fetal transplants rescue axial muscle representations in M1 cortex of neonatally transected rats that develop weight support. *J Neurophysiol.* 80:3021-3030.

74. Bennett, D.J., H. Hultborn, B. Fedirchk, M. Gorassini (1998) Synaptic activation of plateaus in hindlimb motoneurons of decerebrate cats *J. Neurophysiol.* 80:2023-2037.

75. Bennett, D.J., Hultborn, H., Fedirchk, B., and Gorassini, M. (1998) Short-term plasticity in hindlimb motoneurons of decerebrate cats *J. Neurophysiol.* 80:2038-2045

76. Kim, D., V. Adipudi, M. Shibayama, S.F. Giszter, A. Tessler, M. Murray, and K.J. Simansky (1999) Direct agonists for serotonin receptors enhance locomotor function in rats that received neural transplants after neonatal spinal transection. *J. Neurosci.* 19(1):6213-6224.

77. Noga, B.R., P.A. Fortier, D.J. Kriellaars, X. Dai, G.R. Detillieux, and L.M. Jordan (1995) Field potential mapping of neurons in the lumbar spinal cord activated following stimulation of the mesencephalic locomotor region. *J. Neurosci.* 15:2203-17.

78. Godwin, D.W., D. Che, D.M. O'Mallet, and Q. Zhou (1997) Photostimulation with caged neurotransmitters using fiber optic lightguides. *J. Neurosci. Methods* 73:91-106.

5 How to Use Nerve Cuffs to Stimulate, Record or Modulate Neural Activity

Joaquín Andrés Hoffer and Klaus Kallesøe

CONTENTS

5.1   INTRODUCTION

A nerve cuff is an insulating sleeve that can be placed around a length of nerve without compromising its integrity and used to make external connections to the nerve. New clinical and research applications are enabled by using nerve cuffs as permanent electrochemical interfaces with nerves. Cuffs can be used to selectively activate nerves and muscles, record the electrical signals that travel along nerves, or modulate nerve functions with pharmacological agents. This chapter provides a practical introduction of what is needed to make a permanent, safe and reliable connection to a peripheral nerve, cranial nerve or spinal root.

Cuff electrodes placed on nerves are less likely to disturb the normal function of neurons than microelectrodes chronically implanted near the nerve cell bodies in the spinal cord or the brain (discussed in Chapter 6). This is because cuffs remain outside the blood–brain barrier, far away from the sites where synaptic integration and action potential generation take place within the central nervous system. However, nerves are mechanically delicate structures that often must bend or stretch considerably in the course of normal movement. Nerves can suffer severe damage and may cease to function if subjected to bending torques, compression, or disruptions to their blood supply. Thus, the cuff dimensions, shape and flexibility, the routing and mechanical properties of the leads, and the method of surgical installation are all critically important determinants for the successful functional survival of instrumented nerves.

A nerve cuff that is placed around a nerve provides a simultaneous electrical interface with many, possibly all, of the axons in the nerve. In contrast, intrafascicular electrodes that are inserted in a nerve may contact small groups of adjacent axons within a nerve fascicle. Implanted microelectrodes may some day allow connections with individual axons, but problems of poor signal longevity remain to be solved. Each of these approaches offers distinct advantages and limitations with respect to signal source specificity, noise rejection, signal stability with movement, and the safety, efficacy and stability of the interface over long periods of implantation (reviewed by Hoffer and Haugland,[1] Hoffer et al.[2] and Kallesøe[3]). Of these alternatives, nerve cuffs have emerged as the method of choice for establishing permanent connections with nerves and for recording stable signals indefinitely. By virtue of its geometry, a properly installed nerve cuff cannot migrate and will remain connected to the same neuronal population. First used clinically in the 1960s, stimulating cuffs implanted on phrenic nerves to control breathing have kept individual patients alive for decades.[4] Recording cuffs were introduced as a research tool in the 1970s.[5-8] Their long-term reliability in animals led to the first human trials in the 1990s. Sensory nerve recording cuffs were implanted in paralyzed legs or hands of several volunteers, from one of which signals have now been recorded for nearly five years.[9]

Nerve cuffs share three basic elements: an insulating wall made of biocompatible material, internal electrodes (or fluid ports) connected to insulated lead wires (or catheters), and some method to allow controlled cuff installation and closing. Several evolutions in nerve cuff design and methods of fabrication have taken place. We will summarize the key features found in various types of nerve cuffs and highlight their relative advantages and disadvantages.

Initially, nerve recording cuffs provided a "single channel" of information based on the aggregate potentials generated by all active nerve axons in a nerve. However, for many applications it may be necessary to monitor several distinct signal sources present in a nerve. The same rationale applies to separate stimulation of nerve subpopulations jointly present within a single cuff. This chapter introduces advanced multichannel cuff designs that allow simultaneous monitoring of several nerve signal sources or independent stimulation of several functionally distinct sensory or motor nerve subpopulations that may coexist in a single nerve trunk.

5.2 ANATOMICAL AND SURGICAL CONSIDERATIONS FOR CHRONIC NERVE CUFF IMPLANTS

Nerve cuffs should be implanted in locations far away from joints, where the nerve is not branching or receiving local blood supply and can be safely freed from its surrounding tissues over several cm. The shape, size, thickness, orientation and flexibility of a nerve cuff and its associated leadout cables must be carefully matched to the anatomical site. It is important to emphasize that an inappropriate site selection, incorrect surgical technique or a flawed cuff or lead design is almost guaranteed to cause some form of nerve damage. On the other hand, if basic surgical precautions are followed and the cuff is correctly designed, dimensioned and placed so that the nerve is neither pulled nor torqued by the leadout cables, the long-term prognosis for the nerve after a cuff is implanted is generally excellent.

Immediately following any surgical intervention, two things typically happen. First, all affected tissues, nerves included, undergo some degree of swelling. Second, as fibrocytes converge in regions where surgery was performed, connective tissue layers quickly envelope and penetrate cavities in any foreign objects. As a nerve cuff becomes ensheathed, the mechanical association between cuff and nerve becomes increasingly stable and the likelihood of a displacement of the cuff diminishes. As a bonus, with fibrous tissue replacing fluid inside the cuff over several weeks the longitudinal impedance inside the cuff rises and recorded nerve signal amplitudes can increase significantly.[10] Excessive scar tissue formation outside the cuff, however, is undesirable, as it could adhere the cuff to surrounding structures and reduce nerve mobilization.

To prevent compression neuropathy associated with post-surgical edema,[6,11,12] the cuff lumen must accommodate some nerve swelling. Nerve compression, when it occurs, affects most severely the largest diameter axons in a nerve and causes a reduction in axonal conduction velocities.[13,14] In cat femoral nerves that were implanted for several months with cuffs of inside diameter 20% larger than the nerve diameter, the axonal conduction velocities measured for Group Ia[15,16] as well as alpha-motoneuron axons[17] remained in the normal ranges, supporting the notion that appropriately sized nerve cuffs do not cause compression damage. Quantitative morphological analysis of median nerves of cats implanted for 6–12 months showed that the average myelin thickness in axons located near the outer perimeter of the cuffed nerves was slightly reduced, but the total numbers of axons inside the cuffs were the same as in the contralateral control nerves. No measurable functional deficit could be found in these animals.[18,19]

5.3   HOW TO RECORD NERVE ACTIVITY DURING BEHAVIOR

Nerve recording cuffs are designed to pick up extracellular potentials that propagate along axons. To understand how a cuff works, it is useful to first consider the traditional acute electrophysiological procedure of lifting a section of a nerve into air or mineral oil using a metal hook electrode (Figure 5.1). With the electrode connected to a suitable recording amplifier, a signal can be recorded whenever an action potential propagates along the nerve (Figure 5.1A). The amplitude (V) of the recorded potential is a function of the extracellular action current amplitude, its wavelength (λ) and the length (2d) of the nerve portion that is lifted into air or oil (Figure 5.1B). In analogous fashion, nerve potentials can be recorded if a length of nerve is encompassed by an insulating cuff with electrodes placed inside the cuff (Figure 5.1C). The amplitude of the recorded signal depends non-linearly on the length of the insulated portion (Figure 5.1D). To obtain maximal signal amplitudes the length of a nerve cuff should approximate λ to the extent possible. There is no further advantage to having the cuff length exceed λ. For large myelinated axons, λ ranges between 30 and 40 mm.[20,21]

Beyond these basic tissue geometry considerations, the aggregate electroneurographic (ENG) activity recorded from a nerve depends on the number of active fibers, is usually dominated by the activity of the largest axons and is biased in favor of superficial axons. In 1–4 mm diameter nerves, action potential amplitudes recorded from deep axons can be attenuated 2- or 3-fold.[15,20,22]

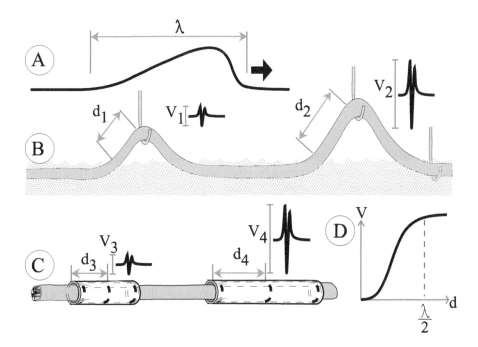

FIGURE 5.1 A: Diagram of nerve action potential wave of wavelength λ traveling in direction of arrow. B: Schematic diagram of two hook electrodes used to lift two segments of a nerve out of saline and into air or mineral oil in a typical acute experiment. The nerve signal amplitudes that are recorded (V) depend non-linearly on the length of nerve between electrode and ground reference (d). C: Chronically implanted cuffs replace the function of mineral oil to provide insulation to segments of nerve. Recorded signals (V) depend on cuff length (2d). D: The nerve signal amplitude (V) depends non-linearly on the length of a cuff (d). Maximal amplitudes can be expected from cuffs having lengths comparable to the wavelength λ of action potential sources.

5.3.1 Cuff Impedance

The total impedance pathway through which extracellular currents must flow determines the voltage signal amplitudes that are resolved with a cuff.[22] Assuming that there are no leaks anywhere along the cuff wall (a condition that, as we will see below, has been difficult to meet), the cuff impedance depends on the longitudinal resistivity of the neural tissue and any fluid present inside the cuff, the length of the cuff and its cross-sectional area.

Cuff electrode impedances are measured using a sinusoidal test signal (usually 1 kHz) and typically range between 2 kΩ and 50 kΩ. It is important to note that a DC meter should *never* be connected to a nerve cuff, because continuous current flow can cause irreversible nerve damage.

Figure 5.1D indicates that shorter cuffs provide smaller signals. In addition, since the cuff impedance is inversely related to its cross-sectional area, larger signal amplitudes can be expected from smaller diameter, snugly fitting cuffs (as long as the nerve is not compressed, of course).

As a rule, adequate signals are recorded during behavioral tasks when the cuff length is about 10 times greater than the cuff inside diameter.[23] For limb nerves, the ENG amplitudes recorded during walking can range from about 5 µV (peak-to-peak) for the cat sciatic nerve using 4-mm I.D., 30-mm long cuffs,[24] to up to 90 µV (peak-to-peak) for the much finer rabbit tenuissimus nerve using 0.3 mm I.D., 5-mm long cuffs.[6]

5.3.2   CUFF SEAL

An essential prerequisite for recording nerve activity is to use an insulating cuff. A porous cuff wall or an incompletely sealed cuff will allow nerve currents to leak out and, additionally, signals generated by structures outside the cuff will leak into the cuff and contaminate the recordings with unwanted noise. Therefore, cuff recording electrodes will not be able to resolve nerve potentials from the noise unless the cuff is well sealed along its entire length.

5.3.2.1   Cuff Amplification Effect

The containment of neural signals by the insulating cuff wall creates an amplification effect that is essential for resolving extraneural potentials. To illustrate how it works, a nerve cuff encompassing a peripheral nerve can be viewed as a tunnel encompassing a noisy road. If a car blows its horn inside the tunnel, the perceived sound is much louder inside the tunnel than the same event taking place outside the tunnel. In a similar manner, the signal intensities that electrodes can detect inside a nerve cuff are much higher than what similar electrodes could detect from the same nerve at locations outside the cuff.

A drawback of this amplification effect is a reduction in the listener's ability to localize the source of the signal. Inside a busy tunnel it is hard to distinguish which car sounded its horn. Similarly, inside a nerve recording cuff, an electrode that detects the ensemble potentials from active axons in the nerve has very limited selectivity for determining exactly which axons contributed to the extraneural potential. For this reason, unless special provisions are taken (discussed below), a traditional nerve recording cuff can only be expected to provide a single channel of information that reflects the aggregate contributions from all the active axons in the nerve.

5.3.3   ELECTRODE CONFIGURATION

Besides the roles of the insulating wall, the quality of recordings obtained with cuffs depends critically on the electrode configuration. Of importance are the number of electrodes, their placement relative to one another, and the size and shape of each electrode.

5.3.3.1   Number and Placement of Electrodes

The simplest recording electrode configuration is to place one conductive lead inside the cuff as the recording electrode and a reference electrode outside the cuff. This "monopolar" configuration can work well in acute experiments with no movement, but not in awake recordings, as it is very susceptible to large noise pickup from electromyographic (EMG) sources as well as other signals originating outside the cuff and movement-related artifacts.

Bipolar electrode configurations help reduce noise pickup. Differential recording between two electrodes placed in relatively close vicinity inside the cuff increases the rejection of common mode noise originating from sources outside the cuff.

A tripolar configuration can further reduce noise pickup. If the cuff wall is truly well sealed any noise from sources external to the cuff can only enter through the ends of the cuff. Placing a reference electrode near each end of the cuff and a central recording electrode midway between the two outer electrodes creates a balanced tripolar configuration. The middle electrode must be symmetrically placed between the outer electrodes in order to maximize the common mode rejection of external noise.[6,7]

The amplitude of externally generated noise that can squeeze into a cuff through the ends depends on the ratio between the parallel internal and external longitudinal impedances. As a further step towards reducing noise pickup, grounding electrodes placed near the ends of the cuff can shunt away much of the extraneural noise.[6]

5.3.3.2   Electrode Geometry

In its simplest form, an electrically conductive surface suitable for recording potentials can just be the stripped end of an insulated wire placed inside the cuff. More complex, two-piece, separate electrode and lead structures can also be employed, although at the expense of increased fabrication complexity and additional failure modes.

Early recording cuffs contained either platinum foil electrodes or deinsulated sections of Pt-iridium wire electrodes that were sewn onto a fairly thick and rigid wall structure made of commercially available silicone tubing.[6,10] The recording electrode was attached along the inner circumference of the cuff and the insulated lead wire was adhered to the outside of the cuff. Starting in the early 1980s, Teflon coated, multi-stranded stainless steel wire (e.g., AS 631, Cooner Wire Co.) was often used instead of Pt-Ir wire for the electrodes and leads. This recording electrode design and method of construction remained basically unchanged for the following 20 years and was used in a wide variety of long-term studies in animals and humans.[1,12,15-17,24-34]

Alternatively, recording cuffs can be fitted with metal foil electrodes that surround the nerve over most of its circumference.[6,35,36] Foil electrodes require a separate lead that must be spot-welded or attached by another method. The joint is often a weak mechanical point. If two dissimilar metals are used, the presence of junction potentials can cause corrosion and further weakening. Alternative designs and methods of construction of nerve cuffs and electrodes are described in greater detail in Section 5.6.

5.3.3.3   Electrode Asymmetries

Until recently, the electrode dimensions, orientations and placements inside recording cuffs could not be accurately controlled because the cuffs were largely fabricated by hand. Mismatches in distances between recording and indifferent electrodes and mismatches in the contact impedances of individual electrodes caused degradation of common mode noise rejection properties. To partially counterbalance these short-comings, Thomsen et al.[37] placed a reference electrode inside a second, larger diam-eter, "dummy" cuff around the nerve cuff in order to record with it approximately the same noise created by external sources, but no nerve signals. With this "dual-cuff" the nerve signal is recorded monopolarly but artifacts are recorded bipolarly and may be rejected using differential recording. The cost, however, is a considerably larger cuff.

Other approaches that may improve common mode noise rejection are to not tie together the two indifferent electrodes and balance out any impedance mismatch using external resistors[6] or, alternatively, employ dual channel differential amplifiers to record separately from each indifferent electrode with respect to the central recording electrode.[38] These measures require adding an extra lead and two additional differential amplifiers per tripolar cuff; electronic compensation for inhomogeneities in hand-fabricated cuffs requires considerable fine-tuning and is limited by the precision in electronic components. Fortunately, the need for such compensatory measures will become obsolete with the availability of more precisely dimensioned, laser-fabricated nerve cuffs and electrodes (described in Section 5.6).

5.3.4   MULTICHANNEL RECORDINGS WITH NERVE CUFFS

Recording electrodes do not need to cover the total inner circumference of the cuff. They can be limited to span half the cuff, a segment of the cuff, or even a point-sized area determined, for example, by the cross-section of the wire used to form a recording surface. When electrode surface areas are reduced, more selective record-ings of activity in local regions close to each electrode may be obtained. Placing a number of punctual electrodes around a nerve inside a cuff may allow for selective recordings of activity arising from various different fascicles of the nerve.

Initial studies that tested the recording selectivity of multiple small electrodes inside cuffs found only modest differences in their recorded signals.[39,40] To interpret these findings, it is important to recall the "tunnel effect" that occurs inside a cuff. Note that these studies used cuffs with smooth inside walls where considerable transverse shunting of signals probably occurred and therefore the impedance between pairs of electrodes was quite low. The selectivity of multiple electrodes inside a cuff can be increased if the continuity of the space between the nerve and the cuff wall is broken up so that the "tunnel" is compartmentalized into several separate "chambers" around the nerve that are isolated from each other and separate electrodes are placed inside each chamber as described below.

5.3.4.1   Multichambered Cuffs

The differences among the signals recorded by a set of small electrodes distributed around a nerve can be enhanced by changing the surface geometry of the inside wall of the cuff to create locally insulated compartments along the surface of the nerve. This is the principle behind the multichambered, multichannel recording nerve cuff.[41-43] Multiple chambers or compartments, each containing a bipolar or tripolar recording electrode configuration, are electrically isolated from each other by longitudinal insulating ridges that just make contact with the nerve. The parallel ridge geometry is ideally suited for peripheral nerves and greatly enhances the recording selectivity achieved from electrodes in separate regions of the nerve trunk.[41] In addition, using signal identification techniques it is possible to extract signal source information from the ensemble of signals recorded by several electrodes inside multichambered cuffs and identify, for example, which of the five digits is being contacted by a test object.[44,45] In future clinical applications it may be possible to implant one or two multichannel cuffs on proximal nerve trunks, far away from moving joints and close to telemetry units, and obtain sensory information that would have otherwise required implanting several small cuff electrodes on distal, more fragile nerve branches.

5.3.5   SIGNAL AMPLIFICATION REQUIREMENTS

The selection of appropriate and efficient nerve signal amplification methods is not a trivial task. Peripheral nerves generate signals with very small amplitudes that must often be recorded in electrically noisy environments. The muscles that surround nerves elicit 1000-fold larger potentials. The body as a whole may be subjected to significant electric field pickup from external sources such as power lines or electrical stimulators used for FES.

The most important aspect of selecting a nerve cuff recording technique is to obtain a reasonable signal-to-noise ratio. The ENG signals generated by mammalian peripheral nerves are in the $10\,\mu V$ range with most of the energy between 1 kHz and 3 kHz. The electromyographic (EMG) noise originating from nearby muscles is in the low mV range, two or three orders of magnitude larger than the ENG signal. To record ENG signals under these conditions is the equivalent of trying to understand quiet speech (50 dB) within 1 meter of a loud horn (80 dB). An even worse problem can be the influx of external electric noise, i.e., power-line noise (~1 V range), equivalent to replacing the horn with a jackhammer (110 dB). Obviously, this type of background noise can severely contaminate ENG recordings unless special precautions are taken.

What makes ENG recordings feasible despite such noisy environment is the fact that the frequency spectra of the EMG and power-line noise are predominantly below 1 kHz. The EMG power spectrum is centered around 250 Hz but reaches up to 3 kHz, whereas power-line noise is located at 50 or 60 Hz. Thus, both sources of noise can be reduced using filtering techniques. Common-mode noise is further reduced by using a balanced tripolar electrode configuration and differential recording amplifiers with high common mode rejection (typically $\geqslant$ 90 dB) and high input impedance.

If the nerve recording amplification circuitry can be located outside the body, amplifiers can be readily assembled using commercially available components. Externally located amplifiers, however, require long leads that must course sub- and transcutaneously. Long cables increase the risk of wire breakage and the added resistance and capacitance contribute to signal shunting and greater pick-up of stimulation artifacts and power-line influx noise. The best way to eliminate these noise problems is to implant an amplifier in close proximity to the nerve cuff and to use a telemetry system for transcutaneous transmission of the recorded neural signals. However, the requirements for fully implanted nerve signal amplifiers are far more stringent than for external amplifiers.[46] Implantable amplifiers suitable for nerve signal recordings are not yet commercially available.

5.3.6 DAY-TO-DAY STABILITY AND REPRODUCIBILITY OF RECORDED SIGNALS

A nerve cuff that is correctly constructed, sized and installed cannot drift, so it will always record from the same population of neurons and the recorded signals will remain stable indefinitely.[12,24] Cases where the nerve is damaged during surgery soon become evident because the nerve signal deteriorates very rapidly.[47,48] In correctly installed cuffs, the recorded signals tend to be very reproducible for matched testing conditions over different days with the exception of some drift that may occur from day to day in the overall gain scaling factor.[49-52]

5.4 HOW TO STIMULATE NERVES WITH CUFF ELECTRODES

Whereas cuffs are essential for recording potentials from the surface of nerves, the same is not true for stimulation applications. All that may be needed to stimulate a nerve or a muscle are a pair of properly placed electrodes and sufficient current delivered through them. We will first review the properties of nerve stimulation when cuffs are not used, and subsequently describe the advantages of using cuffs for specific applications.

5.4.1 NERVE STIMULATION WITHOUT NERVE CUFFS

Research experience with electrical stimulation of nerves goes back three centuries to Benjamin Franklin and the first commercial clinical stimulation systems go back at least a century (reviewed by Hambrecht[53]). Considerable information is available on appropriate materials for stimulation electrodes and on safe and effective electrical stimulation parameters (see excellent review by Agnew and McCreery[54]).

To stimulate superficially located structures the electrodes may sometimes be placed outside the skin. However, when skin surface stimulation is used, the current requirements are much larger, the selectivity is lower, unwanted side effects such as pain may occur, the stability and reliability of stimulation is lower, and electrode or lead breakage is more frequent than when implanted electrodes are used.

Implanted electrodes can be as simple as the bared surfaces of otherwise insulated leads. Because of their simplicity of insertion, such "lead" electrodes can be effective in clinical applications that require fairly widespread, nonspecific stimulation, for example for pain suppression using dorsal column stimulation with intrathecally placed lead electrodes.[55] Another type of clinically used stimulation electrode consists of an open helical coil wound around a nerve.[56-58] Electrodes of this type have been implanted around the vagus nerve of over 3,000 epileptic patients and are stimulated following preprogrammed protocols that prevent or reduce the onset of seizures.

The main drawbacks of using nerve stimulating electrodes of either the lead or open spiral type are their higher current requirements, greater drift, and insufficient selectivity of stimulation. Neither electrode type may be well suited for applications where it is necessary to stimulate only one structure (for example, a specific muscle) without cross-stimulation of other, neighboring structures. In these applications nerve cuffs can provide more focused and effective stimulation.

Cuff-type stimulation electrodes have been used clinically for the past four decades, although thus far in fewer patients than lead or open spiral type stimulation electrodes. An early development for gait assistance in foot drop patients (Neuro Muscular Assist by Medtronic in 1971[59]) was not commercially successful. Applications for respiratory control using phrenic nerve electrodes were successful only after the cuff was reduced in size and made more planar.[60] Approximately 1,500 phrenic nerve patients have been implanted since 1968. A main reason for the slow acceptance of stimulation cuffs was the large size and stiffness of the cuffs that were developed in the 1960s and 1970s (reviewed by Hambrecht and Reswick[61] and Naples et al.[62]). The advent of the self-spiraling cuff type of stimulation electrode[63,64] represented a significant improvement in the mechanical properties of stimulation cuffs. However, the difficulty of installation of self-spiraling cuffs may have hindered the general acceptance of this type of electrode for clinical applications.

The recent development of the nerve recording field has brought about further advances in cuff design and fabrication methods (reviewed in Section 5.6 of this chapter). These new cuff technologies can be equally applied to fabricate better stimulation cuffs. Nevertheless, cuffs are not the final answer for all stimulation needs. In this section we review the advantages as well as the limitations of using nerve cuffs as the implanted interfaces for stimulation purposes.

5.4.2 ADVANTAGES OF USING CUFFS TO STIMULATE NERVES

A first advantage in using a cuff that contains the stimulating electrodes is that the locations of the electrodes with respect to the nerve become stabilized and the electrodes are less likely to drift during movement or over extended periods of time. A second advantage in using an insulating cuff is a major reduction in stimulus current requirements because little or no current needs to flow out of the cuff and through low-impedance surrounding tissues. An order of magnitude reduction in required stimulus current can be expected when a nerve cuff electrode is used instead of lead electrodes, open spiral stimulating electrodes wrapped around a nerve, or epimysial electrodes sutured over the muscle nerve entry point.[62,65] This reduction can have important, desirable consequences on battery life, on electrode corrosion and on the tissue damage that can potentially occur when high charge intensities are used. A third advantage is that unwanted side effects, such as current spillover causing stimulation of other nerves located near the electrodes, can be eliminated or markedly reduced when a cuff is used.

5.4.3 ELECTRODE CONFIGURATIONS

Stimulating cuffs can differ from recording cuffs in three ways. First, the electrode configurations, sizes and materials can be different. Second, stimulating cuffs need not be as well sealed as recording cuffs. Third, stimulating cuffs need not be as long as recording cuffs.

The simplest configuration is to place one stimulating electrode inside a cuff and a reference electrode outside. If there are several stimulating cuffs, a single common reference electrode may often be used. The effective surface area of stimulating electrodes should be as large as possible and the contact impedance as low as possible to minimize the charge density required for excitation of neural tissue. Nerve stimulation typically requires application of negative pulses of brief duration (10–200 ms) delivered from a constant current source, preferably optically isolated. Each stimulus pulse should be followed by a positive pulse with about 10 times longer duration and 10 times lower amplitude, which by itself will normally not cause stimulation but will balance the ionic flow and reduce electrode polarization.[66] The amplitude of stimulation required depends on the spectrum of axonal diameters in the nerve, the cross-section of the nerve, the impedance of the cuff and the effective current density achieved with stimulation. It is important to be aware of the current density levels that can cause damage to tissues or corrosion to the electrodes.[67] For typical applications in peripheral nerves of mammals, threshold current amplitudes for stimulating large motor axons are in the order of 50–200 μA. Avoid using a voltage source for nerve stimulation, because its effective current output would be unknown and would be prone to fluctuate with any changes in the electrode–tissue interface impedance. This danger is greatest when using skin stimulation electrodes or implanted wire electrodes that may move with respect to the stimulated nerve. Cuff electrode impedances are far less prone to drift, but caution is necessary nevertheless.

Monopolar stimulation as described above is not optimal, because the stimulus current must leak out of the cuff to reach the return electrode. This current spillover can cause side effects such as unwanted stimulation of other nerves located near the electrodes and if this happens, the benefits of using a cuff would not be achieved. Spillover can be eliminated or markedly reduced if both the current source and current return electrodes are placed inside the same cuff, creating a bipolar stimulation cuff. Bipolar cuffs are more efficient and contribute to longer battery life and lower electrode corrosion.

Beyond simple bipolar stimulation cuffs, several electrode configurations have been described that can be used, for example, to recruit small motor axons before large axons[68] to elicit unidirectional propagation of action potentials in stimulated axons[69] or to "steer" a current field in preferred directions within a cuff.[70]

5.4.4 CUFF SEAL

Whereas a good electrical seal is essential for recording nerve activity with a cuff, this requirement can often be relaxed for nerve stimulation purposes. A poorly fitting or poorly sealed cuff may still stimulate adequately for some applications even if some current is wasted. Nevertheless, it is advisable to use cuffs that fit properly around the nerve in order to stabilize the position of the stimulating electrodes during movement and to avoid causing mechanical damage to the nerve.

5.4.5 CUFF LENGTH

Nerve cuffs for stimulation need not be as long as recording cuffs. Cuffs measuring 5 mm, containing bipolar electrodes separated by 2–3 mm are usually sufficient to stimulate myelinated axons in nerves of up to 3 mm in diameter. If a longer cuff can be used, however, the current requirements will be lower and there will be less spillover.

5.4.6 CLINICAL EXPERIENCE WITH STIMULATION CUFFS

The history of development and uses of cuff electrodes for nerve stimulation and previous clinical experience was well reviewed by Naples et al.[62] and Grandjean.[65] In spite of concerns about the lack of documentation on the safety of early types of nerve cuffs used for stimulation, nerve cuffs implanted on phrenic nerves for diaphragm pacing or peroneal nerves for foot drop correction have performed well for several years and even decades in limited numbers of patients. The continued survival of hundreds of patients with respiratory paralysis attests to the long-term efficacy and stable performance of at least some types of nerve cuff stimulating electrodes in humans (reviewed by Glenn et al.,[4] Hambrecht and Reswick,[61] McNeal and Bowman[71]).

5.4.7 MULTICHANNEL NERVE CUFFS FOR STIMULATION

The 1990s witnessed advances in methods for selective stimulation of multiple muscle groups using multi-electrode nerve cuffs that were placed in cat hindlimb nerves[72-74] and in human limbs.[75] The multichannel stimulating cuff approach has the advantage of centralizing the location of stimulation electrodes in one surgically more accessible site that is mechanically stable and safer than placing individual stimulating electrodes at the various distal sites near the nerve entry points of the target muscles.

5.4.7.1 Multichambered Nerve Cuffs for Stimulation

The experiments mentioned above were done using variants of smooth-walled, self-spiraling cuffs[63] containing several punctual electrodes disposed along the inside wall of the cuff. When self-spiraling cuffs are used it is necessary to have a snug fit to the nerve in order for different electrodes to recruit different muscle groups. However, snug-fitting cuffs are not recommended for long-term implantation as they can cause nerve compression.

Multichambered cuffs increase the stimulation selectivity of small stimulating electrodes. The longitudinal ridges isolate the space around the nerve into separate chambers where different stimulation electrodes can be placed. This results in a spatial partitioning of stimulation currents and enhanced local effects of stimulation. The increase in selectivity is similar to what can be obtained when multichambered cuffs are used for recording purposes.[43]

5.5   HOW TO MODULATE NERVE ACTIVITY
WITH NERVE CUFFS

Cuffs can be equipped with electrodes, catheters, or both. A nerve cuff that is equipped with a catheter can be used as a receptacle for focused delivery of pharmacological agents. For example, controlled paralysis of a peripheral nerve and its associated muscles can be achieved using a blocking nerve cuff into which small doses of local anesthetic are infused via a percutaneous catheter to transiently and reversibly block all neural transmission through the cuff at selected times.[28,52]

Selective blockage of only small diameter fibers (e.g., pain fibers, fusimotor neurons) can be achieved with infusion of dilute lidocaine, which can be reversed within minutes with saline infusion. In this way, the activity of large-diameter fibers can be studied in awake animals performing normal movements alternatively in the presence and absence of activity in small fibers.[26,76]

A simple system for chronic, controlled delivery of a pharmacological substance to a nerve consists of a nerve cuff equipped with a catheter that is connected to an implanted minipump. This method has been used, for example, to study factors that affect axonal transport and regeneration.[77]

Care must be taken to prevent the growth of connective tissue that may obstruct the flow in the catheter lumen and the exit port into the cuff lumen. When the cuff is connected to an external catheter it is necessary to infuse on a daily basis a sufficient amount of heparinized saline (1%–5% heparin solution) to fill the catheter and cuff lumen, and to flush the catheter in a similar way after any infusion procedure.[23,28] When the cuff is connected to a minipump that provides a constant flow of fluid, this flow usually prevents the formation of excessive connective tissue that could block the flow.

Cuffs intended for fluid delivery should fit loosely around the nerve and provide a sufficiently large reservoir around it for fresh solution to be contained. Though it would seem desirable to hermetically contain the delivered solution inside the cuff by having tightly fitting rings at the ends of the cuff, such end rings must be avoided, for two reasons. First, no transverse mechanical elements must contact the nerve, as these could constrict longitudinal flow in axons and vessels and interfere with normal nerve function. Second, if a cuff reservoir is too well sealed, high pressure might be required to infuse fresh solution and such rise in pressure could itself compromise the nerve. Therefore, it is prudent to provide outlets for fluid outflow at the ends of the cuff. In general, it may be desirable to fit more than one catheter to a nerve cuff and use additional catheters to drain excess fluid as fresh solution is being administered. The use of draining catheters also allows for the disposal of excess solution in an appropriate location, e.g., the abdominal cavity, rather than have solution linger in the vicinity of the cuff and possibly other nerves on which the administered drug might cause unwanted effects.

Cuffs equipped with catheters may find clinical applications in the treatment of pain of neuropathic origin, in promoting regeneration of damaged nerves and in other areas that may require administration of substances directly and exclusively to peripheral or cranial nerves (see Table 5.1 on page 169).

5.6   NERVE CUFF DESIGNS AND METHODS OF FABRICATION

Cuff-like stimulation electrodes that were used in pioneering clinical applications had insulating walls and lead wire cables made of thick, stiff materials (reviewed by Naples et al.[62]). Early nerve stimulation cuffs consisted of two pre-formed halves[78] or a single strip of insulating material bent into a U-shape[59,79] that contained either disc or foil electrodes. The cuff was placed around the nerve and sutured to itself or to nearby tissue. The cuff lumen was not well sealed and sometimes was intentionally made much larger than the nerve diameter to lessen the risk of compression neuropathy. These early cuffs demonstrated that nerve stimulation could be adequately provided, but the electrode fabrication technologies then available were unsatisfactory. Nevertheless, Medtronic continued to provide Neuro Muscular Assist cuff electrodes through the 1980s and 1990s for various research uses (in particular for stimulation of the latissimus dorsi nerve in cardiomyoplasty applications[80]) until this product was discontinued in 1998.

With the advent of recording cuffs the design requirements became more stringent. Not only must recording cuffs be longer than stimulation cuffs, and well sealing, but recording cuffs can also provide direct information on the amplitude of nerve action potentials and therefore on the health of the nerve following cuff implantation (reviewed by Hoffer[23]). In contrast, it may not always be possible to assess the degree of damage caused by an implanted stimulation cuff. Muscle output upon stimulation can mask long-term damage caused to a nerve, since surviving motor axons tend to sprout and reinnervate denervated muscle fibers.[12] The lessons learned from developing nerve recording cuffs in the past 25 years are of benefit to the stimulation field as well, and improved stimulation cuff designs have now become available.

When selecting a nerve cuff, all the materials used must, of course, meet chemical biocompatibility requirements. In addition, it is important to avoid mechanical failure modes that have been associated with previous cuff designs. Mechanical failure modes can include nerve damage, electrode failure, or both. Failures can be caused by (1) the wall structure, (2) the opening or closure method or (3) the electrodes or lead-out cables.

5.6.1   Cuff Wall Structure

The cuff wall has two primary functions: it provides insulation and mechanical support for the electrodes that are attached or embedded in it. Wall materials and design shapes must be mechanically strong and stiff enough to support conductive leads or catheters, but also flexible enough to let the cuff "float" with the natural movements of the nerve.

Ideally, the cuff will appear "transparent" to the neural tissue, i.e., its presence will pose no mechanical, electrical or chemical load to the nerve. In practice, it is imperative to use flexible materials. The shape and size of the cuff must match as closely as possible the dimensions of the target nerve and allow for the expected swelling that follows the surgical implantation procedure.

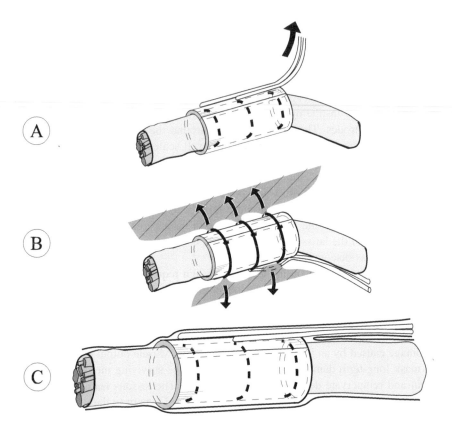

FIGURE 5.2 Examples of failure modes that can be caused by excessive mechanical loads on nerves. A: A cable pulled in direction of arrow can cause bending of the nerve around the edge of the cuff and may lead to compression neuropathy. B: Excessive connective tissue adhesions to closure sutures, cable or other elements on the surface of a cuff can cause the nerve to be pulled along by moving muscles. C: Design features that reduce mechanical loads include an absence of mechanical complexities on the cuff surface and routing of cables longitudinally along the nerve.

The lead-out cables or catheters that emerge from the cuff should be routed longitudinally along the nerve in order to minimize transverse loads on the cuff. Transverse loads can create mechanical stress points on the nerve or on the cables (Figure 5.2A). It is also imperative to avoid tissue adhesions through loops or growth into materials such as braided suture, which could pull or kink the nerve (Figure 5.2B). The ideal cuff is therefore smooth and flexible and the lead-out cables or catheters are routed longitudinally along the nerve (Figure 5.2C).

The lumen of the cuff must be large enough to accommodate any post-surgical swelling of the nerve in order to avoid causing compression damage (Figure 5.3A).

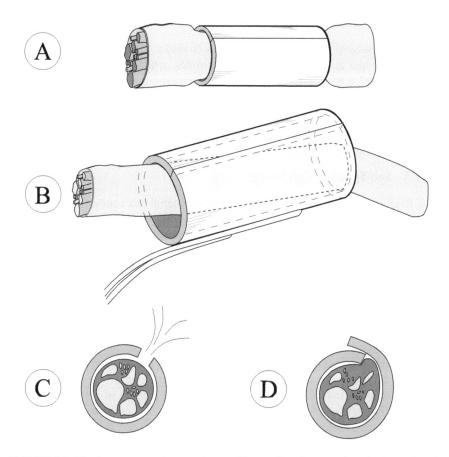

FIGURE 5.3 The importance of correctly matching cuff and nerve sizes is shown by the following examples. A: A cuff that is too small will cause compression neuropathy. B: A cuff that is too large and fits too loosely will produce smaller signals (due to signal shunting by lower resistivity saline) and, if misaligned, can cause the nerve to bend and suffer compression damage. C: An electrical leak across the opening of a poorly sealed cuff will degrade the signal produced by the nerve and allow greater noise pickup by the recording electrodes. The signal-to-noise ratio will be degraded by both of these effects. D: If a cuff is closed too tightly around a nerve, compression neuropathy will occur.

5.6.1.1   Cuff Wall Made of Commercial Tubing

For the first twenty years (1974–1994), recording cuffs were almost exclusively made from commercially available silicone tubing.[5,6,17,81] The walls were fairly thick and stiff, which presented sharp mechanical discontinuities to the nerve at each end of the cuff. Improper cuff alignment could cause the neural tissue to suffer compression or bending damage at these points[12] (Figure 5.3B). Nevertheless, with this cuff design it was possible to record good signals over months or years from many nerves in animal preparations as well as human volunteers.[1,12,15-17,24-34]

5.6.1.2   Cuff Wall Formed *In Situ*

Julien and Rossignol[82] placed wire electrodes around a nerve and made a "cuff" by pouring elastomer around the nerve and electrodes, which then cured *in situ*. This approach provided adequate recordings in acute experiments and has some attractive features. The safety and viability of this method for chronic recording experiments was demonstrated by Milner et al.[30] who used a similar approach to construct a cuff around the median nerve of a monkey and obtained useful recordings of sensory signal of cutaneous origin that lasted for several weeks.

5.6.1.3   Self-Spiraling Cuff Wall

Thin, flexible sheeting has been used as the wall material in a number of more recent designs. Some thin sheeting materials that can readily be curled into a cylindrical shape, like Teflon®, mylar or polyimide, are nevertheless very stiff and have sharp edges. Such materials should, for this reason, either be clad with more compliant materials along the edges, or avoided altogether. Fabrication of cuffs using such thin film substrates has not been successful thus far. Loeb and Peck[83] summarized the problems encountered with five recent cuff designs that included flexible polyimide and polyesterimide self-coiling film substrates with printed electrodes.

Cuffs made from thin silicone sheeting, on the other hand, are very flexible and have soft edges, making this an excellent choice of material for the wall structure of nerve cuffs. Naples et al.[63,64] described the methods of fabrication of silicone self-spiraling cuff electrodes suitable for nerve stimulation. When flexible thin sheeting is used as the substrate, however, the attachment and support of electrodes, which are generally made of much stiffer metallic materials, has continued to be a challenging problem. No commercial product has yet been based on the self-spiraling cuff wall approach.

5.6.1.4   Double Interlocking Cuff Wall

Loeb and Peck[83] presented a novel solution based on two interlocking slit open cylinders, one fitted around the other. The inner cylinder is made on a mandrel and has sufficient thickness to support circumferential wire electrodes that are also formed around the mandrel. The outer cyclinder snaps around the inner cylinder. This approach requires that both inner and outer elements have sufficient spring-like stiffness to maintain their shape and lock onto each other mechanically once assembled around the nerve.

5.6.1.5   Compliant Cast Silicone Cuff Wall

Loeb and Peck,[83] Haugland,[35] Kallesøe[3] and Kallesøe and Hoffer[84] introduced several methods for fabrication of thin, compliant cuff walls cast of silicone elastomer. This method can, in principle, produce thinner, more flexible nerve cuff walls than when the thinnest (125 μm) commercially available silicone sheeting is used. As with self-coiling silicone cuff designs, however, two remaining problems must be solved in order to use such thin cuff walls. The first is a method of attachment of generally stiffer electrodes. The second is a method for sealing the cuff. New solutions for these needs are presented in following sections.

5.6.1.6   Cuff Wall Made of Shape-Memory Alloy

A novel approach to cuff construction and installation was explored by Crampon et al.[85] They used a shape-memory alloy to automatically close the cuff when a certain temperature was reached. No description of the sealing properties of these cuffs was provided. The biocompatibility of the material used is also an issue that will require clarification.

5.6.2   Cuff Opening and Closure Methods

The cuff wall must provide an effective insulating barrier against both the influx of currents from external noise sources and the outflow of currents generated by the nerve axons inside the cuff. A key challenge has been to create a cuff wall that is thin, flexible and easy to open at the time of surgical installation, and that can also be securely sealed after the nerve is placed inside the cuff.

The recording cuffs used in the first 20 years were closed by encircling the cuff with several sutures that were partially attached to the external wall (Figure 5.4B). The sutures were used to pull the cuff open during the installation and, once tied together, to keep the cuff tightly closed for implant periods that usually lasted several months.[23] However, either the knots or the suture material itself sometimes yielded and some cuffs were partially opened by the end of the experiment. In general, the sutures could not usually prevent a tongue of connective tissue from growing into the cuff through the longitudinal slit (Figure 5.3C). These failures to maintain the cuff edges tightly closed led to degradation of the electrical and mechanical isolation of the interior of the cuff.

A drawback to the suture closure method is that the cuff edges must be pushed together by the sutures. If the cuff edges are not stiff enough or do not squarely meet one another the cuff cross section can sometimes be distorted into the shape of a spiral rather than a circle. This can cause serious compression damage (Figure 5.3D).

5.6.2.1   Self-Coiling Cuffs

An alternative closure method that does not require sutures implements two laminated flexible silicone sheets to form a self-coiling cuff.[63] In this design the sheets are differentially stretched in the lamination process to bias the passive stretch forces of the material so that the device tends to curl itself around the enclosed nerve. The cuff is presumed to retain its shape and remain in a closed and sealed state indefinitely after its implantation.

The self-coiling cuff design eliminated the use of external sutures to close the cuff. Practical problems, however, exist with this closure mechanism. First, it lacks a simple, convenient method for opening the cuff during surgical conditions. This will pose a risk to the nerve if the cuff snaps itself shut at the wrong time. Second,

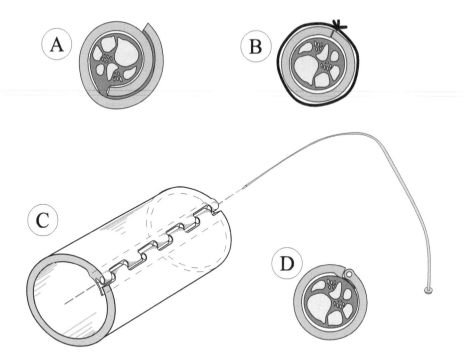

FIGURE 5.4 Comparison among the three main methods for closing nerve cuffs. A: Self-coiling cuffs have a "preferred" lumen size that may or may not be appropriate for the nerve size and may turn out to provide either too tight or too loose a fit. In addition, the external flap is only loosely wrapped around the cuff and can allow for the formation of a connective tissue bridge and a leakage pathway for nerve signals as well as EMG pickup. B: Cuffs made of relatively stiff silicone tubing can be reasonably well sealed with sutures tied around the outside, but the sutures tend to become sites for excessive connective tissue formation, adhesions, and growth of infectious agents. C: An inside-sealing flap together with a piano-hinge interlocking system provides both a good seal and a known lumen for the cuff. A monofilament nylon suture can be used as flexible closure element for rapid closure and a permanent seal with minimal adhesions. After the suture is passed through the closure elements it is trimmed and sealed with local heat application. The closing suture can be cut and easily removed, if needed. D: An inside flap provides a better seal for the cuff opening than an outside flap (not shown).

there may be no effective seal between the interior and the exterior (Figure 5.4A). Finally, there is no method to prevent the cuff from opening following implantation.

By design, self-coiling cuffs are intended to increase their lumen if the nerve swells and reduce their lumen again when the swelling recedes. To accomplish this, the natural resting diameter must be about 0.9× the nerve diameter so the cuff fits snugly at all times.[21,64] However, a snug fit around the nerve implies that mechanical pressure is constantly being exerted on the nerve. This design therefore has inherent potential for causing nerve compression damage. Furthermore, the curled inside edge of the cuff wall can sometimes penetrate and divide the nerve, as was determined histologically.[21]

Once the cuff is on the nerve, the lack of closure devices such as sutures theoretically allows the cuff to expand if the nerve swells. However, this lack also makes it possible for connective tissue to grow into the seam starting from either the interior or exterior of the cuff. This may eventually provide an unwanted current shunting path and possibly even pry the cuff open.[21] As discussed earlier, any breakdown of the cuff seal is undesirable, as it will harm the electrical and mechanical isolation of the interior or the cuff from the exterior.

5.6.2.2   Double Interlocking Cuffs

Loeb and Peck,[83] after reviewing the above mentioned suture and self-spiraling closure mechanisms, introduced a new closure method based on two interlocking slit open cylinders, one fitted around the other. This can in theory create a very efficient seal, but it is again a device that relies on spring properties in the cuff wall that has no apparent provisions to prevent the cuff from collapsing and constricting the nerve. If the fit is too loose, connective tissue will grow in through the opening. Loeb and Peck[83] therefore introduced a modified double cylinder design in which the two halves were adhered together with silicone adhesive applied *in situ* after the nerve was loaded in the cuff. Whereas a good cuff closure may be obtained with this design, the use of silicone adhesives near nerves during a surgical procedure cannot be recommended for human implants unless the adhesive can be proven not to be toxic.

5.6.2.3   Interlocking Hinge Method of Cuff Closure

An alternative to the above mentioned closure methods, all of which have the potential for unknown changes in lumen (intentional or unintentional), is to use an interlocking "piano-hinge" type of closure.[86] Matched sets of interdigitated tubes attached along the edges of the cuff opening (Figure 5.4C) are used to draw the two edges together. This system removes most structural strength requirements from the cuff wall and allows it to be made of thin, flexible silicone. To load the nerve, the cuff is positioned under the nerve and easily opened by pulling apart on two temporary sutures passed through the closure tubes at either edge of the cuff. The nerve is allowed to gently drop into the open cuff without any need for gripping, holding or pushing the nerve. Once the nerve is inside the cuff, the temporary sutures are removed and the closure tubes are aligned. A flexible closure element made of monofilament nylon or equivalent is passed through the closure tubes (Figure 5.4C). A tapered leading portion of the closure element facilitates its insertion and allows it to be easily pulled through the interdigitated closure tubes. When the cuff is very flexible it can be difficult to push a closure member through an interdigitating closure. It is typically much easier to pull a gradually tapering closure member into place. After insertion, the ends of the closure element are trimmed off and they can be quickly and safely sealed with focally applied heat. This method ensures that the cuff has invariant lumen dimensions and therefore will not exert compressive forces on the nerve if correctly sized at the time of installation. The closure mechanism is also easily reversible and little or no connective tissue will adhere to the cuff since the surface is very smooth. It is therefore possible to remove a cuff and replace it, even months or years later, with minimal consequences to the nerve.

5.6.3   CUFF ELECTRODES AND LEAD-OUT CABLES

Three main types of electrodes have been used in nerve cuffs. These will be described in the following sections.

5.6.3.1   Wire Electrode Sewn to Cuff Wall

Most recording cuffs used until 1995 had de-insulated stainless steel or platinum-iridium wire electrodes that were sewn into silicone tubing wall structures (Figure 5.5A). The exposed surface of the wire was placed along the inner circumference of the wall and the insulated leadout wires were attached to the outside wall. These electrodes typically encompassed the nerve more or less completely.

The drawback of this electrode design is that the electrodes are not likely to conform to the inside wall of the cuff. Connective tissue will invariably grow and attach to these electrodes, making it virtually impossible to safely remove the cuff. Hand sewn wire electrodes also pose a risk of cutting into the nerve because deformation of the cuff wall can create taut strands that project inside the lumen of the cuff (Figure 5.5B). Furthermore, the manufacturing method is laborious with variable results. The wires are prone to misalignment, splaying and unbalanced inter-electrode distances. Splaying of strands can occur when the individual strands are not twisted tightly enough. This defect causes a diminished temporal resolution of the recorded signal. Misaligned electrodes contribute to distorting the shape of the recorded signal.[3,18,83]

5.6.3.2   Foil Electrode Incorporated in Cuff Wall

Platinum foil electrodes were used in early nerve stimulation cuffs.[59,60] Naples et al.[63,64] described methods for including Pt foil electrodes inside laminated silicone walls. Haugland[35] more recently described a method of casting a silicone cuff wall around pre-aligned Pt foil electrodes disposed on a mandrel that achieved more precise and secure electrode attachment (Figure 5.5C). This cuff fabrication method provides a flexible wall with large surface area electrodes, ideal for stimulation. The foil electrode method, however, has some inherent drawbacks. A weld is required between the foil and the lead-out cable and this junction of dissimilar metals becomes a point of potential electrode failure. Similarly, bending of the foil sheet creates mechanical stress points where electrode breakage can be expected to occur. Finally, when the electrodes nearly encircle the nerve there may be no room for expansion of the tissue unless the inner diameter of the cuff is deliberately oversized.

5.6.3.3   Coil Electrode Embedded in Cuff Wall

A novel alternative to traditional electrodes mentioned above is a flexible coil electrode that is made from the same wire used as the lead-out cable, for example Cooner AS 631. The coil is partially embedded in the cuff wall during the casting

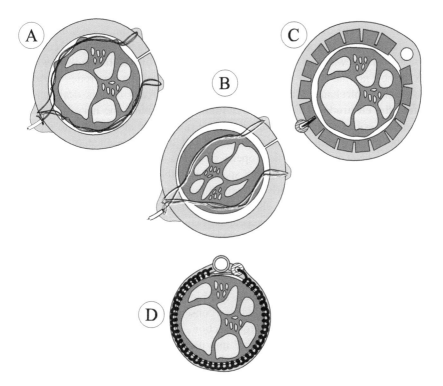

FIGURE 5.5 Alternative methods for installing circumferential electrodes inside a nerve cuff. A: The majority of nerve recording cuffs that have been used in animal research and also in initial patient trials had de-insulated wire electrodes that were individually sewn to the cuff wall made of relatively stiff silicone tubing. Problems with this method included asymmetrical placements of electrodes, frayed or broken strands of wire and excessive connective tissue adhesions to the wire electrodes and the nerve that prevented the safe removal of a malfunctioning cuff some weeks or months after cuff implantation. B: A further problem with sewn wire electrodes was the difficulty in controlling the length of wire inside the cuff wall. Repeated opening and closing of a cuff could cause tightening of the wire electrode spans inside the cuff lumen, with the consequence that wire segments could constrict the nerve and cause local compression. C: Metal foil electrodes can provide larger surface areas but may constrict the nerve circumferentially, are difficult to attach without causing deformation of the electrode or the wall, and the joints to lead-out wires may corrode or break. D: Circumferential, optimally coiled electrodes can be embedded in and are as flexible as thin silicone walls, provide increased electrode surface area and, if correctly recessed into the wall, will not provide a good substrate for connective tissue adhesion. The coiled wire segments that come in contact with the nerve are oriented nearly longitudinally and therefore cause minimal or no damage to the nerve.

process (Figure 5.5D). Dissimilar metal junctions are eliminated by maintaining continuity between the insulated lead-out cable and the de-insulated electrode wire. By coiling the de-insulated wire to an intermediate coil spacing, the electrode is given high mechanical flexibility. This property allows the electrode to become incorporated and move together with very thin silicone cast wall structures, without distorting the shape of the cuff wall.[3,84] The lead-out wires run longitudinally to the nerve and are also incorporated within the cuff wall during the casting process. Cuffs fabricated in this way are so flexible that they can be flipped inside-out and return back to their normal shape without any damage to either the electrodes or the wall.

A further mechanical advantage of coiled electrodes of this design is that, whereas the electrode is disposed circumferentially around the nerve, each of the segments of coiled wire that contact the nerve runs essentially parallel to the nerve. Thus, the nerve is not transversely constricted by the presence of the coil. In microscopic-level analogy to the property of longitudinal ridges along the cuff wall, the nerve can expand as needed into spaces available between individual loops of the coil, thus allowing for a fairly snug fit without causing compression to the nerve. Finally, the coiled nature of the electrodes reduces the likelihood that the electrodes will fatigue or break as a result of repeated flexing of the walls of an implanted cuff.

5.6.4 EVOLUTION OF CUFF WALL AND CLOSURE SYSTEMS

Figure 5.6 summarizes major evolutionary milestones in design of cuff wall and closure systems over the past 30 years, and highlights the progress made in efforts to reduce the cuff wall dimensions and, as a consequence, increase the cuff flexibility.

Five basic cuffs are shown in cross-section. Throughout Figure 5.6, the dimensions of cuffs appropriately sized for 2 mm nerves are compared. Note that all the cuffs described are largely made of silicone elastomer.

The diagram at the top represents the NeuroMuscular Assist bipolar stimulation cuff electrode that was commercially made by Medtronic from circa 1971 until 1998. Other neuromuscular stimulation cuffs developed in the late 1960s (e.g., phrenic stimulation cuffs[60]) and early 1970s (e.g., Avery cuff electrodes[78,79]) shared similar general dimensions. Note that although the cuff wall, made of silicone, was approx. 1 mm thick, the long tabs contributed considerable additional cross-sectional area, mass and rigidity. The method of closure using sutures could not ensure as good a seal as would be needed for recording nerve signals.

The diagram immediately below represents the first nerve recording cuffs introduced in the mid-1970s by Hoffer et al.[5,6] and Stein et al.[7,10,81] which consisted of a length of longitudinally slit commercial silicone tubing that was closed as tightly as possible with sutures. Note that although the wall thickness was still 75% of the stimulation cuff wall thickness, the overall cross-sectional area of the first recording cuffs was only 20% of the stimulation cuff cross-sectional area. Recording cuffs of essentially similar design and dimensions continued to be used until the mid-1990s.

The next major milestone occurred in 1986 when the self-spiraling cuff design (center of Figure 5.6) was introduced by Naples et al.[63] By reducing the need for structural rigidity in the cuff wall and introducing the self-coiling concept, Naples et al.[64] were able to reduce the wall thickness to 300 μm. However, the requirement to have the cuff wall wrap around itself at least 1.5 times, or as much as 2.75 times,[87] means that the effective cuff wall is greater than a single layer of laminate. The introduction of silicone self-spiraling cuffs represented an overall improvement in wall thickness (by 20%) and cross-section (by 40%). Although these cuffs were

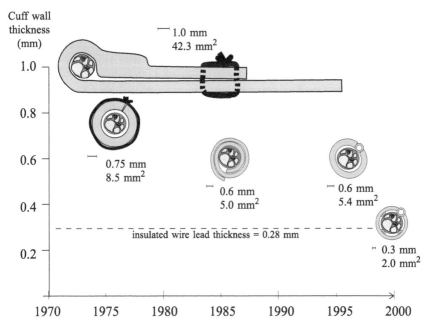

FIGURE 5.6 Diagrammatic representation of evolution of nerve cuff wall and closure systems in the past 30 years. Ordinate: average cuff wall thickness in mm. Abscissa: time in years. Five basic cuffs are shown in cross-section with a 2-mm diameter nerve represented inside the lumen. The coordinates represent the year of introduction and the wall thickness for each cuff. The wall thickness as well as the wall cross-sectional area are indicated next to each cuff. Further description in text.

developed primarily for stimulation purposes, where the quality of cuff closure is not as important as it is for recording self-spiraling cuffs have been used recently for recording from 2.5 mm thick hypoglossal nerves in dogs.[87]

The next milestone in cuff design was the development of a secure, "piano-hinge" closure system by Kallesøe et al.[86] This system eliminated the need for tight closing sutures and, consequently, for rigid cuff walls. Initially, cuffs made with this closing system still used a commercial tubing wall, although somewhat thinner than in earlier cuffs. The overall wall thickness and cross-sectional area were equivalent to the self-spiraling cuffs.

The most recent improvement in cuff design has resulted from the combination of three elements: a piano-hinge closing system, a very thin, flexible cast wall, and a coiled electrode with sufficient flexibility to be incorporated in the flexible wall.[84] The cuff wall dimensions were significantly reduced: the wall thickness was reduced by 50% and the wall cross-sectional area was reduced by 63%. These gains have very significant implications on the flexibility and overall biocompatibility of this new generation of cuff electrodes. The new cuffs are suitable for recording, stimulation or modulation and may incorporate either circumferential or multichannel electrode geometries.

To further understand the significance of the cuff wall thickness, the thickness of the insulation of Cooner AS 631 wire is shown in Figure 5.6 by a horizontal dotted line for comparative purposes. Note that the cuff wall thickness is now approximately equal to the thickness of an insulated wire lead (Cooner AS 631, which consists of 10 strands of 25-μm stainless steel wire plus four coats of Teflon insulation for a total thickness of 280 μm). Whereas thinner lead wire or insulation may be available, there are practical lower limits to the surface area of nerve cuff electrodes, which may preclude the reduction of lead wire dimensions much below what is currently used. The cuff wall thickness, however, may continue to be reduced beyond the lead wire thickness in future cuffs as long as the wall will not tear with normal use.

5.7 RESEARCH AND CLINICAL USES OF NERVE CUFFS

A variety of uses of nerve cuffs in research and clinical applications was reviewed by Hoffer[23] and Hoffer et al.[2] Sinkjær et al.[88] recently reviewed uses of recording cuffs in animals and humans, and Hoffer and Kallesøe[89] reviewed applications of nerve cuffs for nerve repair and regeneration.

5.7.1 Typical Research Uses of Nerve Cuffs

We present in this section examples of uses of nerve cuffs that have been successfully implemented in animal experiments.

Figure 5.7 shows typical implant locations for three types of nerve cuffs in a cat hindlimb and how these cuffs may be used for studies of nerve signal selectivity and stability as a function of time. Under deep halothane anesthesia (see Hoffer[23] for general anesthetic and surgical protocols), four single-channel tripolar cuffs are implanted in a first surgical procedure, one on the sciatic nerve proximal to the hip joint and one on each of three distal branches of the sciatic nerve: the sural nerve (not shown in Figure 5.7), the superficial peroneal nerve (also not shown) and the tibial nerve. In a second surgery that typically takes place 2–4 weeks later, after the animal has fully recovered from the first surgery and the nerve signals have stabilized, an 8-channel multichambered cuff is placed on the sciatic nerve, between the hip and the knee. Figure 5.7B illustrates the conformal fit to the natural cross-section of the sciatic nerve in the thigh region that can be obtained with a two-piece, hinged, flexible multichambered nerve cuff.[43]

At periodic intervals following the first surgery, under general gas anesthesia, each of the tripolar cuffs can be stimulated and compound action potentials (CAPs) are recorded from the proximal sciatic cuff. After the multichannel cuff has also been implanted, the stimulation and recording protocols are repeated at periodic intervals. If downward shifts in the peak-to-peak amplitude and increases in the latency of the CAPs occur, these may reflect damage to the nerve caused by the test cuff. The long-term safety and efficacy of devices implanted in a forelimb or a hindlimb have been monitored in this way for at least six months per animal in our laboratory.[42,90]

In addition, at periodic intervals and again under gas anesthesia, sensory neural traffic signals can be simultaneously recorded by the recording electrodes in the

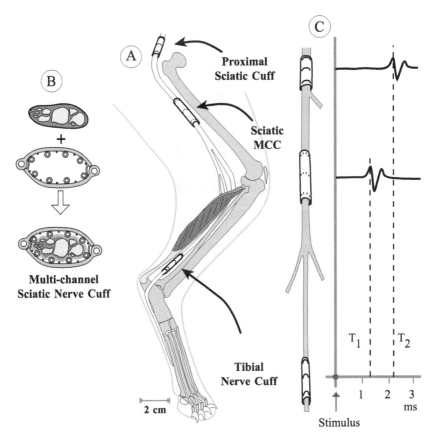

FIGURE 5.7 Diagram showing implant locations of nerve cuffs in a cat hindlimb to study nerve signal selectivity and stability as a function of time. A: Single-channel tripolar cuffs placed on sciatic nerve proximal to hip joint and on tibial nerve. In a second surgery, an 8-channel multichambered cuff is placed on the sciatic nerve between hip and knee. B: Conformal fit to the shape of the sciatic nerve in the thigh region is obtained with two-piece, hinged, flexible multichambered cuffs. C: At periodic intervals following the first surgery, under gas anesthesia, each tripolar cuff is stimulated and compound action potentials (CAPs) are recorded. After the multichannel cuff is implanted, major shifts in the peak-to-peak amplitude or in the latency of the CAPs can indicate that damage to the nerve was caused by the test cuff.

multichannel cuff during passive manipulations of the ankle joint, digits and skin. Similar recordings can be performed in the awake animal while walking on a treadmill[91,92] or performing other tasks such as reaching.[93] Signal processing techniques can be applied to identify the sources of peripheral input from the multichannel recording array. The stability of the recorded signals can thus be assessed during the duration of the implant. If EMG electrodes are also implanted in selected limb muscles, the selectivity for individual muscle recruitment with stimulation of the

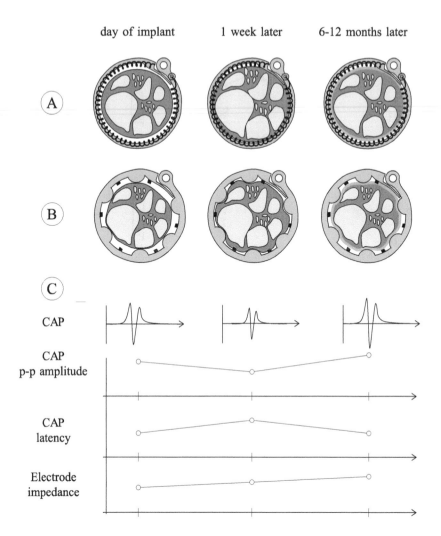

FIGURE 5.8 Schematic representation of typical sequential changes to nerves after cuff installation. Further description in text.

array of electrodes in the multichannel cuff can also be assessed and the stability of recruitment can be monitored as a function of time.

The sequential changes that are typically seen in nerves after cuff installation are represented schematically in Figure 5.8. The nerve status is shown at three time points: day of implant, one week later and 6–12 months later. Figure 5.8A shows a diagram of a cross-sectional view through a peripheral nerve inside a single-channel tripolar cuff with flexible wall and a flexible circumferential electrode. The day of implant, the nerve circumference is carefully measured with a flexible ruler and a cuff with inside circumference 10% larger than the nerve circumference is selected. The cuff thus fits loosely around the nerve. One week later, the nerve has swollen and almost fills the cuff lumen. The coil electrode elements that project into the cuff lumen may contact and indent the nerve during this period but cannot be expected to penetrate the nerve sheath. Six to 12 months later, the nerve has regained its original size. A thin (about 100 μm; light gray) connective tissue sheath has formed around the nerve inside the cuff, but the lumen is otherwise filled with clear fluid, not with connective tissue, and there are few or no adhesions to the electrodes.

Figure 5.8B shows a cross-sectional view through a peripheral nerve inside a multichambered cuff with ridged walls and six punctual electrodes. A cuff with inside circumference equal to the nerve circumference is selected. The cuff is made to fit snugly, with its ridges just touching the nerve. One week later, the nerve has swollen and almost fills the chambered spaces between the ridges. Six–12 months later, the nerve has regained its original size. A thin (about 100 μm; light gray) connective tissue sheath has formed around the nerve inside the cuff, but the chambered spaces are filled with clear fluid, not with connective tissue, and there are few or no adhesions to the electrodes.

Figure 5.8C shows examples of electrophysiological data that are typically recorded from either single-channel or multichambered nerve cuff electrodes. Top trace: Compound action potentials (CAP) recorded as was shown in Figure 5.7C. Note that the CAP amplitude tends to decline somewhat in the first week after cuff implant (second trace) and the CAP latency tends to increase by 10–30% (third trace). By about a month after implant the CAP values typically recover and after 6–12 months the CAP amplitudes can be somewhat larger than control values, which can be attributed to increases in cuff impedance after connective sheath formation (bottom traces), and the CAP latencies have returned to original control values.

Other applications are possible using the same implanted cuffs. For example, any of the cuffs may be fitted with catheters and can be used to transiently block activity in either the whole nerve inside that cuff, or the activity in just the small-diameter (gamma-range and smaller) motoneurons and sensory fibers by anesthetic infusion into the cuff. These procedures have allowed "us," for example, to study aspects of movement control in the temporary absence of some of the active motor elements.[26,76,92] Similarly, cuffs equipped with catheters can be used to treat axons in selected nerves by selective infusion of pharmacological agents that affect axonal regeneration or repair.[89]

5.7.2 EMERGING CLINICAL USES FOR NERVE CUFFS

Heart pacemakers, the first electrical stimulation systems to gain wide acceptance, have been implanted in millions of patients in the past 30 years. By comparison, widespread clinical uses of implanted neuroprosthetic systems have lagged behind. Among the currently more advanced neuroprostheses are cochlear multichannel stimulating arrays that can restore hearing to the profoundly deaf. Commercial versions became available in the 1990s after experimental multichannel cochlear implants were shown to function far better than previous generations of implanted cochlear stimulators.

In the past three years, pioneering FES systems became commercially available for epilepsy suppression (Model 101/300 system, Cyberonics Corp.), paralyzed hand control (FreeHand, NeuroControl Corp.) and bladder control (Vocare, NeuroControl

Corp.). As a consequence of the total time required for development, clinical testing and obtaining FDA approvals, these newly approved systems use electrodes that were developed in the 1970s and 1980s to stimulate cranial nerves, muscle nerves or nerve roots. Whereas these systems are now providing life-saving functions for seriously affected patients, the latest advances in electrode technology will enable the development of 21st century neuroprosthetic systems; provide greater stimulation selectivity, fewer unwanted side effects, longer battery life, higher reliability, and improved control through the use of natural feedback signals; and restore lost functions that cannot be addressed with existing commercial systems. Because of the pioneering efforts of the developers of recently approved systems, the FDA approval process can be expected to be faster for future systems in similar fields of use.

Advanced nerve cuffs that provide multichannel stimulation, reliably record natural feedback signals or focus the delivery of pharmacological substances to targeted nerves can be expected to become integral components in a variety of new neuroprosthetic systems.

Two traditional barriers to more widespread use of nerve cuffs have been (1) valid concerns about safety, efficacy and technical unreliability of early versions of nerve stimulating cuffs and (2) previous commercial unavailability of advanced nerve cuffs. The first concerns are being addressed by an increasing body of data available from animal and human uses of nerve cuffs. Acceptance and further development of cuff-based technologies will hinge on the emergence of commercially available nerve cuffs.

A synopsis of emerging clinical applications of nerve cuffs is presented in Table 5.1. Nerve stimulation, recording and modulation applications are separately listed, and applications are further sorted by the location of cuffs on peripheral nerves, cranial nerves or roots. References are given to related clinical implementations when these exist or to research that has demonstrated the feasibility of the proposed approach.

It is apparent that a wide variety of serious disorders can be addressed through the use of nerve cuff electrodes and drug delivery systems. This list is by no means exhaustive; it merely reflects applications for which initial clinical experience or available research results already suggest feasibility of clinical implementation. We trust that a good number of the applications listed in Table 5.1 will become clinically available in the next several years, and more applications will be added to a growing list of emerging applications for neuroprostheses on the basis of future research results.

ACKNOWLEDGMENTS

Funded in part by a research contract from NIH (NINDS-NO1-NS-6-2339) and by NeuroStream Technologies, Inc. We are grateful to our many collaborators who over the years contributed to works referred to in this study and in particular to M. Barú, A. Caputi, Y. Chen, P. Christensen, D. Crouch, E. Heygood, M. Hansen, M. Haugland, C. Kamimura, F. Maloufi, W. Ng, D. Popović, H. Qi, T. Sinkjær, R. B. Stein, K. Strange and I. Valenzuela.

TABLE 5.1

Emerging Clinical Applications for Nerve Cuffs

Location of Cuffs	Disorder	Restorable Function	Refs.
		1. Stimulation Cuffs	
Peripheral nerve stimulation			
Motor nerves in upper limb	Spinal cord injury	Activate grasp, wrist and elbow in high-level quadriplegia	94
Motor nerves in both lower limbs	Spinal cord injury	Activate paralyzed leg muscles for transfers and standing	95
Motor nerves in lower limb	Stroke	Activate foot dorsiflexor muscles for walking	71
Lattissimus dorsi nerve in chest	Cardiac	Stimulate latissimus dorsi muscle nerve for cardiomyoplasty	80
Ventral root stimulation			
Sacral motor roots	Spinal cord injury	Bladder and bowel voiding	61
Sacral motor roots	Spinal cord injury	Erection and ejaculation	61
Cranial nerve stimulation			
Vagus nerve	Epilepsy	Suppress onset of seizures	58
Phrenic nerve	Brainstem	Control respiration/cough assistance	4
Hypoglossal nerve	Apnea	Control tongue position to prevent apnea	36
Optic nerve	Blindness	Restore vision	96
		2. Modulation Cuffs	
Peripheral nerve modulation			
Affected peripheral nerve	Chronic pain	Pharmacological control of pain	55
Affected peripheral nerve	Nerve trauma	Promote nerve repair & regeneration	77
Cranial nerve modulation			
Affected cranial nerve	Chronic pain	Pharmacological control of pain	55
Affected cranial nerve	Nerve trauma	Promote nerve repair & regeneration	77
		3. Recording Cuffs	
Peripheral nerve recording			
Sensory nerves in upper limb	Spinal cord injury	Record touch, slip, joint position information and use to control paralyzed arm/hand with FES	21
Severed motor nerves in upper limb	Amputee	Record command signals from severed motor nerves and use to control prosthetic hand, wrist, elbow	47
Sensory nerve in lower limb	Stroke	Monitor plantar pressure, joint position and use for feedback control of paralyzed leg with FES	31
Dorsal root recording			
Sacral dorsal roots	Spinal cord injury	Record from bladder pressure afferents and use to control bladder voiding	21
Cranial nerve recording			
Hypoglossal nerve	Apnea	Monitor tongue position, use as feedback to control apnea	36

REFERENCES

1. Hoffer, J. A, Haugland, M. K., Signals from tactile sensors in glabrous skin: recording, processing, and applications for the restoration of motor functions in paralyzed humans, in *Neural Prostheses: Replacing motor function after disease or disability*, Stein, R. B., Peckham, P. H., Popovic, D. P. Eds., Oxford University Press, 99, 1992.
2. Hoffer, J. A., Stein, R. B., Haugland, M. K., Sinkjær, T., Durfee, W. K., Schwartz, A. B., Loeb, G. E., Kantor, C., Neural signals for command control and feedback in functional neuromuscular stimulation: a review, *J. Rehabil. Res. Dev.*, 33, 145, 1996.
3. Kallesøe, K., *Implantable transducers for neurokinesiological research and neural prostheses*, Ph.D. Thesis, Simon Fraser University, Feb 1998, ISBN 0612377180, Burnaby, Canada, pp. 196, 1998.
4. Glenn, W. W. L., Phelps, M. L., Diaphragm pacing by electrical stimulation of the phrenic nerve, *Neurosurg.*, 17, 974, 1985.
5. Hoffer, J. A., Marks, W. B., Rymer, W. Z., Nerve fiber activity during normal movements, *Abstr. Soc. Neurosci.*, 258, 1974.
6. Hoffer, J. A., *Long-term peripheral nerve activity during behaviour in the rabbit: the control of locomotion*, Ph.D. Thesis, Johns Hopkins Univ., Nov. 1975. Publ. No. 76-8530, University Microfilms, Ann Arbor, Michigan, 124 pp., 1976 .
7. Hoffer, J. A., Marks, W. B., Long term peripheral nerve activity during behavior in the rabbit, *Adv. Behav. Biol.*, 18, 767, 1976.
8. Stein, R. B., Charles, D., Davis, L., Jhamandas, J., Mannard, A., Nichols, T. R., Principles underlying new methods for chronic neural recording, *Can. J. Neurol. Sci.*, 2, 235, 1975.
9. Struijk, J. J., Thomsen, M., Larsen, J O., Sinkjær, T., Cuff electrodes for long-term recording of natural sensory information, *IEEE Eng. Med. & Bio.*, May/June, 91, 1999.
10. Stein, R. B., Charles, D., Gordon, T., Hoffer, J. A., Jhamandas, J., Impedance properties of metal electrodes for chronic recording from mammalian nerves, *IEEE Trans. BME* 25: 532, 1978.
11. Aguayo, A., Nair, C. P. V., Midley, R., Experimental progressive neuropathy in the rabbit, *Arch. Neurol.*, 24, 358, 1971.
12. Davis, L. A., Gordon, T., Hoffer, J. A., Jhamandas, J., Stein, R. B., Compound action potentials recorded from mammalian peripheral nerves following ligation or resuturing, *J. Physiol. (Lond)*, 285, 543, 1978.
13. Sunderland, S., *Nerve and Nerve Injuries.* 2nd ed., Churchill Livingstone, Edinburgh, London and New York. 1978.
14. Gillespie, M. J., Stein, R. B., The relationship between axon diameter, myelin thickness and conduction velocity during atrophy of mammalian peripheral nerves, *Brain Res.*, 259, 41, 1983.
15. Hoffer, J. A., Loeb, G. E., Pratt, C. A., Single unit conduction velocities from averaged nerve cuff electrode records in freely moving cats, *J. Neurosci. Methods*, 4, 211, 1981.
16. Loeb, G. E., Hoffer, J. A., Pratt, C. A., Activity of spindle afferents from cat anterior thigh muscles. I. Identification and patterns during normal locomotion, *J. Neurophysiol.*, 54, 549, 1985.
17. Hoffer, J. A., Loeb, G. E., Marks, W. B., O'Donovan, M. J., Pratt, C. A., Sugano, N., Cat hindlimb motoneurons during locomotion. I. Destination, axonal conduction velocity and recruitment threshold, *J. Neurophysiol.*, 57, 530, 1987.
18. Crouch, D., *Morphometric analysis of neural tissue following the long-term implantation of nerve cuffs in the cat forelimb*, M.Sc. Thesis, Simon Fraser University, 1997, ISBN 0612241130, Burnaby, Canada, pp. 118, 1997.

19. Crouch, D., Strange, K. D., Hoffer J. A., Morphometric analysis of cat median nerves after long-term implantation of nerve cuff recording electrodes, *IFESS/Neural Prostheses V Int'l. Conf.*, Vancouver, BC, pp. 245, 1997.
20. Marks, W. B., Loeb, G. E., Action currents, internodal potentials, and extracellular records of myelinated mammalian nerve fibers derived from node potentials, *Biophys. J.*, 16, 655, 1976.
21. Grill, W. M. and Mortimer, J. T., Neural and connective tissue response to long-term implantation of multiple contact nerve cuff electrodes, *J. Biomed. Mater. Res.*, 50, 215–226, 2000.
22. Stein, R. B., Oguztöreli, M. N., The radial decline of nerve impulses in a restricted cylindrical extracellular space, *Biol. Cybernetics*, 28, 159, 1978.
23. Hoffer, J. A., Techniques to record spinal cord, peripheral nerve and muscle activity in freely moving animals, in *Neurophysiological Techniques: Applications to Neural Systems. NEUROMETHODS, Vol. 15,* Boulton, A. A., Baker, G. B. and Vanderwolf, C. H., Eds., Humana Press, Clifton, NJ, pp. 65, 1990.
24. Gordon, T., Hoffer, J. A., Jhamandas, J., Stein, R. B., Long-term effects of axotomy on neural activity during cat locomotion, *J. Physiol..* 303, 243, 1980.
25. Hoffer, J. A., O'Donovan, M. J., Pratt, C. A., Loeb, G. E., Discharge patterns of hindlimb motoneurons during normal cat locomotion., *Science,* 213, 466, 1981.
26. Hoffer, J. A., Loeb, G. E., A technique for reversible fusimotor blockade during chronic recording from spindle afferents in walking cats, *Exp. Brain Res.* (Suppl. 7), 272, 1983.
27. Krarup, C., Loeb, G. E., Conduction studies in peripheral cat nerve using implanted electrodes: I. Methods and findings in control, *Muscle and Nerve,* 11, 922, 1987.
28. Hoffer, J. A., Leonard, T. R., Cleland, C. L., Sinkjær, T., Segmental reflex action in normal and decerebrate cats, *J. Neurophysiol.,* 64, 1611, 1990.
29. Sinkjær, T., Hoffer, J. A., Factors determining segmental reflex action in normal and decerebrate cats., *J. Neurophysiol.,* 64: 1625, 1990.
30. Milner, T. E., Dugas, C., Picard, N., Smith A. M., Cutaneous afferent activity in the median nerve during grasping in the primate, *Brain Res.,* 548, 228, 1991.
31. Sinkjær, T. Haugland, M. K., Haase, J., Neural cuff electrode recordings as a replacement of lost sensory feedback in paraplegic patients, *Neurobionics,* 267, 1993.
32. Popovic, D. B., Stein, R. B., Jovanovic, K. L., Dai, R., Kostov, A., Armstrong, W. W., Sensory nerve recording for closed-loop control to restore motor functions, *IEEE Trans. BME,* 40, 1024, 1993.
33. Haugland, M. K., Hoffer, J. A., Sinkjær, T., Skin contact force information in sensory nerve signals recorded by implanted cuff electrodes. *IEEE Trans. Rehab. Engng.,* 2, 18, 1994.
34. Haugland, M. K., Sinkjær, T., Cutaneous whole nerve recordings used for correction of footdrop in hemiplegic man, *IEEE Trans. Rehab. Eng.,* 3, 307, 1995.
35. Haugland, M. K., A flexible method for fabrication of nerve cuff electrodes, The 18th Annual International Conference of the IEEE Engineering in Medicine and Biology Society, Amsterdam, 31 Oct.–3 Nov., 1996.
36. Sahin, M., Haxhiu, M. A., Durand, D. M., Dreshaj, I. A., Spiral nerve cuff electrode for recordings of respiratory output, *J. Appl. Physiol.,* 83, 317, 1997.
37. Thomsen, M., Struijk, J. J., Sinkjær, T., Signal, noise, and artifacts in dual-cuff nerve recordings, *IEEE Trans. Rehab. Eng.,* provisionally accepted, 1999.
38. Pflaum, C., Riso, R. R., Performance of alternative amplifier configurations for tripolar nerve cuff recorded ENG, *Proc. 18th Annual Meeting IEEE/Eng Med & Bio. Sac.,* Amsterdam, 1996.

39. Lichtenberg, B. K., De Luca, C. J., Distinguishability of functionally distinct evoked neuroelectric signals on the surface of a nerve, *IEEE Trans. BME*, 26, 228, 1979.

40. Struijk, J. J., Haugland, M. K., Thomsen, M., Fascicle selective recording with a nerve cuff electrode, The 18th Annual International Conference of the IEEE Engineering in Medicine and Biology Society, Amsterdam, 31 Oct.–3 Nov., 1996.

41. Chen, Y., Christensen, P. R., Strange, K. D., Hoffer, J. A., Multichannel recordings from peripheral nerves: 2. Measurement of selectivity, *IFESS/Neural Prostheses V Int'l. Conf.*, Vancouver, BC, pp. 241, 1997.

42. Hoffer, J. A., Strange, K. D., Christensen, P. R., Chen, Y., Yoshida, K., Multichannel recordings from peripheral nerves: 1. Properties of multi-contact cuff (MCC) and longitudinal intra-fascicular electrode (LIFE) arrays implanted in cat forelimb nerves, *IFESS/Neural Prostheses V Int'l. Conf.*, Vancouver, BC, pp. 239, 1997.

43. Hoffer, J. A., Chen, Y., Strange, K., Christensen, P. R., *Nerve Cuff having One or More Isolated Chambers*, United States Patent No. 5,824,027, October 20, 1998.

44. Christensen, P. R. , *Sensory source identification from nerve recordings with multichannel electrode arrays*, M.A.Sc. Thesis, Simon Fraser University, 1997, ISBN 0612241084, Burnaby, Canada, pp. 105, 1997.

45. Christensen, P. R., Chen, Y., Strange, K. D., Hoffer, J. A., Multichannel recordings from peripheral nerves: 4. Evaluation of selectivity using mechanical stimulation of individual digits, *IFESS/Neural Prostheses V Int'l. Conf.*, Vancouver, BC, pp. 217, 1997.

46. Charles, D., Neural and EMG bioetelemetry implant for control of powered prostheses and functional electrical stimulation. *Biotelemetry X, Proc. X Intl. Symp. Biotelemetry*, C.J. Amlaner, Ed., Univ. Arkansas Press, 544, 1989.

47. Stein, R. B., Charles, D., Hoffer, J. A., Arsenault, J., Davis, L. A., Moorman, S., Moss, B., New approaches to controlling powered arm prostheses, particularly by high-level amputees, *Bull. Prosth. Res.*, 17, 51, 1980.

48. Krarup, C., Loeb, G. E., Pezeshkpour, G. H., Conduction studies in peripheral cat nerve using implanted electrodes. II. The effect of prolonged constriction on regeneration of crushed nerve fibers, *Muscle and Nerve*, 11, 933, 1988 .

49. Hoffer, J. A., Haugland, M. K., Li, T., Obtaining skin contact force information from implanted nerve cuff recording electrodes, *Proc. IEEE Eng. in Med. & Biol. Soc. Intl. Conf.* 11, 928, 1989.

50. Haugland, M. K., Hoffer, J. A. Sinkjær, T. , Skin contact force information in sensory nerve signals recorded by implanted cuff electrodes, *IEEE Trans. Rehab. Eng.*, 2, 18, 1994.

51. Kostov, A., Fuhr, B., Strange, K., Hoffer, J. A., Potential role of afferent recordings as a sensory feedback in movement control systems: animal model, *Proc. IEEE — EMBS '98 Conference*, Hong Kong, pp. 2536, Oct. 1998.

52. Strange, K., Hoffer, J.A., Restoration of use of paralyzed limb muscles using sensory nerve signals for state control of FES-assisted walking, *IEEE Trans. Rehab. Engineering*, 7, 289, 1999.

53. Hambrecht, F. T., The history of neural stimulation and its relevance to future neural prostheses, in *Neural Prostheses: Fundamental Studies*, Agnew, W. F., McCreery, D. B., Eds., Prentice Hall, Englewood Cliffs, NJ, 1990, chap. 1.

54. *Neural prostheses: Fundamental studies*, Agnew, W. F., McCreery, D. B., Eds., Prentice Hall, Englewood Cliffs, NJ, 1990.

55. *Medtronic Lead Implant Manual*, Medtronic, Inc., 1990.

56. Bullara, L. A., *Implantable electrode array*, United States Patent No. 4,573,481, March 4, 1986.

57. Bullara, L. A., *Bidirectional helical electrode for nerve stimulation,* United States Patent No. 4,920,979, May 1, 1990.

58. Maschino, S. Ross B.G. Jr., *Circumneural electrode assembly,* United States Patent No. 5,531,778, Jul. 2, 1996.

59. Waters, R. , Electrical stimulation of the peroneal and femoral nerves in man, in *Functional Electrical Stimulation, Application in Neural Prostheses,* Hambrecht, F. T., Reswick, J. B., Eds., Biomed. Eng. and Instrum. Ser. 3, Marcel Dekker, Inc., New York and Basel, 1977, 55.

60. Glenn, W. W. L., Holcomb, W. G, Hogan, J. F., Kaneyuki, T., Kim J., Long-term stimulation of the phrenic nerve for diaphragm pacing, In *Functional Electrical Stimulation, Application in Neural Prostheses,* Hambrecht, F. T., Reswick, J. B., Eds., Biomed. Eng. and Instrum. Ser. 3, Marcel Dekker, Inc., New York and Basel, 1977, 97.

61. *Functional Electrical Stimulation, Application in Neural Prostheses,* Hambrecht, F. T., Reswick, J. B., Eds., Biomed. Eng. and Instrum. Ser. 3, Marcel Dekker, Inc., New York and Basel, 1977.

62. Naples, G. G., Mortimer J. T., Yuen, G. H., Overview of peripheral nerve electrode design and implantation, in *Neural Prostheses: Fundamental Studies,* Agnew, W. F., McCreery, D. B., Eds., Prentice Hall, Englewood Cliffs, NJ, 1990, chap. 5.

63. Naples, G. G., Sweeney, J. D., Mortimer, J. T., *Implantable cuff, method of manufacture, and method of installation,* United States Patent No. 4,602,624, Jul. 29, 1986.

64. Naples, G. G., Mortimer, J. T., Scheiner, A., Sweeney, J. D., A spiral nerve cuff electrode for peripheral nerve stimulation, *IEEE Trans. BME,* BME-35, 905, 1988.

65. Grandjean, P., Electrical stimulation of skeletal muscles, in *Cardiomyoplasty,* Carpentier, A., Chachques, J. C., Grandjean, P., Eds., Futura Publishing Company, Inc., New York, 1991, chap. 4.

66. Mortimer J. T., Electrical excitation of nerve, in *Neural Prostheses: Fundamental Studies*, Agnew, W. F., McCreery, D. B., Eds., Prentice Hall, Englewood Cliffs, NJ, 1990, chap. 3.

67. Robblee, L. S., Rose, T. L., Electrochemical guidelines for selection of protocols and electrode materials for neural stimulation, in *Neural Prostheses: Fundamental Studies*, Agnew, W. F., McCreery, D. B., Eds., Prentice Hall, Englewood Cliffs, NJ, 1990, chap. 2.

68. Fang, Z. P., Mortimer, J. T., Selective activation of small motor axons by quasitrapezoidal current pulses, *IEEE Trans BME,* 38, 168, 1991.

69. Sweeney, J. D., Mortimer, J. T., An asymmetric two electrode cuff for generation of unidirectionally propagated action potentials, *IEEE Trans BME,* 33, 541, 1986.

70. Sweeney, J. D., Ksienski, D. A., Mortimer, J. T., A nerve cuff technique for selective excitation of peripheral nerve trunk regions, *IEEE Trans. BME,* BME-37, 706, 1990.

71. McNeal, D. R., Bowman, B. R., Selective activation of muscles using peripheral nerve electrodes, *Med. Biol. Eng. Comput.,* 23, 249, 1985.

72. Rozman, J., Trlep, M., Multielectrode spiral cuff for selective stimulation of nerve fibers, *J. Med. Eng. & Tech.,* 16, 194, 1992.

73. Grill, W. M., Mortimer, J. T., The effect of stimulus pulse duration on selectivity of neural stimulation", *IEEE Trans. BME,* 43, 161. 1996.

74. Grill, W. M., Mortimer, J. T., Stability of the input-output properties of chronically implanted multiple contact nerve cuff electrodes, *IEEE Trans. Rehab. Eng.,* 6, 364, 1998.

75. Boznjak, R, Vinko, V. D., Kralj, A., Biomechanical response in the ankle to stimulation of lumbosacral nerve roots with spiral cuff multielectrode, *Neurol Med Chir (Tokyo),* 39, 659, 1999.

76. Loeb, G. E., Hoffer, J. A., Activity of spindle afferents from cat anterior thigh muscles. II. Effects of fusimotor blockade, *J. Neurophysiol.*, 54, 565, 1985.

77. Brown, J. C., Bouldin, T. W., Goodrum, J. F., Myelination of regenerating rat sciatic nerve occurs in the absence of cholesterol reutilization, *Soc Neurosci Abstr*, 24, 616.19, 1998.

78. Avery, R. E., Wepsic, J. S., *Implantable electrodes for the stimulation of the sciatic nerve*, United States Patent No. 3,738,368, June 12, 1973.

79. Avery, R. E., *Implantable nerve stimulation electrode*, United States Patent No. 3,774,618, Nov. 27, 1973.

80. *Cardiomyoplasty*, Carpentier, A., Chachques, J. C., Grandjean, P., Eds., Futura Publishing Company, Inc., New York, 1991.

81. Stein, R. B., Nichols, T. R., Jhamandas, J., Davis, L., Charles, D. Stable long-term recordings from cat peripheral nerves. *Brain Res.*, 128, 21, 1977.

82. Julien, C., Rossignol, S., Electroneurographic recordings with polymer cuff electrodes in paralyzed cats, *J. Neurosci. Methods*, 5, 267, 1982.

83. Loeb, G. E., Peck, R. A., Cuff electrodes for chronic stimulation and recording of peripheral nerve activity, *J. Neurosci. Methods*, 64, 95, 1996.

84. Kallesøe, K., Hoffer, J. A., *Flexible implantable electrochemical interfaces*, Submitted for US patent, Feb, 24, 2000.

85. Crampon, M. A., Sawan, M., Brailovksi, V., Trochu, F., New easy to install nerve cuff electrode using shape memory alloy armature, *Artif. Organs*, 23, 392, 1999.

86. Kallesøe, K., Hoffer, J. A., Strange, K. and Valenzuela, I., *Implantable Cuff having Improved Closure*, United States Patent No. 5,487,756, January 30, 1996.

87. Sahin, M., Durand, D. M., Haxhiu, M. A., Chronic recordings of hypoglossal nerve activity in a dog model of upper airway obstruction, *J. Appl. Physiol.*, in press.

88. Sinkjær, T., Haugland, M. K., Struijk, J., Riso, R., Long-term cuff electrode recordings from peripheral nerves in animals and humans, submitted to Springer-Verlag as a contribution to *Modern Techniques in Neuroscience*, 1999.

89. Hoffer, J. A., Kallesøe, K., Nerve cuffs for nerve repair and regeneration, *Prog. Brain Res.*, 128, 121, 2000.

90. Strange, K. D., Christensen, P. R., Chen, Y., Yoshida, K., Hoffer, J. A., Multichannel recordings from peripheral nerves: 3. Evaluation of selectivity using electrical stimulation of individual digits, *IFESS/Neural Prostheses V Int'l. Conf.*, Vancouver, BC, 243, 1997.

91. Strange, K. D., Hoffer, J. A., Gait phase information provided by sensory nerve activity during walking: applicability as state controller feedback for FES, *IEEE Trans. BME*, 46, 797, 1999.

92. Strange, K., Hoffer, J. A., Restoration of use of paralyzed limb muscles using sensory nerve signals for state control of FES-assisted walking, *IEEE Trans. Rehab. Eng.*, 7, 289, 1999.

93. Hansen, M., Hoffer, J. A., Strange, K. D., Chen, Y., Sensory feedback for control of reaching and grasping using functional electrical stimulation, *IFESS/Neural Prostheses V Int'l. Conf.*, Vancouver, BC, pp. 253, 1997.

94. Kilgore, K. L., Peckham, P. H., Thorpe, G. B., Keith, M. W., Gallaher-Stone, K. A., Synthesis of hand grasp using functional neuromuscular stimulation, *IEEE Trans. BME*, 36, 761, 1989.

95. Cybulski, G. R., Penn, R. D., Jaeger, T. J., Lower extremity functional neuromuscular stimulation in cases of spinal cord injury, *Neurosurgery*, 15, 132, 1984.

96. Veraart, C., Raftopoulos, C., Mortimer, J. T., Delbeke, J., Pins, D., Michaux, G., Vanlierde, A., Parrini, S., Wanet-Defalque, M. C., Visual sensations produced by optic nerve stimulation using an implanted self-sizing spiral cuff electrode, *Brain Res.*, 813, 181, 1998.

Part 2

Brain Control of Neural Prostheses

6 Designing a Brain-Machine Interface for Neuroprosthetic Control

Karen A. Moxon, James Morizio, John K. Chapin,
Miguel A. L. Nicolelis, and Patrick D. Wolf

CONTENTS

0-8493-2225-1/01/$0.00+$.50

6.1 INTRODUCTION

A central issue that arises from the previous chapters (describing neuroprosthetic devices that stimulate nerves and muscles) is how to control these neuroprosthetic devices. One solution, described in the next three chapters, is to use command signals recorded directly from electrodes implanted into the brain. While practical, clinical applications of this method are just beginning to be feasible (see Chapter 7 for a case study), no device exists for creating a command signal to control fine motor movements continuously in time. This device will require recording neural signals from many neurons across several motor areas of the brain. In this chapter, we will describe some of the issues and problems associated with producing brain derived command signals to control fine-grained temporal and spatial movements of a prosthetic device and, where applicable, point out some of the crucial design issues and trade-offs involved in developing such a device.

There are roughly four major subassemblies of a brain derived neuroprosthetic control device: (1) the electrodes subassembly, (2) signal conditioning subassembly, (3) signal acquisition subassembly, and (4) transmitter subassemblies (Table 6.1). Notably absent from these subassemblies is a device to generate the command signal. This issue will be addressed in Chapter 8. Each of these subassemblies has been studied in detail and some of the major advances are presented below. However, each of these steps requires major technological advances in order to produce a device for clinical use and we will attempt to point out some of the possible solutions.

Simultaneous recordings from large numbers (>64) of single neurons are routinely recorded from mammals in laboratory settings.[1-19] The development of multichannel recording systems provides a blueprint for how to develop an interface for clinical use. This interface must include brain tissue that is receptive to the electrode and a biocompatible electrode to interface with the neural tissue. The electrode will have to be integrated with appropriate signal conditioning electronics to extract useful data from the raw neural signals and a processing subassembly to combine the neural signals into a command signal (Figure 6.1). Lastly, a telemetric subassembly will likely be needed to tune the processing subassembly and transmit the command signal to the neuroprosthetic device. In each of these areas there is a need for improvement in order to build a device for clinical applications.

TABLE 6.1
Neuroprosthetic Control Device Subassemblies

Electrode Subassembly	Signal Conditioning	Signal Acquisition	Transmitter Subassembly
1. Neural Tissue Interface Section 6.2	1. Pre-amplifier Circuits Section 6.4.1	1. Mixed Signal VLSI Section 6.5.1	1. Bandwidth Concerns Section 6.6.1
2. Electrode Interface Section 6.3	2. Noise Sources Section 6.4.2	2. Device Packaging Constraints Section 6.5.2	2. Power Source Constraints Section 6.6.1
	3. Analog to Digital Conversion Section 6.4.3		

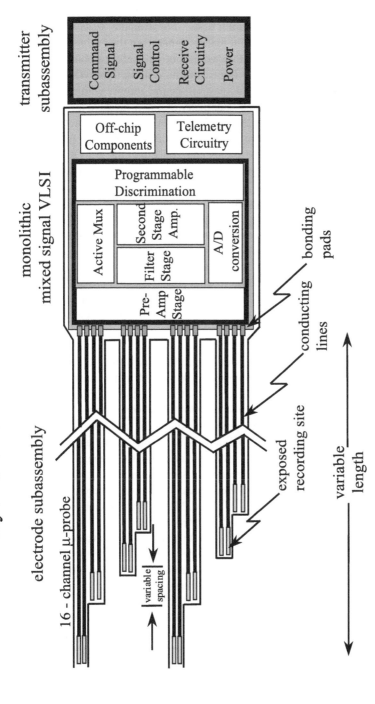

FIGURE 6.1 Schematic representation of a neuroprosthetic control device. The recording sites, conducting lines and bonding pads that define the electrode are patterned onto a wafer using photolithography. The electrodes are bonded to a hybrid circuit device containing mixed-signal VLSI for signal conditioning, amplification, multiplexing and telemetry circuits. The signals would be transmitted to a remote device. Ideally, the system would include on-chip templates for discriminating individual action potential waveforms.

While current research is also pursuing non-invasive techniques to acquire signals from the brain, the results of these studies suggest that it may not be possible to acquire the appropriate spatial and time resolution that can be achieved with electrodes implanted directly into brain tissue. For example, EEG signals and field potentials have been used to extract a binary signal from the brain to register on/off signals (see Chapter 2). However, this would not be useful to control fine-grained temporal and spatial movements of limbs continuously in time. While non-invasive techniques may be able to pick up increases in neural activity within the appropriate brain region for a particular limb, it is unlikely that non-invasive techniques will be able to resolve neural signals at the single neuron level or interpret brain derived commands for fine-grained control of movement. The information presented in this chapter is therefore a sampling of issues and concerns related to the development of a brain derived command signal that combine signals from large numbers of individual neurons.

These issues include minimizing the damage from invasive microelectrodes, which will require re-engineering the response of neural tissue to the introduction of the electrode. Recent studies[20] of the response of neural tissue to stab wounds have shown the responses to be stereotypic. There appears to be a largely trophic, early phase and a more adverse late phase. The underlying causes of progressive degeneration of neural tissue leading to glial scar formation are still under investigation. Section 6.2 presents some recent studies of the molecular biology of brain injury to provide insights into how the brain's response to the electrode might be re-engineered to produce a limited response to the initial insult and to stabilize the brain's interface with the electrode.

Section 6.3 presents an overview of electrodes used to record neural signals. As demonstrated in the previous chapters and presented in the next three chapters, microwires have been the most successful electrode for recording brain derived signals and stimulating nerve cells and muscles. However, it is rare for these signals to be maintained for more than one year and usually the signal lasts for only several months. In addition, the volume of brain tissue displaced for each recording site (i.e., each wire) is high and will result in more neural tissue damage when compared with electrodes that have multiple recording sites. Thin-film technology holds out the promise of highly compatible electrodes with several recording sites while displacing relatively small volumes of tissue. Manufacturing techniques and examples of current thin-film electrodes will be presented.

Issues surrounding the acquisition of single neuron signals in a device for clinical use are presented in Sections 6.4 and 6.5. These issues include ultra-miniature on-board electronics for signal processing and direct output to the prosthetic device. The on-board signal processing includes subassemblies to amplify and filter the neural signals, digitize them, and combine them to generate a command signal. Packaging and bandwidth constraints limit the ability to process analog neural signals from large numbers of channels, and issues involving on board digitizing and multiplexing the signal are explored. Lastly, Section 6.5 presents some of the issues involved in transmitting these signals via telemetry. There are limited examples of most of the subassemblies necessary for such a device and no examples of a complete and functional device for routine clinical use. The goal of this chapter, then, is to describe some of the currently available technology and discuss where major improvements in subassembly technology are necessary. How to use these signals to actually create a command signal is addressed in Chapter 8.

6.2 ENGINEERING THE NEURAL INTERFACE

6.2.1 RE-ENGINEERING THE GLIAL RESPONSE

Recent studies examining the effects of acute neural injury suggest that the damage to the brain caused by inserting a microelectrode can be controlled to maximize the beneficial aspects of the response and minimize or eliminate adverse responses.[21] Inserting the electrode into the brain tissue can cause severe damage and result in glial scarring preventing the neural signals from being recorded. In general, the neural response to acute injury has two parts, a short- and a long-term response. If the short-term response can be controlled, it appears likely that the long-term response, glial scarring, can be prevented.

Acute injury to the brain triggers a large network of morphological and metabolic changes that play a role in two crucial physiological processes: protection against infectious agents and repair of the damaged tissue.[22-23] In response to injury, the damaged tissue assumes a state of emergency, rapidly changing its gene expression and stimulating nearby microglia and astrocytes. This neuroglial response is a graded, stereotypic response that is readily observed during insult to the brain such as the introduction of a recording electrode. This stereotypic response consists of a largely beneficial initial phase and a more adverse long-term response that is dependent on the extent of the injury. Because the initial response is stereotypic across a wide variety of injuries, it may be possible to understand the signals that lead to neuroglial activation and create a targeted intervention to control the response, reduce the adverse nature of the response and maintain an ideal environment for the brain–electrode interface.

By re-engineering the brain's response to the introduction of electrodes through controlled molecular engineering, it may be possible to find a balance that minimizes any adverse responses and enhances the trophic responses. When the electrode is introduced to the brain, regardless of the method of introduction (see Section 6.3 below for details), neural processes will be damaged and likely severed from the cell body and some neurons will die from the electrode penetration. The blood brain barrier will be disrupted and tiny blood vessels will be torn all along the shaft of the electrode.[24] The brain's response to this insult includes a response by microglia, astrocytes, macrophages, leukocytes, and expression of signaling molecules. Each of these cellular responses plays an important function in protecting the brain from infectious disease and further injury.

The initial responses of microglia to brain insult are likely beneficial and include production of neurotrophic substances and cell adhesion molecules, which support injured neurons.[25] Activated cells express glia filament asssociated proteins (GFAP) near the site of injury and into surrounding tissue and intermediate filament associated proteins (IFAP) at sites near tissue damage.[26] Expression of these proteins appears necessary for restorative events to take place. Molecules that may aid in controlling the microglia response include nerve growth factor[27] and transforming growth factor (TGF), which strongly inhibit proliferation of astrocytes and may play a role in tissue repair.[28] A second, beneficial molecule is protein-S (PS) which has been shown to markedly suppress proliferation of astrocytes and thereby reduce gliosis. When PS is contacted by tyrosine-3, the combined system stops further gliosis in later stages of wound healing after CNS injury.[29]

These events allow the microglia to efficiently take up ions that diffuse from the injured cells. If neurons die during the implantation of the electrodes, the microglia will transform into phagocytotic cells and remove the neural debris.[30-31] However, if the injury is severe, which is likely the case with the introduction of electrodes, the reactive astrocytes can create a glial scar effectively producing a physical barrier between the electrode and healthy cells.[22] Although the extent of scarring depends on the severity of the brain damage, a minimal response around affected neurons is always observed after a mild, indirect trauma or axotomy. This glial scarring is composed of a dense network of hypertrophic astrocytes with thick interdigitating processes and associated extracellular matrix. This buildup glial scar can prevent contact of the electrode with healthy tissue. However, it possible for phagocytotic microglia to remove cellular debris within a few days, detach from the microglial nodules and down regulate their activation markers which instigated the process. The goal, then, for re-engineering the glial response is two fold: first, to ensure that the damage to neural tissue is minimized during implantation of the electrode (see Section 6.3 below) and second, to ensure that the neuroglia response is limited.

Activation of leukocytes, like that of microglia and astrocytes, also seems to have two phases: an early largely beneficial stage and a second stage that is dependent on the extent of the injury. During the early stages, leukocytes aggregate around phagocytotic microglia and provide a structural base for efficient antigen presentation.[32] However, later stages involving long-term presences of lymphocytes may cause a secondary, antigen-mediated neurotoxicity.[33-34]

In addition to microglia and leukocytes, if the blood brain barrier (BBB) is damaged, macrophages from the blood stream can enter the neural tissue and become activated.[35] Some microglia that migrate from surrounding tissue can also become macrophage-like when activated. Leakage from the BBB has been shown to be necessary but not sufficient to activate the macrophage response to toxic levels that are inhibitory to the healing process in the brain. Introduction of activated macrophages into neural tissue can result in a cascade of events that leads to severe necrosis and cavitation around the point of entry.[22] Generally, these events are associated with an increase in cytokines that may produce a secondary inflammation that enhances the formation of a glial scar.

The absolute number of activated macrophages and concentrations of cytokines appear to be the major pathological correlate of disease severity in response to this type of activation. Some cytokines are restricted to severe brain injury while others serve different functions depending on the type of injury. Many of the cytokines including IL-6 promote the release of neurotrophic factors such as neural growth factor (NGF).[36] However, the effects of NGF are also controversial and any neuro-protective benefit is limited to a period just after injury. Prolonged exposure to NGF may induce β-amyloid precursor protein messenger ribonucleic acid (mRNA) formation, which can lead to plaques.[37]

Both pro- and anti-inflammatory cytokines have been shown to regulate glial scar formation.[38-41] Studies using transgenic and knockout models have shown that cytokines act as molecular signals, which orchestrate the graded cellular response in the damaged brain. This work suggests molecular targets for possible pharmacological intervention that could be used to manipulate the magnitude of the glial response and improve neurological recovery after introduction of an electrode.

Reducing the number of microglia/macrophages leads to an attenuation of astrocytic activation.[42-43] While it is likely that the microglial buildup in response to brain injury alone may be relatively harmless, the inhibition of molecular markers on activated glia strongly reduces post-injury tissue loss and enhances the protective effects of the glial response. Re-engineering this response to maximize the ability of the brain's immune system to screen for possible pathogens while preventing excessive glial activation appears to be possible. This function is assisted by gradual transformation of microglia into antigen presenting cells expressing a large set of immunoreactive molecules.

Another method for controlling the severity of the neural glia response[21,44-46] suggests that the endothelin response may also be controlled. For example, endothelins can act as trophic factors causing the beneficial proliferation of astrocytes during initial phases of brain injury. By introducing an endothelin receptor antagonist, astrocyte proliferation was attenuated. The content of endothelins is increased at brain-injured sites and are derived from blood components and brain cells.[47-48] There is extensive coupling of endothelia and astrocytes in the maintenance of the blood brain barrier. Controlling the endothelins' response can benefit repair of the BBB after insertion of the electrode.

Microglia adhere closely to the neuronal cell bodies, placing them in a strategic position to remove degenerating neural processes and prevent debris or toxic material from leaking out of the damaged neurons. However, this process must be stopped before astrocytes are able to produce dense glial scars and wall off the electrode from the rest of the brain tissue. The knowledge that both trophic and adverse pathways can coexist suggests that future research that delineates the mechanisms that regulate these pathways might yield optimal brain responses to the introduction of the electrode.

These new ideas from the study of brain injury and especially stab wounds provide several methods for modifying the neural response to insertion of the electrode. It is clear from these studies that the initial response of microglia is beneficial and should be allowed to take place. However, the response needs to be controlled to ensure that excessive proliferation of activated microglia and macrophages does not induce secondary inflammation and lead to cavitation, necrosis and glial scarring.

6.2.2 STABILIZING THE RECORDINGS OVER LONG PERIODS

In order to ensure that the initial injury to the brain is minimized and, therefore, glial scarring is minimized, the electrode must be properly introduced to the brain. Proper introduction would minimize damage due to the initial insult to the brain and maximize the ability of the electrode to remain in the brain without creating further damage. This would require little or no damage be done to capillaries, axons and dendritic processes throughout the brain and that cell death be minimzed. In order to achieve this it is probably best to introduce the electrode to the brain by cutting through the brain tissue, not tearing the tissue.[24] Tearing neural tissue can result in stress and damage through large volumes of tissue compared with damage created by cutting through the tissue. In addition, with what is currently known about the glial response to brain damage (Section 6.2.1) it is clear that minimizing the damage is essential and that the brain is actually very efficient at removing small amounts of damaged neural tissue and protecting against infection. Therefore, an ideal electrode would a be smooth, ultra-sharp electrode with a tip that will slide past or sever blood vessels and neural processes without distorting or tearing cell walls. Excessive tearing could lead to an increased amount of bleeding which will increase the response of paravascular macrophages.

An additional question raised by the issue of electrode shape is how the electrode should be inserted. It appears that slowly lowering the electrode into the tissue produces the least amount of stress and minimizes tissue tearing. This is especially true if the electrode is to access deep brain nuclei. Then, once the electrode has been introduced into the brain, further damage needs to be minimized by eliminating the movement of the electrode against the brain tissue. One way to achieve this is by anchoring the electrode to the brain tissue. Novel substrates for electrodes designed with an active matrix that includes neuro-protective molecules in the matrix may be an effective way to minimize damage. Later, if neural processes grow into the matrix it would keep the electrode from moving within the local brain tissue and prevent continued damage to the neural tissue. For example, a form of silicon with a porous network has been suggested. If this network was treated with anti-inflammatory agents and attached to the electrode it could be used to minimize the toxic glial response and to anchor the electrode in the neural tissue, preventing further move-ment and eliminating further irritation of the neural tissue.

Another question related to the stability of recordings is whether the electrode should be anchored to the skull or left floating in the brain. Since, especially in humans, the brain moves a significant amount within the skull cavity it has been suggested that if the electrode is connected to the skull it will continually cut through the brain tissue when the brain moves. Two ingenious methods have been developed to circumvent this problem. One, developed by the University of Michigan Center for Sensor technology is a silicon based ribbon cable.[49] By using boron doped silicon to define a small (approximately 5 micron diameter) cable between the electrode and the connector, the electrode can essentially move with the brain tissue. The silicon ribbon is flexible and deforms with the movement of the brain. A second method was developed by the Univeristy of Utah[50] (also see Section 6.3 below) and is an array of electrodes etched out of a block of silicon such that the pad holding all of the electrodes floats on the surface of the brain. Liu et al.[51] reported stability of recording signals in cats for up to 250 days using an iridium wire electrode array that floats on the pia surface. Their work suggests that tissue growth around the wires helps to stabilize the electrode after several days, allowing for consistent neural recordings for the next few months. However, because the array is allowed to float on the pia, they also suggest this neural growth gradually pushes the microelectrode down into the tissue or out of the tissue of interest resulting in the loss of the targeted neural signals. While each of these anchoring methods holds out promise, none has shown consistent recordings of single neurons for years.

6.2.3 ROLE OF NEURAL PLASTICITY

A final issue that relates to the ability of the brain to maintain an effective interface with the electrode is neural plasticity. These issues are addressed in detail in the subsequent three chapters. However, a long history of examining neural plasticity in primates suggests that the somatosensory response properties of single neurons undergo short- and long-term plastic changes.[52-55] There is reason to believe that the motor response properties of cells can also undergo plastic changes[56] (also see Chapters 7 and 8). This plasticity will be essential for long-term success of brain derived neural signals for the control of prosthetic devices. Because patients who would require devices to control their prosthetics will likely have a long recovery period before a device is implanted, the response properties of neurons will change and likely no longer perform the functions they performed before the injury or disease. However, these plasticity changes will continue after the neural implant is complete and motor signals to control the prosthetic device can be tuned to maximize control (see Chapter 7 for an example).

6.3 ENGINEERING THE ELECTRODE INTERFACE

While the above section has demonstrated that it is possible to affect the way the brain responds to the electrode, it is also necessary to engineer the electrodes to maintain effective contact with neural tissue and minimize adverse reaction by the brain. Since signals from large numbers of neurons from several different regions of the brain will be required to control fine motor movements, it will be necessary to maximize the number of recording sites per volume of neural tissue displaced (RS/NTD ratio). Increasing the number of active recording sites increases the likelihood of getting and maintaining appropriate neural signals for neuroprosthetic control.[56] As outlined in Section 6.2 above, reducing the amount of brain trauma reduces the likelihood of glial scarring. Microwire array electrodes that record chronic neural signals from multiple channels are available for recordings in mammals including humans (see Chapter 7 for a case study). However, recent advances in thin-film technology have shown that it is possible to produce single microelectrodes as small as a single microwire with as many as 16 recording sites.[57] These thin-film arrays have a much greater RS/NTD ratio when compared to traditional microwire electrode. Both types of electrodes will be reviewed below.

Combining the spatial and temporal activity of large numbers of neurons' individual action potentials into a command signal to control a prosthetic device will require implanting several electrodes, each with several recording sites. The probability of recording at least one neuron per channel is a function of the quality of the electrode interface with the neural tissue. At this interface, a double layer of electrical charge is formed, creating a high impedance interface.[58] The electrodes are described in the following sections and the subsequent processing of the signals is discussed in Sections 6.4 and 6.5.

6.3.1 MICROWIRES

To date, insulated metal wires have made the most successful, stable chronic recording electrode.[59-60] Each of the neuroprosthetic devices presented in this book was successful using some type of wire electrode to record from or stimulate the neural tissue. Willliams et al.[8] presented data suggesting that single unit recordings can be maintained routinely for three months. The wire must have a small diameter while maintaining adequate stiffness and tip shape to penetrate the dura and traverse the tissue with minimal bending and mechanical disturbance to surrounding tissue. The wire must be insulated with a material that is biologically inert in order to minimize the reactivity with the tissue. If the impedance of the electrodes is low with minimal noise, it is more likely to record multiple neurons per wire, thus greatly increasing the yield of these electrode arrays. These electrodes have generally been produced within the lab by paying particular attention to the tips.

The benefits of the 25–50 micron wires include being strong enough to go through the dura of rodents and rigid enough to be lowered into the correct position. Wires smaller than 25 microns have difficulty penetrating the dura and can be diverted by a fiber bundle when lowered into the brain. Therefore, they often do not end up in the intended brain structure.

The tips of these electrodes can be carefully shaped to enhance the neural signal. The electrode tips can be shaped by chemical etching or beveling. Current theory suggests that a fine-pointed tip, with insulation covering the wire as close to the tip as possible, is the most ideal. The smaller the exposed surface area, the higher the input impedance and therefore the better the signal-to-noise characteristics of the electrode. However, suitable recordings have been acquired from blunt tip electrodes.[61] The divergent theories regarding the effect of tip shape arise because the resulting interface with the brain has not been sufficiently studied.

The two most popular types of wires are tungsten[16] and platinum,[62] though carbon has been used with success.[63] Several types of insulating material and pre-insulated wires are now readily available. Two commonly used insulating materials are polyimide or polydicholoroparaxylene[64] and Teflon.[65] The integrity of the insulation is very important. If the insulation is breached and the wires come in contact with each other or with surrounding tissue, the recordings will be compromised.

A distinct advantage to using arrays of microwires is that the wires can be shaped at the time of implant.[59-66] Unique geometric configurations of electrodes can be used to compensate for the complex structure of the brain. These electrode arrays can be shaped and positioned to record from multiple dimensions. They can be configured to follow the curvature of a structure within the brain and lowered during surgery at various angles to ensure proper placement in the brain area of interest or avoid trouble spots such as the ventricles.

There are three main disadvantages to using microwires for neural control devices. The first is the low RS/NTD ratio; the second is the difficulty of integrating on-board electronics. It is important to integrate minimal signal conditioning electronics as close to the electrode as possible to reduce noise (see Section 6.4 below). Lastly, there has been no method yet developed to produce quality microwire electrodes for neural recording using batch processing. Generally these electrodes are made by hand and the result is considerable variation in the recording characteristics of each electrode tip. As discussed below (Section 6.3.2), since on-board electronics must be included, a wide range of electrical characteristics of the electrode results in an inability to properly match impedances.[67] This variability can be overcome with the development of thin-film electrodes described in the following section. Thin-film electrodes are alternative electrodes that have the potential to maintain the brain–electrode interface using a high RS/NTD ratio using a substrate material and insulation that is strong and resistant to the brain microenvironment. Novel methods for applying the insulation can create an electrode that will last in the saline microenvironment of the brain for decades. These issues and manufacturing methods will be described below.

6.3.2 THIN-FILM ELECTRODES

While there are examples of thin-film electrodes to record chronic neuron activity,[2,9,12,68-70] to produce an electrode subassembly for a neural control device will require recordings that are stable for a period much longer than the current state-of-the-art and on-board electronics to reduce noise.[71-74] Since the hostile response of the brain microenvironment is a function of the severity of the insult, the electrode must also be engineered to limit a negative reaction and enhance the signal recorded from the brain. In addition, the electrode will have to be manufactured on a supporting substrate that is resistant to the background neural activity so that cross talk between recording sites is minimized.[57] The electrode will also require an insulating material capable of protecting the electrode from the corrosive brain environment for decades.

Recent advances in thin-film technology provide a method to produce a more ideal electrode interface for a neuroprosthetic control device.[75] There are several distinct advantages of thin-film electrodes over microwires. First, the technology allows these electrodes to be very small compared with microwires. In addition, each electrode can have multiple recording sites, thereby greatly reducing the RS/NTD ratio. The devices can be produced using precision techniques, in large quantities that ensure the same recording characteristics for ideal impedance matching, precise inter-electrode spacing and on-board electronics. Thin-film electrode designs have been produced since the early seventies.[2,12,58,68-70] Thin-film technology is the process used to create the recording sites, bonding pads and signal conditioning circuitry by applying thin films of metal or dopants to a substrate (Figure 6.2). Traditionally integrated circuits were built on semiconductors such as silicon and germanium. Wafers were grown with a crystalline structure. In their pure state, these semiconductors make poor conductors, but can be "doped" with a material such that the conductivity of the silicon can be controlled. The dopant either lends an electron to the crystalline structure or takes an electron away, leaving a hole, so that the material becomes conductive. For example, boron doped silicon is a good conductor

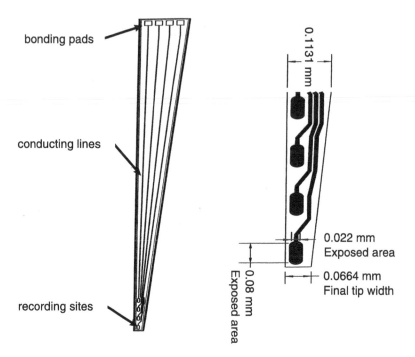

FIGURE 6.2 Drawing of a four-site thin-film electrode. A. Recording sites are defined at the tip of the electrode. Conducting lines transmit the recorded signal to bonding pads at the opposite end of the electrode. B. The size of the electrode is dependent on the smallest feature size available. The drawing presented here shows recording sites 0.08 × 0.02 mm. Each conducting line is 0.10 mm wide and the spacing between features is 0.10 mm. The design shows a blunt tip but a sharp tip for cutting though neural tissue could also be designed.

because boron removes an electron from the silicon creating a "hole" that now easily moves through the material.[72] By patterning the wafer, holes and extra electrons can be precisely added to the silicon substrate to define circuits. In addition to doping the material to define devices such as amplifiers and filters, thin-film technology can be used to put precisely defined lines of metal on the wafer by the process of metallization. These metal lines define the recording sites, bonding pads and conducting lines (refer to Figure 6.2). Thin films of metals can be patterned onto the surface of a wafer with feature sizes as small as 5 microns.[57]

These processes are the same processes used to pattern integrated circuits onto the wafer. In fact, one of the distinct advantages of using thin-film technology and starting with a semiconductor wafer is the idea that integrated circuits can be patterned onto the wafer along with the design of the electrode. Having signal-conditioning circuitry close to the source of the recording site reduces noise in the signal. When designing integrated circuits, several processes including etching, doping, and chemical vapor deposition are used to define the circuits. The process basically entails using photolithography (described below in Section 6.3.2.1) to transfer the design from a mask that defines the electrode to a substrate that acts as the carrier for the electrode.

6.3.2.1 Photolithography

The photolithography process uses a mask to define a pattern in a photosensitive material, the photoresist. For this process, a mask is created that defines the pattern to be transferred to the substrate. The mask design and production is a relatively expensive process, and generally one mask is produced and used repeatedly to make thousands of components. Once a mask is produced that defines the pattern it can then be used to transfer this pattern to the substrate by creating structures in the photoresist on the wafer. During development, the unexposed resist is removed, exposing the substrate in locations where further processing is to be performed. The resist left behind protects the covered substrate from further processing steps. This process creates resist structures on the wafer that define the electrode. The photolithography process begins with the substrate first being cleaned and prepared for processing. To prepare the photoresist for application to the substrate, the photoresist is dissolved in a solvent. The photoresist materials are light-sensitive polymers. When the material is exposed to light (generally ultraviolet light [UV]), the resist turns from developable to non-developable (positive resists) or vice versa (negative resists).[76] The resist is applied to the substrate by spin coating to ensure that the resist forms a uniform layer. This is a critical step and many factors influence the proper coating of the substrate with resist, including the viscosity of the photoresist and the amount deposited on the wafer. The dissolved photoresist is applied to the surface of the wafer and the wafer spun at high speeds to create the uniform coating of resist over the wafer. Other methods for applying the photoresist include spraying for liquid crystal device manufacturing and meniscus coating for organic films such as polyimide, but these processes are generally not performed during standard integrated circuit production. The wafer is then soft baked to evaporate some of the solvent from the resist film. This process increases the ability of the resist to adhere to the wafer. Because the resist will define the pattern applied to the wafer, it is essential that it adheres properly to the wafer.

After the wafer is properly coated with the photoresist, the mask is laid over the substrate and the photoresist is exposed using UV light and developed to create the resist structures. The resolution of the resist structures is a function of the wavelength of light used to expose the photoresist (Equation 6.1). By using shorter wavelengths of light, the resolution of the pattern can be increased, allowing smaller feature sizes. This is due to the relationship between resolution and wavelength:

$$\nu_o = \frac{2NA}{\lambda} \qquad (6.1)$$

where: ν_o = frequency cutoff
NA = numerical aperture
λ = wavelength in millimeters

The numerical aperture (NA) is a function of the optical system used to produce the light. Therefore, the resolution ($1/\nu_o$) can be reduced by reducing the wavelength. In addition to the resolution and resist thickness, the size of the wafer influences the resolution of the features. The size of the wafer is important because it is difficult to apply uniform light intensity as the wafer size increases. The solution to this problem is to use a step-and-repeat process. For this process, only a small section of the wafer is exposed at a time. The mask is then moved and the process is repeated. This ensures uniform light exposure across the entire wafer.

Another important factor, especially for the metallization process, is the integrity and shape of the sidewalls. Many factors contribute to the integrity of the sidewalls including the thickness of the photoresist applied, the wavelength of light used to expose the photoresist, the size of the original pattern (step-and-repeat process) and the process steps that define the angle of the sidewalls. Often a negative photoresist is used because the development time between the exposed and unexposed resist is much greater than with a positive resist.[77] This ensures better definition of the pattern. With a negative photoresist, after exposure, the resist is changed from an unpolymerized state to a polymerized state. When the substrate is developed, the developing chemical rapidly dissolves the unpolymerized region and the polymerized region is dissolved much more slowly. After development of the substrate, the result is that the resist is removed from the wafer in all the places that the mask defined for the application of metal. At this point, the wafer then undergoes a hardbake to ensure good adhesion of the remaining photoresist to the wafer by evaporating the remaining solvent out of the photoresist. The exposed surface of the wafer is ready for further processing. If silicon or another semiconductor is used, the exposed surface of the wafer generally requires an etching step to remove oxides that may have formed on the surface of the wafer. The patterned wafer is now ready for processing using standard procedures that include doping to create n or p type regions for integrated circuits (ICs) or deposition of thin films of metal. The metal can be applied to the wafer using sputtering, evaporation, chemical vapor deposition (CVD), or ion-beam assisted deposition (IBAD).

6.3.2.2   Metallization

To create the metal recording sites, conducting lines and bonding pads that define the electrode, a thin film of metal is applied to the substrate after the resist features are created during the photolithographic process. Photomasks are made that define the pattern for the metallization. Photoresist is applied, exposed to UV light and developed. For metallization, care must be taken to ensure that the pattern defined by the photoresist will produce a continuous line. Any cracks or breaks in the lines destroy the integrity of the electrode. On the other hand, the smaller the feature or line size, the more recording sites can be fit onto the electrode. There are limits to the size of any feature that can be patterned. Ten micron features are standard and easy to create. Features smaller than 10 microns require special processing steps. Typically, to define metal electrodes, the metal is applied by evaporation or sputtering. When the metal is applied to the wafer it covers the remaining photoresist as well as the exposed wafer. Layering metals onto the substrate is a multistep process similar to the process used to dope a semiconductor.

Because the thickness of photoresist exposed is a function of the length of time the UV light penetrates the resist, the resist farthest away from the wafer (on top)

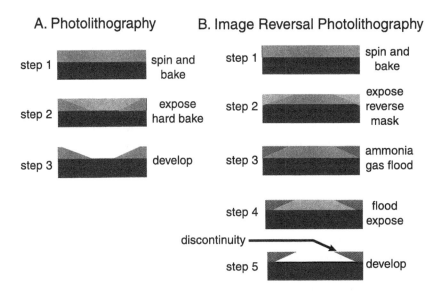

FIGURE 6.3 Basic step in the photolithography process to deposit photoresist structures onto a wafer. A. Demonstration of a positive sidewall angle. The photoresist (grey) is spun onto a substrate (black), usually silicon or ceramic. A mask defining the pattern is placed over the substrate and the substrate is exposed to UV light. The photoresist is converted (dotted pattern) and can subsequently be removed during development to define areas where metal can be applied to the substrate. During development the resist closest to the wafer which was not completely exposed is not removed. The resulting positive sidewall angle is ideal for etched doping masks because it avoids sharp corners. B. When image reversal photolithography is used, a negative mask is applied to the photoresist and the initial exposure to UV converts photoresist where no metal is wanted. The substrate is then flood exposed with ammonium gas to transform the converted photoresist to carboxylic acid, making it resistant to further exposure. The substrate is then flood exposed with UV light, converting the remaining photoresist, which is subsequently developed away. This process of reverse photolithography produces a negative sidewall angle. During subsequent metallization steps, this negative sidewall angle ensures good adhesion of the metal to the substrate because the base is broader than the top of the metal structure. In addition, the discontinuity in the photoresist structure ensures ideal lift-off conditions (refer to Figure 6.4).

is exposed to the light longer than the resist closest to the wafer. When the resist is developed, the sidewall angle produced is generally a positive angle, wider at the top, narrower at the bottom because the light does not always penetrate to the bottom of the photoresist (Figure 6.3A). Therefore, thinner photoresist produces a sidewall angle closer to 90 degrees. The sidewall angle can be varied depending on the process to be performed. For example, a positive sidewall angle is used when etched doping masks are used to avoid sharp corners. However, for metallization processes, a negative sidewall angle is preferable.

To produce a negative sidewall angle, an effective procedure is reverse photolithography.[78] In this case, a reverse mask is used. The process allows for precise control of the resist wall angles and improves the resolution of the features applied to the substrate. For this process, the initial mask is the reversal of the usual positive mask. The photoresist exposed is the area where the metal is to adhere to the substrate. Following UV light exposure, the wafer is flooded with an amine vapor that binds to the exposed resist, neutralizing it by forming a carboxylic acid that is highly insensitive to development. The UV light can no longer affect the previously exposed resist. The entire wafer is now exposed to UV light and the reverse image from the original mask is now polymerized and accessible to subsequent development. The photoresist exposed in this last step is removed and the pattern, with a negative sidewall angle, is produced (Figure 6.3B). An added benefit is that the widths of the features to be patterned are now well defined down to the level of the substrate.[79]

Once the wafer is prepared with appropriate resist features that define the electrode pattern, the metal is applied to the wafer, completely covering the exposed regions and the resist features. There are many choices for the metal conductor, including nickel, stainless steel, tungsten, gold or platinum. Gold can cause difficulties in the manufacturing process because it is very soft. When the lift-off or etching process is conducted, it may be difficult to leave behind a quality metal line. Platinum has been shown to cause the least histological damage for long-term implants.[80] However, because of the nature of the thin-film process, one conductor can be used for the circuit while a final layer of the material most desirable as an electrode–biological interface can be sputtered only on the recording sites.

Processes for metal deposition include sputtering, evaporation and ion beam deposition. Where the resist has been developed away, the metal resides directly on the wafer surface. The negative sidewall angle results in a thinner layer of metal at the step from wafer surface to resist (Figure 6.4). The unwanted metal along with the underlying photoresist is then removed using a lift-off procedure. The wafer is submersed in a photoresist stripper, and the photoresist expands, cracking the surface of the metal at the step and allowing the stripper access to the phototresist. The remaining photoresist is "lifted off" along with the unwanted overlying metal. The metal left behind defines the recording sites, conducting lines and the bonding pads of the electrode (Figure 6.5).

6.3.2.3 Bio-Resilient Features

In addition to general issues of biocompatibility, described in Section 6.2 above, there are several issues that must be addressed regarding the ability of a multisite probe to remain effective for long periods of time. The electrode must be small to minimze neural damage and strong to maintain its integrity for decades. The most popular trend in thin-film electrodes has been to use silicon as a substrate. However, since silicon is a semiconductor, when it is introduced into the brain microenvironment, it will conduct neural signals and must be completely insulated. The surrounding neural tissue (including tissue in contact with the back side) and the recording sites patterned onto the substrate must be isolated from substrate in order for the individual recording sites to maintain their integrity in the electrically noisy brain. A thorough study of this phenomena[57] showed that when a dielectric layer (either SiO_2 or Si_3N_4) is used to insulate the silicon substrate, a coupling capacitance is

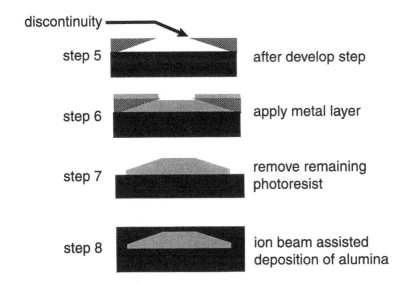

FIGURE 6.4 Example of metal applied over the step of a photoresist feature. The negative sidewall angle and discontinuity ensure a thin layer of metal over this step. During the lift-off step, the substrate is immersed in acetone to dissolve the remaining photoresist. The discontinuity allows easy access to the photoresist for better defined metal structures, which subsequently allow smaller feature sizes. During the insulation step (step 8), the insulating layer can better encase the metal when there is a negative sidewall angle than if there is a positive sidewall angle, which often leads to discontinuities in the insulation.

introduced between the silicon semiconductor and the metal of the electrode with the insulation acting as the dielectric. This capacitance between electrode and recording site is in addition to the coupling capacitance between conducting lines on the electrode. The conductivity of the silicon substrate is increased if the silicon is doped with ions such as boron, effectively increasing the coupling capacitance. This coupling capacitance can be eliminated if an insulating substrate, such as ceramic, is used instead of more traditional semiconductors.[79]

In addition to the issue of using an insulating substrate, it is important that the electrode be strong. Both silicon and ceramic are strong, with ceramic probably being slightly stronger due to increased bond strength.[81] However, the thin substrates used in the electrode manufacturing process produce substrates that are quite fragile regardless of the material used. However, certain advantages can be leveraged depending on the material. For example, a silicon based electrode is extremely flexible.[81] This may be advantageous under certain circumstances, for example a cochlear implant that requires the electrode to bend along a curved structure.[72] However, for recording or stimulating in deep structures it is preferable to have a stiff electrode that will penetrate the brain and maintain a straight course through the brain tissue to the desired target structure.[16] Ceramic, even for relatively thin substrates (<45 microns), maintains its stiff nature and is ideal for reaching deep structures in the brain.

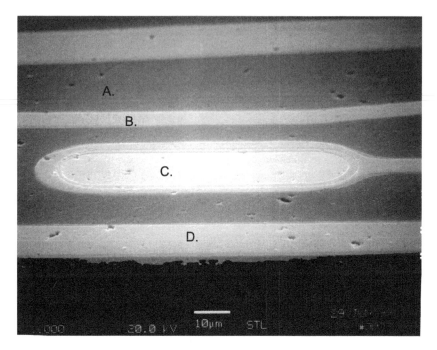

FIGURE 6.5 Scanning electron micrographs of platinum electrodes patterned onto a ceramic wafer. The drawings in Figure 6.2 were used to design this electrode. A. The ceramic substrate. During subsequent processing steps, the electrodes will be cut from this supporting ceramic structure. B. A conducting line from an electrode whose recording site is to the left of the figure. C. One of the four recording sites. The center is the platinum surface of the electrode that is used to conduct electrical signals from the brain. The immediately surrounding edge indicates where insulation has been added to the electrode. This insulation covers the conducting line indicated by B. D. Guide lines used to cut the electrode from the wafer. There is an additional guide line at the top of the figure. These electrodes are cut from the wafer using a guided laser.

A final characteristic of electrodes for a neuroprosthetic control device is the size. In light of data presented in Section 6.2, it is clear that by displacing the least amount of brain tissue, the smallest potentially negative brain response is produced. The University of Michigan center for sensor technology has developed a unique solution to this problem.[79] By performing a boron etch step that dopes the silicon with boron ions before patterning the electrode, the final shape of the electrode can be defined by the amount of boron diffused. Therefore, by carefully controlling the boron diffusion process, multisite electrodes (up to 16 sites) that are 15 microns thick have been manufactured. This represents two orders of magnitude improvement in the RS/BTD ratio when compared with traditional microwires. However, as mentioned above, boron doped silicon is more conductive than natural silicon and the coupling capacitance is increased.

For chronic recording, a more flexible cable is required. Several different kinds of flexible cables are available. Hetke et al.[49] have developed a silicon ribbon cable that is extremely flexible and can be from 2 to 20 microns in thickness. The advantage is that the same process that is used to make the electrodes is used to make the cable. Therefore, by applying appropriate masks, the cable and electrode can be integrated during fabrication, eliminating the need to bond the electrode to the cable.

Bionic Technologies provides a very unique electrode design. The arrays consist of up to 100 sharpened needles that project from a silicon substrate.[82-83] The needles are electrically isolated with a nonconducting glass and the tips are coated with platinum to make them electrically active recording sites. The tips are typically 50 to 100 microns in length and have electrode impedances from 0.1 to 0.2 megaohms. The needles are either 1.0 or 1.5 mm long. This limits their use in *in vivo* brain preparations to cortical regions. However, they are also ideal for spinal cord, peripheral tissue, and brain slices. For acute recordings all 100 sites can be connected to a thin-film card for connection to second stage electronics. The high density of recording needles makes it difficult to implant the arrays. They provide a pneumatically actuated impulse inserter that implants the array at high velocity with little tissue damage. For chronic recordings, they provide a twelve lead connector to an electrode cap. The number of active sites is currently limited by the electronics of the cap. The twelve leads provide for 11 active recording sites and one reference wire.

6.3.2.4   Insulation

A crucial element of an electrode interface subassembly is the insulation used to protect the metal conducting lines from the brain microenvironment. Any breach in the insulation along the length of an electrode will destroy the integrity of the recording site by shorting it out. This is especially true if the underlying substrate has any conductive properties, such as silicon. The only exposed sites on the electrode should be the recording sites. The exposed metal surface of the recording site becomes a conducting medium and if there is any breach in the insulation exposing a conducting line, there will be cross-talk noise in the signal from this damaged site. The effects of the brain microenvironment on the integrity of the insulation is also an important consideration. The electrode is essentially soaking in a salt solution. Very few insulating materials are impermeable to the extracellular fluid, and this is an important problem for long-term recording devices.

Because of the close proximity of the metal lines that carry the signal from the recording sites to the bonding terminals, any breach in the insulation will greatly increase noise, cross talk and distortion in the signal. Some materials that have been known to be protective against fluid penetration in other applications are not suitable for microelectrodes. The general measure of the insulating property of a material is its pinhole density. The thinner the layer of material, the greater the pinhole density in many cases. While insulating materials such as Parylene® C and Silastic® have been used to insulate and encapsulate microelectronic implants and cardiac pacemakers, the films are relatively thick compared to thin-film application of insulators for microelectrodes.[84] To date, silicon nitride[85] is the most commonly used insulating material.

Silicon nitride as an insulating material does not provide a sufficient barrier to the saline environment of the brain. During application, pinholes develop in the thin-film layer. The silicon nitride becomes permeable to water, which then increases the cross talk between recording sites. Blum et al.[9] used polyimide as an insulator for solid state electrodes and found it to be a better insulator than silicon nitride. However, the long-term integrity of this material is also questionable when it is subjected to the brain's saline environment for long periods of time. Recent work in our lab suggests that alumina (Al_2O_3) may be a superior insulating material.[79] When the starting substrate is also derived from a ceramic or alumina wafer, the recording sites can be patterned and then an insulating layer of alumina can be applied using ion-beam assisted deposition. In this procedure, the alumina is ionized using an electron beam. The alumina can be patterned just like the metallization layer, exposing the recording sites but completely covering the conducting lines. This effectively encases the conducting lines in ceramic (refer to Figure 6.4). In addition, the dielectric properties of alumina are lower than silicon nitride, reducing the capacitive coupling between conducting lines and allowing the space between features to be reduced.

The insulation procedure is generally similar to the metallization procedure. A second photomask is produced that leaves only the recording sites and the bonding terminals exposed. Photoresists are applied over the entire circuit and developed using this photomask so that the terminals and bonding pads are protected. The entire substrate is then layered with an insulator and finally the resist over the recording sites and bonding terminals is removed, simultaneously removing the insulator.

The individual electrodes are then released from the substrate by sawing or laser cutting or through an etching process. Either of these processes exposes the sides of the electrode to the brain microenvironment, further exemplifying the need for an inert, biocompatible substrate. If the original substrate is appropriately prepro-cessed to control the etching process, back etching of the substrate can produce very thin electrodes. The Michigan center uses this method. The electrode shape is patterned into the substrate using deep boron diffusion to heavily dope the intended electrode area before metallization. The temperature and time of this diffusion determine the final thickness of the probe. The final process is to subject the substrate to an ethylenediamine-pyrocatechol-water etch that separates the electrodes and stops at the boron boundary.

To complete the assembly, the bonding terminals are then attached to second stage recording equipment. The integrity of these bonds affects the recording capa-bility of the entire electrode. The small size of the electrodes makes them difficult to manipulate and most of the yield is lost during this processing step. A common bonding procedure is ultrasonic wire bonding that can be accomplished with gold or platinum bonding terminals. For acute recordings, the electrodes are generally connected to a larger PC board that may or may not contain other integrated circuits and sets of connectors to a third stage of recording equipment. Also refer to examples for connectors discussed in Section 6.3.2.3.

6.4 CONDITIONING THE NEURAL SIGNAL

A critical component of a neuroprosthetic control device for clinical applications includes an integrated circuit to interface with the recording electrodes. The initial design challenge begins at the source of the signal. The quality of the neural signal is poor.[86] At the electrode–brain interface, the signals recorded from single neuron action potentials are small, typically less than 100 µVolts in peak to peak amplitude. The critical frequency band for these signals is from 0.5 to 5 kHz. These signals are corrupted with low and high frequency common mode noise sources that are abundant inside and outside the brain tissue. Noise levels can exceed an average of 0.5 millivolts. In addition to these noise considerations, the electrode interface with the neural signals is a high impedance source generally greater than 10 Megaohms. This makes the design requirement for the pre-amplifier extremely low current noise density of less than 5 pA per root hertz, and eliminates the possibility of a simple capacitor coupling stage to filter out DC offsets. In addition to design considerations for signal conditioning, there are significant design considerations for a single very large scale integrated (VLSI) device which contains both analog and digital signals (mixed-signal VLSI) (see Section 6.4.2 below).

The electrode interface consists of a metal in contact with a salt solution. This combination of materials forms a half-cell with a DC offset potential. The magnitude of the half-cell potential is dependent on the metal used in the electrode and the size and shape of the recording site. The potential is typically several hundred millivolts. This half-cell potential is caused by the buildup of a charge layer in the tissue and an adjacent layer on the metal surface. This polarization layer can fluctuate with small movements of the electrode, causing voltage artifacts in the measured signal that can be synchronous with blood pressure, respiration and other physical motion. The brain microenvironment acts like a volume conductor with many electrical sources. Electrical charge in the tissue consists of the flow of ions in this volume conductor. For example, electrical signals from the heart, other neurons and local (facial) muscle activity all contribute to the intrinsic noise sources seen by the electrodes.

Electrodes made from noble metals are typically completely polarizable and the signal is coupled into the metal recording site through the capacitive component of the interface impedance. In order to capture the low frequency neural activity, a pre-amplifier with a suitably high input impedance is required. With both a high impedance source and high pre-amplifier input impedance, the input stage of the pre-amplifier is extremely susceptible to induced noise from both intrinsic and extrinsic sources. To minimize this pickup, the pre-amp should be located as closely as possible to the electrode. Following the pre-amp, the individual neural signals must be amplified and filtered from the background noise. Then they must be combined into an appropriate command signal to control a neuroprosthetic device. Ideally, signal-conditioning circuitry will be designed as a hybrid circuit device small enough to be integrated directly onto the electrode. By designing a thin-film electrode on a stiff, strong substrate, the development of a monolithic hybrid circuit device is feasible. Ideally the amplifier system, together with the pre-amp and buffer circuits, would be integrated onto the electrode array.

6.4.1 PRE-AMPLIFIER CIRCUITS

The complex nature of the electrode–tissue interface, the small magnitude of the neural signals, and the presence of many noise sources combine to make an integrated pre-amplifier design nontrivial. One of the most difficult problems to overcome is the large DC offset potential caused by the electrode half-cell potential. Ideally, AC coupling with a capacitor would be used to eliminate this offset and allow high gain in the first stage. However, capacitors of the size required cannot be manufactured on an IC, thus making a fully integrated design impossible. Switched capacitor circuits, which can be designed to meet the frequency and capacitor size requirements, are extremely difficult to fabricate with noise floors in the microvolt range. One unique solution to this problem is to use the capacitance of the electrode itself to AC couple the signal into the amplifier. A laser trimmed input impedance was used to match the characteristics of the electrode capacitance. This approach is only possible if the characteristics of the electrode are known in advance, and it demonstrates one advantage of designing an integrated electrode-amplifier system with high precision thin-film technology to manufacture the electrodes.

A pre-amplifier stage is required in the hybrid circuit device. The pre-amplifiers must perform an impedance transformation from the complex source impedance to a low impedance output of less than 200 ohms. This transformation is necessary to provide matching impedance for the bandpass filter input stage. At this stage, a high gain is not necessarily needed. Critical requirements include small offset voltages of less than 15 mV and low distortion of -80 dB total harmonic distortion (THD). Typically, the pre-amplifiers are single device amplifiers using junctional field effect transistors (JFETS), nmos or CMOS source follower circuits which have very small input bias current.

After pre-amplifying, the signal needs to be filtered. Since the neural signals are band limited to the audio range, the bandpass filter needs to have a passband frequency range of 500 Hz to 5 kHz. The 500 Hz zero frequency is needed to remove power supply common mode noise at 60 Hz noise from the devices, and low frequency noise sources from natural biological activity such as respiration and heart rate. The high frequency attenuation is required to limit bandwidth dependent noise from the devices and external sources of noise including RF broadcasts. Because of the magnitude of the out-of-band noise, the stop band attenuation should be 40–80 dB/decade. Voltage gain above 10 dB is not a critical requirement for the bandpass filter since a separate gain stage can follow, although higher gain in this stage might limit overall power consumption.

Due to the high impedance, small amplitude and noisy ambient environment of the neuron signals, signal conditioning is a critical component of the neuroprosthetic control device. A -70 dB, signal-to-noise ratio (S/N) over a 2 volt input range is a design target for each analog channel. This becomes a difficult challenge given the noise environment at the electrode interface. Differential signal amplification is not necessarily needed for -70 dB. However, if noise floors below -70 dB to -80 dB are needed, differential signaling within the IC must be considered for the design. Single ended circuits require less power and less integrated circuit area than differential circuits. In addition, single ended, or nondifferential, signal channels are more efficient because every recording site can be used to provide additional useful signals. However, it is not unlikely that some input channels will not have a discriminable single unit. This channel is then ideal as the reference channel. This presumes that the hardware can switch to any channel on demand. When designing a device for clinical use (i.e., small, lightweight), this may be prohibitive. A second design choice would be to predetermine one recorded signal per electrode as the reference for the other signals on the same electrode. This will ensure full differential recording without the need to design switching. However, this does reduce the number of useful signal channels and may be prohibitive. The design decision ultimately depends upon how many electrodes can be safely implanted, how many recording sites can be fit on each electrode and the minimum number of single neuron signals required to provide an adequate command signal. Each of the analog subcomponents found in the hybrid circuit device has distinct features for optimal neural signal conditioning.

If the device is to have a selectable reference channel, a low noise analog selector will be required. This selector will allow for a switchable ground scheme that allows the reference to be selected at any time after the electrodes are implanted. Ideally, this switchable reference channel would be programmable and updated periodically. The selected channel is used as a unipolar reference for all other analog channels. This selectable reference could be changed at any time using a micro dip switch or serial interface.

6.4.2 NOISE SOURCES

The analog signals present in the hybrid circuit device are prone to noise sources classified in two categories — inherent noise and ambient noise. The inherent noise sources are caused by the physical properties of the active and passive electrical components. Inherent noise sources are defined as shot, thermal, flicker, burst and avalanche noise.[88] The ambient or remote noise sources can dominate for single-ended electronics and are defined by common mode noise and substrate noise. All these noise sources are discussed below in order of priority.

1. Common mode noise. Common mode noise is usually most dominant and caused by electromagnetic waves surrounding an analog signal conductor. This noise can be 60 Hz noise from AC power, or AM/FM radio signals from a local radio station or cell phone. It can also be broadband or white noise caused by digital clock signals near the metal conductor lines on an integrated circuit. This induced noise is capacitive or inductively coupled into the signal lines. Ground shielding an analog layout conductor, like a coaxial cable, will help shield the analog line from common mode noise. However, this is sometimes difficult given the limitations of the integration technology.

2. Substrate noise. Since integrated circuits all share a common substrate, minority carrier charge can be induced into the sensitive analog circuits by the conducting substrate from other noise sources near the circuit. If high resistive substrates are used, this becomes less of a problem. Unfortunately, as CMOS technology advances, low resistive substrates are becoming the standard because they are more resistant to the latchup phenomena found in CMOS ICs. Triple Well and Silicon on Insulator (SOI) technologies are being developed to reduce substrate noise.

3. Shot noise. Shot noise is always associated with current flow. It results whenever charges cross a potential barrier, like a pn junction. This charge crossing is a stochastic event; thus, the instantaneous current, I(t), is composed of a large number of random, independent current pulses with an average value, $I_{ave}(t)$. Shot noise is generally specified in terms of its mean-square variation about the average value. It is spectrally flat, is independent of temperature, and has a gaussian shaped probability density function.

4. Thermal noise. Thermal noise is caused by thermal agitation of charge carriers (electron or holes) in a conductor or semiconductor. This noise is present in all passive and active elements. Like shot noise, thermal noise is spectrally flat; however, it is independent of current flow. Thermal noise in a conductor can be modeled as voltage or current using an integral equation of T, absolute temperature; k, the Boltzman constant; and R, the resistance of the conductor. It has units of volts squared per hertz, V^2/Hz, current squared per hertz, or I^2/Hz; hence it has a 1/frequency dependence.

5. Flicker noise. Flicker noise is a 1/frequency noise, is present in all active devices and has various origins. Flicker noise is always associated with DC current and its average mean-square value is the integral of a current constant over frequency. If the current is kept low enough in the device, thermal noise will dominate the other sources.

6. Burst noise. Burst noise, also called popcorn noise, appears to be related to imperfections in the semiconductor material and heavy ion implants. Burst noise makes a popping sound at rates below 100 Hz when played through a speaker. Low burst noise is achieved by using clean device processing.

7. Avalanche noise. Avalanche noise is created when a pn junction is operated in the reverse breakdown mode. Under the influence of a strong reverse electric field within the junction's depletion region, electrons have enough kinetic energy so when they collide with the atoms of the crystal lattice, additional electron–hole pairs are formed. These collisions are purely random and produce random current pulses similar to shot noise, but much more intense.

The level of noise added to the analog signal from these sources varies with the overall design of the hybrid circuit device. For example, in the operational amplifier used in the bandpass filters, burst noise and avalanche noise are normally not a problem. Since noise sources have amplitudes that vary randomly with time, they can only be specified by a probability density function. When multiple noise sources are present in a circuit, the noise signal sources must be combined heuristically to obtain the overall noise signal.

6.4.3 ANALOG TO DIGITAL SIGNAL CONVERSION

The amplified and filtered analog signal must be digitized before transmitting it from the hybrid circuit device to the RF receiver hardware. The amplifier stage will require a low noise, 20-dB or greater amplifier. This stage must increase the peak-to-peak level of the neuron signal from 50 µV to acceptable levels of 400 mV over the noise floor in order to ensure the 10–12 bit analog to digital (A/D) conversion is performed efficiently. It is critical to have low THD of –75 dB and very small offset voltage of no more than 20 mV.[89] The power consumption for the high gain amplifiers is proportional to the bias current of the differential and output gain stages. Typical bias currents of 100–200 µA at 5 V are possible, which correspond to 1 mW power dissipation. Slew rates of the amplifier are not necessarily large because of the limited bandwidth. A slew rate of less than 5 V/µs is adequate assuming the input impedance to the A/D converter is high. It is also important to have the amplifier in close proximity to the signal source. Since common mode noise will be induced all along the front-end device, it is best to amplify the signal as early as possible to maintain the signal-to-noise ratio in the latter stages.

The embedded analog to digital converter (ADC) specifications for most neuron recording applications is around 70 dB signal-to-noise ratio (S/N) or 10–12 bits for a 2 v peak-to-peak input range. Since the neuron input signal frequencies are band limited to the audio range, a 12.5 Khz sampling rate is required. A delta-sigma ADC architecture is optimal to meet these requirements.[90] Delta-sigma ADCs exchange oversampling for precision and employ a switched capacitor architecture.[91] S/Ns are a function of oversampling and modulator order. For the hybrid circuit device, a second order delta-sigma ADC with a 500 kHz oversampling rate is sufficient. Power consumption of less than 25 mW is required for the ADC core since total system power budget is critical. One major disadvantage of the delta-sigma architecture is that a crude digital decimation section is needed to reduce the bandwidth of the digitized output.[92] This decimation section will convert the 500 kHz 1 bit data stream to 12 bit Nyquist rate samples at 12.5 kHz.

Lastly, special considerations must be made for transmitting the digitized signals to a neuroprosthetic device or external computer/robotic device. Standard frequency shift keying (FSK) communication protocol can be implemented using a voltage controlled oscillators (VCO) for the FSK modulator/transmitter and a phase lock loop (PLL) for the FSK demodulator/receiver.[93] (See Section 6.5 for further details.) CMOS VCO carrier frequencies can range from 1 kHz to 1 GHz using current-starved ring oscillator topologies.[94] The RF carrier center frequency is predicated based on the bandwidth requirements of the neural signal. For a single channel, a 12-bit neuron signal at a 12.5 kHz sampling rate, a 150 kbits per second data rate (150 Kb/s) is needed. For 16 parallel receive channels, the RF link data rates are 2.4 Mb/s, which require 16 times the power consumption of a single channel. The frequency range of the FSK channel is not limited by the technology. It is only limited to the available FCC ranges and the power consumption budgets of the device.

In addition to the VCO and PLL circuits, the matching filter and antenna designs of the receiver and transmitter are very important to the RF link.[95] The high quality of these designs will make it possible to achieve low bit-error rates of 1×10^{-15} at 6–10 feet, using 5 mW per channel. The receiver and transmit antenna are typically a 1/8 wavelength dipole layout in the printed circuit board signal layer of the hybrid circuit device. The matching filter utilizes ferrite beads and large capacitors that normally cannot be integrated on the IC of a single VLSI. These components are typically surface mounted onto the system board of the hybrid circuit device. This will increase the size of the front-end system, and careful considerations need to be made to minimize adverse effects on the overall design of the device.

6.5 NEURAL SIGNAL ACQUISITION SYSTEM

A critical component of a neuroprosthetic control device for clinical applications includes an IC device to combine the neural signals recorded from each channel into a command signal. As discussed below (Sections 6.5.2 and 6.6), packaging constraints prevent designing a device to buffer the complete analog signal from all of the recorded channels. Therefore, some processing components must be available to detect transmission of the analog signals from all of the channels and extract the spike times from the neurons recorded from each channel. The trick then is to digitize the analog signals, compare them to on-board templates, and finally convert the spike times and neuron identity to a command signal. On-line discrimination techniques are beyond the scope of this chapter.[96] However, issues related to handling the mixed signal on a single VLSI chip and packaging constraints for large numbers of input channels will be discussed in this section. In systems currently used today to record multiple signal channels from a freely behaving animal, this receiver is normally associated with a multiprocessor computer that is responsible for the subsequent realtime processing and storage of brain derived signals.[97] (Also see Chapters 7 and 8 for examples.)

6.5.1 MIXED SIGNAL VLSI

Developments for an integrated circuit that interfaces with electrodes implanted directly into brain tissue and contains spike sorting processors are being driven by advancements in the mixed-signal VLSI and device packaging technologies.[98-100] Mixed-signal VLSI combines both analog and digital signals on the same chip. Discrete, single tethered channels, mounted on a PCB (printed circuit board), are now being transformed to full custom, multichannel, programmable, wireless, hybrid packages to allow for more accurate and lower power analog receive channels. This migration of the neuron micro-instrumentation electronics into wireless devices is progressing similarly in direction and pace to the analog circuit integration found in low power consumer electronics and personal computers. The advantages of VLSI technology and packaging trends will allow the possibility of integrating more analog macro functions with improved analog performance.[101-103] The discrete, single channel analog functions, which include the preamplifiers, bandpass filters, analog selectors, high gain amplifiers, multiplexors and A/D converters, can now be integrated with more channels, lower noise, lower power, and packaged in smaller and lighter weight hybrid circuit sub-assemblies (Figure 6.6).

There are four major areas of technology migration that will provide advancement in the design of this hybrid circuit sub-assembly. These areas include enhanced circuit design techniques and architectures, improved analog circuit fabrication processes, robust mixed-signal CAD design software, and high-density pin count packaging technology.

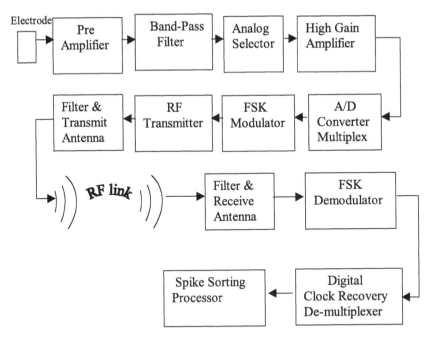

FIGURE 6.6 Illustration of a hybrid circuit sub-assembly to interface with the recording electrodes. The device consists of mixed signal VLSI to condition the neural signals and telemetry transmit devices. The top view illustrates the layout of the ICs and the side view illustrates the interface with electrodes.

Improved mixed-signal circuit architectures, designs and layout techniques are being advanced for both the continuous time domain and the discrete time (switch capacitor) domain. Today's circuits for each of these domains can provide low noise and low power solutions that achieve a –70 dB noise floor and 15 mW per channel signal channel requirement. Advances in micro power and higher order amplifier topologies will continue to drive the low noise and low power evolution of integrated mixed-signal electronics.[104] The low power, high gain, single ended or fully differential operational amplifiers continue to provide the basic building block for most high gain large bandwidth circuits. In addition, new low bias current amplifier circuits[105-106] are now being implemented with low power process technologies that will continue to set the stage for more advances in mixed-signal technology.

Complimentary metal oxide silicon (CMOS) technologies continue to follow the technology migration paths seen over the last three decades. Device gate lengths shrink in size by half every 2–3 years. Additional metal layers are added to enable high density, low resistive signal and power bus routing. In addition, system clock frequencies are doubling every 2–3 years. The 0.5 µm micron, double poly-silicon, and four level metal CMOS and Bi-CMOS technology can offer a viable integration process for most –70 dB noise floors with microvolt input levels used in today's mixed-signal electronics. With the double poly-silicon option, linear capacitors with 0.1% voltage coefficients can be implemented. These devices are needed for most of the mixed-signal analog circuit options. It is because of these improvements in silicon process technologies that Ghz RF links and high bandwidth amplifiers can be implemented with excellent reliability.

Advancements with computer aided mixed-signal design software (CAD) and engineering computer hardware is keeping up with the silicon process improvements. For today's CAD software and methodologies, mixed-signal CAD flows are improving designers' productivity and chance of first pass silicon success. New and improved device models, high-level design languages for modeling and system analysis, and power estimation algorithms are increasing the VLSI designer's level of accuracy and confidence in device operation. As the complexity of VLSI increases due to an increased number of devices, higher circuit density, and faster clocks, greater computer power (MIPs) is needed for the engineering workstation to churn out simulation and verification results. Additional memory and computer horsepower are becoming increasingly valuable for a productive VLSI designer's workstation, and these are available on the market today.

It is best for noise considerations to include the pre-amplifier through A/D converter sub-assemblies together in close proximity to the recording electrode. This layout will minimize the common mode noise in the analog receive channel.[101] However, combining analog and digital signals on the same VLSI device can create difficulties (discussed below in Section 6.5.2). Depending on the integration technology of the analog components, this may be the ideal solution for this application. Separating the analog and digital components is probably not a viable design option because the resulting size and weight of the individual components are prohibitive.

6.5.2 DEVICE PACKAGING

There are significant packaging challenges to be overcome in the design of a neural signal acquisition system. These design challenges are well known[74,107-109] and worth noting since the system specifications of this electrode interface are constrained by these electrical and packaging challenges. The number of electrodes per interface can vary and depends on the technology of the electrode and system requirements. Typically, electrode sub-assemblies have 4–16 channels per electrode, but this is expected to increase to 32 and 64 channels in the next 2–5 years.

These design challenges include limits on the size and weight of the neuron signal acquisition system. These are a function of the size and strength of the mammal whose signals are being monitored.[110] For relatively small mammals like mice and rats, the entire electrode data acquisition system interface has to be extremely light in weight and physically small. Generally speaking, the entire weight of the sub-assembly for 16 receive channels has to be less than 100 grams and smaller than 8 mm by 10 mm by 1 mm in dimension. Hence, packaging requirements for a rat headstage sub-assembly warrants as much device integration and surface mount packaging as possible to guarantee the light weight and compact size. While humans might be able to sustain much greater weights than mice or rats, for aesthetic reasons, the package should remain as small as possible.

Top View

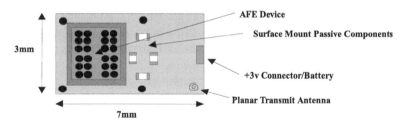

Side View

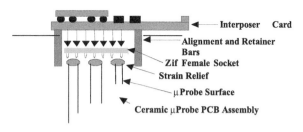

FIGURE 6.7 A telemetric neuron signal acquisition system. The complete systems can be divided into two parts: sub-assemblies before and after the RF link. Sub-assemblies before the RF link include the electrodes, the pre-amplifier, bandpass filter, analog selector, high gain amplifier, A/D converter, frequency shift keying (FSK) modulator, RF transmitter, transmit filter and antenna. The RF link receiver hardware follows the RF link and consists of the receive filter and antenna, FSK demodulator, digital recovery and de-multiplexer components that can reside in a remote backpack to provide signals to the prosthetic device or receive station for tuning parameters

It is for these significant packaging design constraints of the neuron signal acquisition system that novel approaches to design an optimized and efficient hardware system are required. Improved packaging technologies such as direct chip attach, miniature surface mount, tab bonding on flex circuits and multi-chip packaging will increase the number of input/outputs per square millimeter allowed for an integrated circuit device. In addition, reduced pin inductance will improve power supply bounce caused by current surges in the device. These packaging improvements increased the number of channels per area of the hybrid circuit device compared to dual in-line package devices.[110]

Another important feature of the overall package is the ability to telemeter the signal away from the patient (see Section 6.6 below). A telemetric signal acquisition system can be divided into two parts (Figure 6.7): (1) a hybrid circuit sub-assembly for signal conditioning and conversion and (2) RF transceiver hardware. The hybrid circuit device consists of the pre-amplifier, bandpass filter, analog selector, high gain amplifier, A/D converter, FSK modulator, RF transmitter, transmit filter and antenna. This device interfaces directly with the electrodes. Ideally, the hybrid device would be packaged into a VLSI chip and associated passive electronics and be small enough to reside on the patient subdurally. The RF link receiver hardware follows the RF link and consists of the receive filter and antenna, FSK demodulator, and digital recovery and de-multiplexer components that can reside in a remote backpack to provide signals to the prosthetic device or receive station for tuning parameters.

A spike sorting processor that can create templates to discriminate individual neurons on-line in the hybrid circuit device will also be necessary but will not be discussed here. During off-line spike discrimination, the patient would reside near the spike sorting processor. Fully sampled neural signals would be transmitted to the spike sorting processor and templates would be created. These templates would then be downloaded to the hybrid circuit device. The complexity of the receiver depends on the robustness of the transmit communication hardware.[111-113]

6.6 TELEMETRY

In order to discriminate the neurons, the neural signals will have to be transmitted to a remote receive station that electrically connects the implanted electrode sub-assembly to a digital spike sorting processor. Ideally, this should be done telemetrically without the need to attach wires directly to the patient. If there is a percutaneous connector available for physical connection, there is risk of infection and irritation (for an example, see Chapter 2). In addition to accessing the signals for spike discrimination, the subsequent processed command signal for controlling a neuroprosthetic device or an external device will have to be transmitted to that device. Ideally, this would also take place telemetrically via a receiver sub-assembly.

The design must consider many parameters when determining the transmission frequency, the modulation method, and the data-encoding scheme. The overwhelming constraint, as can be seen from the discussion below, is minimizing power consumption in the implanted transmitter. As more sophisticated communication protocols are used in portable devices, there will be increased availability of totally integrated transceivers employing more efficient modulation schemes. These devices will employ new integration technology, which can be exploited to further enhance the neuroprosthetic control device communication interface.

For transmitters with extremely limited power (<1 W), the transmission is not regulated by the Federal Communications Commission (FCC). However, it is necessary to pick a transmission frequency that avoids interference from other commonly used RF devices such as radios, cell phones and telephones. A technique called frequency hopping is used by cell phones to virtually insure an unused frequency for transmission. This system actually searches the allowed spectrum for a channel that is not in use locally.

The general guidelines for picking a transmission frequency for low power short transmission systems are based on the bandwidth or data rate of the channel. A general rule of thumb is to use a transmission frequency at least 20 times greater than the bandwidth. The upper end of the spectrum for frequency selection is generally set by the frequency limitations of the semiconductor technology and by the desired power consumption. Generally, the higher the frequency, the more power is required to generate the signal. One caveat to this general rule is burst transmission. Sometimes greater power efficiency can be obtained by transmitting at higher bit rates for shorter periods of time. However, not all systems can take advantage of data in this format.

Many schemes exist for encoding data for serial transmission on an RF data link. One low power encoding scheme that uses a minimum of off chip components is the digital FM method of frequency shift keying (FSK). In this scheme, the voltage signal of ones and zeros is filtered and used to modulate a carrier frequency. A voltage controlled oscillator is used to generate the modulated signal. On the receive side, a phase locked loop (PLL) is used to track the incoming frequency changes and to demodulate the signal directly. This encoding scheme cannot handle long strings of ones and zeros and thus some encoding of the outgoing data stream is required. Frequently this bit steam encoding is combined with error checking. A sub-system to perform these functions is called a packet formatter. A complementary device is required in the receiver to decode the packet data back into the raw format.

6.6.1 BANDWIDTH

The major determinants of the power consumption in an RF data link are the bandwidth of the signal to be transmitted and the distance over which the link is to function. For the neural signals to be telemetered away from the patient, bandwidth is a major consideration. The electrical signals produced by the neurons have about 5 kHz of signal bandwidth per electrode or channel. While it is possible to transmit each of the signals on individual radio channels, the complexity of designing an integrated low power 32- or 64-channel radio transmitter and receiver system is prohibitive.

If the system is made for digital transmission on a single radio channel and the electrode signals are multiplexed, the bandwidth is closely related to the overall bit rate. By comparing bit rates for various schemes, a relative comparison of the bandwidth and, hence, power required to transmit the digital signals can be made:

$$\text{Bit Rate} = \text{Nchan} \times f \times r \times T \tag{6.2}$$

where: $Nchan$ = the number of recording sites
f = sampling frequency
r = resolution
T = transmission factor

The sampling frequency is the sampling rate for each channel. This rate is typically 2.5 × the neural signal bandwidth or 12.5 k samples per channel per second. The resolution is the number of bits in the analog to digital conversion. Typically this is 12 bits. The transmission factor is a constant that relates the data bit rate to the bit rate required to transmit the bit stream. This constant is related primarily to overhead required to encode the data for transmission. For a typical scheme this factor is about 1.25. Thus, the bandwidth required is approximately 187.5 kbits/second/channel.

If the information content of the signal is examined, it can be seen that there is a lot of bandwidth wasted transmitting the unprocessed signal. For this calculation it is assumed that the information in the signal from each channel can be encoded as the state, on or off, of the neurons being monitored by a given electrode. This information would only be available if on-line spike sorting could be accomplished before the telemetry. If a resolution of 1 ms is presumed adequate and a maximum of 4 neurons per electrode is permitted, then it will require only 5 kbits/second/channel to transmit the spike information directly. This is nearly a 40-fold decrease in the transmission bandwidth.

This reduction in transmission bandwidth does not come for free. The tradeoff here is between the power it takes to process and discriminate the signals and transmit the reduced information and the raw power required to simply transmit the unprocessed signals. A system to perform the task of unassisted spike sorting (refer to Section 6.4) in real time would require a huge development effort as well as requiring significant processing and electrical power at the electrode site.

A solution with intermediate processing complexity and bandwidth requirements is the transmission of a compressed version of the raw signals. While there are numerous digital compression schemes that can offer compression rates of about 2–4:1, more efficient (and complex) schemes exist that take advantage of the characteristics of the signal. Many of these schemes have been developed for electrocardiogram (EKG) recording and offer reduction rates in the range of 10–20:1. Without further research it is difficult to predict the performance of these schemes for neural signals but similar results might be expected.

As an example, for a 64-electrode system with 4 neurons per electrode and 1 ms resolution for spike timing, it would take 9.6 Mbits/sec to transmit the raw data. In contrast, it would take about 3.2 Mbits/sec for a data independent compression scheme, possibly 640 kbits/sec for an efficient data dependent compression scheme, and 256 kbits/sec to transmit the spike firing information directly.

6.6.2 POWER SOURCE CONSIDERATIONS

Any implantable or wearable neuro-telemetry system will need to be powered by a battery. Since battery life and weight are critical, power consumption per channel must be as small as possible.[103] An estimate for power available per channel is 15 mW for the complete full function analog channel found described in Figure 6.7. This is an estimate and depends on the size and strength of the mammal being monitored. It has been shown that in medium size primates like chimpanzees and monkeys, the payload of a neural recording device can be 200 times that of a rat.[93] A device designed for humans will require a 3x to 4x increase in available power.

With the need for recording channels virtually unlimited, the major size constraint for this type of system is likely to be the energy storage capacity. The power budget can be divided into three major areas: the analog signal processing, the digital signal processing for information extraction, and the transmission power. Each of these areas can be optimized for minimal power as the characteristics of the required technology are further defined. Fortunately, the technology associated with the mobile communication and computing industry are driving down the cost and power requirements for the digital signal processing and transmission areas. Also, experience from the implantable pacemaker, defibrillator, and drug pump industry has provided a wealth of experience for low power mixed-signal implantable designs. However, the advances in design that led to the 10-year pacemaker have taken place over 3 decades of intense industry competition and were fueled by a multi-billion dollar market. Similar investments will be required to optimize the function of these neurological interfaces.

Battery technology is one area where progress in the areas of personal communication and portable computing can be applied almost directly to the implantable or wearable neuro-device.[114-115] Batteries convert chemical energy to electrical energy. This process is controlled by the chemical elements used for the anode and cathode inside the device. Each combination of materials produces a unique voltage (V) that is characteristic of a given energy cell. The construction of the cell and its size determines the capacity (Amp hours) of the battery that is characterized as the number of amps a battery can produce for one hour. The total energy (watthours) stored can be calculated as the capacity times the voltage. To compare different battery technologies it is useful to calculate the energy density. This number can be reported as either total energy per cubic centimeter or as total energy per gram.

There are two types of batteries: primary cells and rechargeable cells. Primary cells convert all of their internal chemical energy to electricity and then must be replaced; rechargeable cells use a reversible chemical process that can be repeatedly replenished by a secondary electrical source. The cost for this advantage is in energy density and in circuit complexity. Table 6.2 presents a comparison of the characteristics of the newest technology batteries for both the primary and rechargeable types. This list was compiled from the data sheets of commercially available batteries. There can be some variability in the energy density numbers, in particular in the density by weight. This variability comes primarily from the ratio of the container weight to the weight of the contents, with larger batteries being more efficient.

Battery voltage is an important parameter for the system designer. If, for a given system, a higher voltage is required than can be generated by a single cell, multiple

TABLE 6.2
Battery Technologies

Technology	Voltage	Density by volume (W·h/cc)	Density by weight (W·h/g)
Modern Primary Cell Technologies			
Lithium thionyl chloride	3.5	0.97	0.4
Lithium manganese dioxide	3.0	0.67	0.233
Zinc–air	1.4	0.8	0.5
Modern Rechargable Cell Technologies			
Nickel metal hydride	1.2	0.3	0.05
Lithium ion	3.6	0.3	0.11
Lithium maganese dioxide	3.0	0.23	0.06

cells can be placed in series or electronic voltage converters can be used to generate the higher voltage. One significant advantage of lithium based cells is that they have a working voltage above three volts. This voltage is high enough that it can be used to power digital and analog circuits directly. This advantage can eliminate the inefficiencies of voltage converters or the need for multiple cells. Lithium also has the advantage of being a light metal; thus, its energy density by weight is very high.

One unique battery source that is commonly used in hearing aids is the zinc-air battery. In this cell, oxygen is used as the cathode material. The oxygen is not packaged within the cell but is drawn out of the air. This design is efficient because no storage is needed for the cathode material; however, the battery must be exposed to air in order to function properly and cannot be implanted.

Most implantable commercial devices use primary cells for energy storage. These devices usually use lithium as the anode but vary somewhat in their cathode material based on the rate at which power is consumed. In implantable pacemakers where the current draw is very low (μA) and continuous, lithium iodine cells are used most commonly. For devices such as implantable defibrillators where periodically the current drain can be very high (Amps), lithium silver vanadium oxide batteries provide the highest performance. These cells as well as lithium carbon monofluoride cells are used for continuous medium current (mA) applications such as neurostimulators and drug infusion pumps. All of these battery types have energy densities similar to the lithium thionyl based cell listed in Table 6.2. Commercial implanted device manufacturers usually custom design the battery cells for their devices. This allows them to optimize the voltage, current drain characteristics, and the shape of the battery to make the best use of space in the implanted device.

For a totally implanted device, a primary cell is desirable only if the battery life is in the range of 1000 days. A shorter life span would require replacement surgery at too short an interval. This implant duration requires 24,000 hours of operation. A 10 cc battery with an energy density of 1.1 Wh/cc and a working voltage of 3.3 V could draw a continuous average current of 140 μA for 24,000 hours. For a 64-channel system this would be equivalent to an average current drain of 2 μA per channel. The technology available today draws current several orders of magnitude above this number. Clearly, more efficient batteries and circuits will be required to build a realistic implantable neuroprosthetic control device.

A more realistic short-term solution is a wearable device with either primary or rechargeable batteries. Such a device would necessarily have to deal with the problems of transcutaneous wires to deliver the power to the implant and to transmit out the signals. At an estimate of 10 mW per channel, 64 channels would require 640 mW of continuous power. For a rechargeable lithium ion cell with an energy density of 110 mWh/g, a 100 g battery would last about 17 hours.

6.7 NEW DIRECTIONS FOR A NEUROPROSTHETIC CONTROL DEVICE

The issues and problems associated with the development of a neuroprosthetic control device outlined above suggest the need for a multidisciplinary design team consisting of biomedical, electrical, and chemical engineers as well as neuroscientists and medical professionals. To design an optimized and efficient hardware system to interface with the brain, record neural signals for decades and produce an effective neuroprosthetic command signal will require state of the art technological developments. These developments include improved materials for neural biocompatibility, mixed signal (analog and digital), very large scale integration (VLSI), novel power consumption technologies and telemetric device packaging technologies.

We have recently put together a team (see Chapter 8 for detailed discussion) to address each of these issues and test novel device strategies. We have developed novel hybrid circuit devices using ceramic based multisite recording electrodes that have successfully recorded from multiple neurons for up to three months in the rat. The device includes a neuro-chip that performs on-board signal conditioning using VLSI technology. Currently, the device is tethered to an existing multichannel recording device (Plexon Inc., Dallas TX) for spike discrimination. However, ongoing studies are developing an integrated telemetric sub-assembly using technologies described above, and on chip spike discrimination algorithms are under way. These brain derived command signals can then be used to control the patient's limbs via one of the neuroprosthetic devices described in the previous chapters or an artificial prosthetic device such as a computer (see Chapter 7) or robotic device.[56]

Artificial neural network algorithms[56] that can be downloaded to the neuro-chip are being developed. We expect this device to be small enough to reside subdurally and eventually provide consistent neural signals that can last for decades.

ACKNOWLEDGMENTS

This project was supported by DARPA-ONR grant (N00014-98-1-0679), NINDS grant 1-RO1-NS26722, NINDS contract NO1-NS-6-2352 to Dr. Chapin and DARPA grant DARPA-ONR(N0014-98-1-0676) to Dr. Nicolelis.

REFERENCES

1. Gray, C.M., Maldonado, P.E., Wilson, M., McNaughton, B., Tetrodes markedly improve the reliability and yield of multiple single-unit isolation from multi-unit recordins in cat striate cortex, *J. Neurosci. Methods*, 63, 43, 1995.
2. Drake, K.L., Wise, K.D., Farraye, J., Anderson, D.J., BeMent, S.L., Performance of planar multisite microprobes in recording extracellular single-unit intracortical activity, *IEEE Trans. Biomed. Eng.*, 35, 719, 1988.
3. Jones, K.E., Campbell, P.K., Normann, R.A., A glass/silicon composite intracortical electrode array, *Ann. Biomed. Eng.*, 20, 423, 1992.
4. Nicolelis, M.A., Ghazanfar, A.A., Faggin, B.M., Votaw, S., Oliveira, L.M., Reconstructing the engram: simultaneous, multisite, many single neuron recordings, *Neuron*, 18, 529, 1997.
5. Schmidt, E.M., Bak, M.J., McIntosh, J.S., Long-term chronic recording from cortical neurons, *Exp. Neuro.*, 52, 496, 1976.
6. Villa, A.E.P., Hyland, B., Tetko, I.V., Najem, A., Dynamic cell assemblies in the rat auditory cortex in a reaction-time task, *Biosystems*, 48, 269, 1998.

7. Schmidt, E.M., Bak, M.J., McIntosh, J.S., Thomas, J.S., Operant conditioning of firing patterns in monkey cortical neurons, *Exp. Neurol.,* 54, 467, 1977.
8. Williams, J.C., Rennaker, R.L., Kipke, D.R., Long-term neural recording characteristics of wire microelectrode arrays implanted in cerebral cortex, *Brain Res. Protocols,* 4, 303, 1999.
9. Blum, N.A., Carkhuff, B.G., Charles, H.K., Edwards, R.L., Meyer, R.A., Multisite microprobes for neural recordings, *IEEE Trans. on Biomed. Eng.,* 38(1), 68, 1991.
10. Kruger, J., Bach, M., Simultaneous recording of 30 microelectrodes in monkey visual cortex, *Exp. Brain Res.,* 41, 191, 1981.
11. Reitbock, H.J., Werner, G., Multielectrode recording system for the study of spatiotemporal activity patterns of neurons in the central nervous systems, *Experientia,* 39, 339, 1983.
12. Eichman, H., Kuperstein, M., Extracellular neural recording with multichannel microelectrodes, *J. Electrophysiol. Tech.,* 13, 189, 1986.
13. Kennedy, P.R., The cone electrode: a long-term electrode that records from neurites grown onto its recording surface, *J. Neurosci. Meth.,* 29, 181, 1989.
14. Kennedy, P.R., Bakay, R.A.E., Sharpe, S.M., Behavioral correlates of action potentials recorded chronically inside the cone electrode, *NeuroReport,* 3, 605, 1992.
15. Carter, R.R., Houk, J.C., Multiple single-unit recording from the CNS using thinfilm electrode arrays, *IEEE Trans. Rehabil. Eng.,* 1(3)175, 1993.
16. Jaeger, D., Gilman, S., Aldridge, W., A multiwire microelectrode for single unit recording in deep brain structure, *J. Neurosci Methods,* 32, 143, 1990.
17. Kubie, J.L., A driveable bundle of microwires for collecting single-unit data from freely-moving rats, *Physiol. Behav.,* 32, 115, 1984.
18. Jaeger, D., Gilman, S., Aldridge, J. W., A multiwire microelectrode for single unit recording in deep brain structures, *J. Neurosci. Meth.,* 32, 143, 1990.
19. Sonn, M., Feist, W. M., A prototype flexible microelectrode array for implant-prosthesis applications, *Med. Biol. Eng. Comput.,* 12(6), 778, 1974.
20. Raivich, R., Bohatschek, M., Kloss, C.U.A., Werner, A., Jones, L.L., Kreutzberg, G.W., Neuroglial activation repertoire in the injured brain: graded response, molecular mechanisms and cues to physiological function, *Brain Res. Rev.,* 30, 77, 1999.
21. Koyama, Y., Takemura, M., Fujiki, K., Ishikawa, N., Shigenaga, Y., Baba, A., BQ788, an endothelin ET_B receptor anatagonist, attenuates stab wound injury-induced reactive astrocytes in rat brain, *Glia,* 26, 268, 1999.
22. Fitch, M.T., Doller, C., Combs, C.K., Landreth, G.E., Silver, J., Cellular and molecular mechanisms of glial scarring and progressive cavitation: *in-vivo* and *in-vitro* analysis of inflammation-induced secondary injury after CNS trauma, *J. Neurosci.,* 19(19), 8182, 1999.
23. Fawcett, J.W., Asher, R.A., The glial scar and central nervous system repair, *Brain Research Bull.,* 49(6), 377, 1999.
24. Edell, D.J., Toi, V.V., McNeil, V.M., Clark, L.D., Factors influencing the biocompatibility of insertable silicon microshafts in cerebral cortex, *IEEE Trans. Biomed. Eng.,* 39(6), 635, 1992.
25. Eddleston, M., Mucke, L., Molecular profile of reactive astrocytes, *Neuroscience,* 54, 15, 1993.
26. Yang, H. Y., Lieska, N., Kriho, V., Wu, C., Pappas, G.D., A subpopulation of reactive astrocytes at the immediate site of cerebral cortical injury, *Exper. Neurol.,* 146, 199, 1997.
27. Goss, J.R., O'Malley, M.E., Zou, L., Styren, S.D., Kochanek, P. M., DeKosky, S.T., Astrocytes are the major source of nerve growth factor upregulation following traumatic brain injury in the rat, *Exper. Neurol.,* 149, 301, 1998.

28. Logan, A., Green, J., Hunter, A., Jackson, R., Berry, M., Inhibition of glial scarring in the injured rat brain by a recombinant human monoclonal antibody to transforming growth factor-beta2, *European J. of Neurosci.*, 11(7), 2367, 1999.

29. Tomobe, Y.I., Hama, H., Sakurai, T., Fujimori, A., Ae, Y., Goto, K., Anticoagulant factor protein S inhibits the proliferation of rat astrocytes after injury, *Neurosci. Lett.*, 214, 57, 1996.

30. Moller, J.C., Klein, M.A., Haas, S., Jones, L.L., Kreutzberg, G.W., Raivich, G., Regulation of thrombospondin in the regenerating mouse facial nucleus, *Glia*, 17, 121, 1996.

31. Streit, W.J., Kreutzberg, G.W., Response of endogenous glial cells to motor neuron degeneration induced by toxic ricin, *J. Comp. Neurol.*, 268, 248, 1988.

32. Kurkowska-Jastrzebska, I., Wronska, A., Kohutnicka, M., Calonkowski, A., Czlonkowska, A., The inflammatory reaction following 1-methyl-4-phenyl-1,2,3,6-tetrahydropyridine (MPTP) intoxication in mouse, *Exp. Neurol.*, 156, 50, 1999.

33. McGeer, P.L., McGeer, E.G., Anti-inflammatory drugs in the fight against Alzheimer's diseach, *Ann. N.Y. Acad. Sci.*, 777, 213, 1996.

34. Shalit, F., Sredni, B., Brodie, C., Kott, E., Humerman, M., T Lymphocyte subpopulations and activation markers correlate with severity of Alzheimer's disease, *Clin. Immunol. Immunopathol.*, 75, 246, 1995.

35. Fitch, M.T., Silver, J., Activated macrophages and the blood-brain barrier: inflammation after CNS injury leads to increases in putative inhibitory molecules, *Exper. Neurol.*, 148, 587, 1997.

36. Kossmann, T., Hans, V., Imhof, H., Trentz, O., Morganti-Kossmann, M.C., Interleukin-6 released in human cerebrospinal fluid following traumatic brain injury may trigger nerve growth factor production in astrocytes, *Brain Res.*, 713, 143, 1996.

37. Mobley, W.C., Neve, R.L., Prusiner, S.B., McKinley, M.P., Nerve growth factor increases mRNA, levels for the prion protein and the B-amyloid protein precursor in developing hamster brain, *Proc. Natl. Acad. Sci.*, 85, 9811, 1988.

38. Balasingam, V., Tejada-Berges, T., Wright, E., Bouckova, R., Yong, V.W., Reactive astrogliosis in the neonatal mouse brain and its modulation by cytokines, *J. Neurosci.*, 14, 846, 1994.

39. Balasingam, V., Yong, V.W., Attenuation of astroglial reactivity by interleukin-10, *J. Neurosci.*, 16, 2945, 1996.

40. Kahn, M.A., Ellison, J.A., Speight, G.J., de Vellis, J., CNTF regulation of astrogliosis and the activation of microglia in the developing rat central nerbous system, *Brain Res.*, 685, 55, 1995.

41. Rostworowski, M., Balasingam, V., Chabot, S., Woens, T., Yong, V.W., Astrogliosis in the neonatal and adult murine brain post-trauma: elevation of inflammatory cytokines and the lack of requirement for endogenous interferon-gamma, *J. Neurosci.*, 17, 3664, 1997.

42. Giuliah, D., Chen, J., Ingeman, J.E., George, J.K., Noponen, M., The role of mononuclear phagocytes in wound healing after traumatic injury to adult mammalian brain, *J. Neurosci.*, 9, 4416, 1989.

43. Svensson, M., Eriksson, N.P., Aldskogius, H., Evidence for activation of astrocytes via reactive microglial cells following hypoglossal nerve transsection, *J. Neurosci. Res.*, 35, 373, 1993.

44. Janzer, R.C., Raff, M.C., Astrocytes induce blood-brain barrier properties in endothelial cells, *Nature*, 325, 253, 1987.

45. Pekny, M., Stanness, K.A., Eliasson, C., Betsholtz, C., Hanigro, D., Impaired induction of blood-brain barrier properties in aorti endothelial cells by astrocytes from GFAP-deficient mice, *Glia*, 22, 390, 1998.

46. Tao-Cheng, J.H., Nagy, Z., Brightman, M.W., Tight junctions of brain endothelium *in vitro* are enhanced by astroglia, *J. Neurosci.,* 7, 3293, 1987.

47. McKensie, A.L., Hall, J.J., Aihara, N., Fukuda, K., Noble, L.J., Immunolocalization of endothelin in the traumatized spinal cord: relationship to blood-spinal cord barrier breakdown, *J. Neurotrauma,* 12, 257, 1995.

48. Yamada, G., Hama, H., Kasuya, Y., Masaki, T., Goto, K., Possible sources of endot-helin-1 in damaged rat brain, *J. Cardiovasc. Pharmacol.,* 26, S448, 1995.

49. Hetke, J.F., Lund, J.L., Najafi, K., Wise, K.D., Anderson, D.J., Silicon ribbon cables for chronically implantable microelectrode arrays, *IEEE Trans. Biomed. Eng.,* 41, 314, 1994.

50. Campbell, P.K., Jones, K.E., Huber, R.J., Horch, K.W., Normann, R.A., A silicon-based, three dimensional neural interface: manufacturing processes for an intracortical electrode array, *IEEE Trans. Biomed. Eng.,* 38, 758, 1991.

51. Liu, X., McCreery, D.B., Carter, R.R., Bullara, L.A., Yeun, T.G.H., Agnew, W.F., Stability of the interface between neural tissue and chronically implanted intracortical microelectrodes, *IEEE Trans. Rehabil. Eng.,* 7, 315, 1999.

52. Nudo, R.J., Recovery after damage to motor cortical areas, *Current Opinion in Neurobiology,* 9, 740, 1999.

53. Cohen, L.G., Ziemann, U., Chen, R., Classen, J., Hallett, M., Gerloff, C., Butefisch, C., Studies of neuroplasticity with transcranial magnetic stimulation, *J. Clin. Neuro-phys.,* 15, 305, 1998.

54. Donoghue, J.P., Limits of reorganization in cortical circuits, *Cerebral Cortex,* 7, 97, 1997.

55. Nicolelis, M.A.L., Lin, R.C., Chapin, J.K., Neonatal whisker removal reduces the discrimination of tactile stimuli by thalamic ensembles in adult rats, *J. Neurophys.,* 78, 1691, 1997.

56. Chapin, J.K, Moxon, K.A., Markowitz, R.S., Nicolelis, M.A. Real-time control of a robot arm using simultaneously recorded neurons in the motor, *Nature Neurosci.,* 2, 664, 1999.

57. Najafi, K., Ji, J., Wise, K.D., Scaling limitations of silicon multichannel recording probes, *IEEE Trans. Biomed. Eng.,* 37(1), 1, 1990.

58. Wise, K.D., Angell, J.B., Starr, A., An integrated-circuit approach to extracellular microelectrodes, *IEEE Trans. Biomed. Eng.,* 17(3), 238, 1970.

59. Moxon, K.A., Multichannel elctrode design: considerations for different applications, in *Methods for Neural Ensemble Recordings*, Nicolelis, M.A.L., Ed., CRC Press, Boca Raton, FL, 1999, 25.

60. Schmidt, E.M., Electrodes for many single neuron recordigs, in *Methods for Neural Ensemble Recordings*, Nicolelis, M.A.L., Ed., CRC Press, Boca Raton, FL, 1999, 1.

61. Strumwasser, F., Long-term recording from single neurons in brain of unrestrained mammals, *Science,* 127, 469, 1958.

62. Sonn, M., Feist, W. M., A prototype flexible microelectrode array for implant-pros-thesis applications, *Med. Biol. Eng. Comput.,* 12(6), 778, 1974.

63. Fox, K., Armstrong-James, M., Millar, J., The electrical characteristics of carbon fiber microelectrodes, *J. Neurosci. Meth.,* 3, 37,1980.

64. Hahn, A. W., Yasuda, H. K., James, W. J., Nichols, M. F., Sadir, R. K., Sharma, A. K., Pringle, O. A., York, D. H., Charlson, E. J., Glow discharge polymers as coatings for implanted devices, *Biomed. Scis. Instrum.,* 17, 109, 1981.

65. Nichols, M. F., Hahn, A. W., Electrical insulation of implantable devices by composite polymer coatings, *ISA Trans.,* 26, 15, 1987.

66. McNaughton, B. L., O'Keefe, J., Barnes, C. A., The stereotrode: A new technique for simultaneous isolation of several single units in the central nervous system from multiple unit records, *J. Neurosci. Meth.,* 8, 391, 1993.

67. BeMent, S.L., Wise, K.D., Anderson, D.J., Najafi, K., Drake, K.L., Solid-state electrodes for multichannel multiplexed intracortical neuronal recording, *IEEE Trans. Biomed. Eng.*, 33(2), 230, 1986.

68. Prohaska, O.J., et al., Thin-film multiple electrode probes; possibilities and limitations, *IEEE Trans. Biomed. Eng.*, 33, 223, 1986.

69. Hoogerwerf, A.C., Wise, K.D., A three dimensional microelectrode array for chronic neural recording, *IEEE Trans. Biomed. Eng.*, 41, 1136, 1994.

70. Rousche, P.J., Norman, R.A., Chronic recording capability of the Utah intracortical electrode array in cat sensory cortex, *J. Neurosci. Methods*, 82, 1, 1998.

71. Akin, T., Ziaie, B., Nikles, S.A., Najafi, K., A modular micromachined high-density connector system for biomedical applications, *IEEE Trans. Biomed. Eng.*, 46, 471, 1999.

72. Anderson, D.J., Najafi, K., Tanghe, S.J., Evans, D.A., Levy, K.L., Hetke, J.F., Xue, X.X., Zappia, J.J., Wise, K.D., Batch-fabricated thin-film electrodes for stimulation of the central auditory system, *IEEE Trans. Biomed. Eng.*, 36, 693, 1989.

73. Maynard, E.M., Nordhausen, C.T., Normann, R.A., The Utah intracortical electrode array: a recording structure for potential brain-computer interfaces, *Electroencephalogr. Clin. Neurophysiol.*, 102, 228, 1997.

74. Najafi, K., Wise, K.D., An implantable multielectrode array with on-chip signal processing, *IEEE J. Solid-State Circ.*, 21, 1035, 1986.

75. Najafi, K., Wise, K.D., Mochizuki, T., A high-yield IC-compatible multichannel recording array, *IEEE Trans. Elect. Dev.*, 32, 1206, 1985.

76. Elliot, D.J., *Microlithography: Process Technology for IC Fabrication*, McGraw-Hill, New York, 1986.

77. Van Zant, P., *Microchip fabrication: A practical guide to semiconductor processing*, Semiconductor Services, San Jose, CA, 1984.

78. Alling, E., Stauffer, C., Image reversal photoresist, *Solid State Tech.*, 6, 37, 1988.

79. Moxon, K.A., Nicolelis, M.A., Adler, L.E., Gerhardt, G.A., Chapin, J.K, Ceramic based multisite electrodes for multiple single unit recording, *IEEE Trans. Biomed. Eng.*, submitted.

80. White, R. L., Gross, T. J., An evaluation of the resistance to electrolysis of metals for use in biostimulation microprobes, *IEEE Trans. Biomed. Eng.*, BME-21, 487, 1974.

81. Najfi, K., Hetke, J.F., Strength characterization of silicon microprobes in neurophysiological tissues, *IEEE Trans. Biomed. Eng.*, 37, 474, 1990.

82. Nordhausen, C.T., Maynard, E.M., Normann, R. A., Single unit recording capabilities of a 100 microelectrode array, *Brain Res.*, 726, 129, 1996.

83. Rousche, P.J., Petersen, R.S., Battiston, S., Giannotta, S., Diamond, M.E., Examination of the spatial and temporal distribution of sensory cortical activity using a 100-electrode array, *J. Neurosci. Methods*, 90, 57, 1999.

84. Pickard, R. S., A review of printed circuit microelectrodes and their production, *J. Neurosci. Meth.*, 1, 301, 1979.

85. James, K. J., Normann, R. A., Low-stress silicon nitride for insulating multielectrode arrays, *Proc. 16th Ann. Int. Conf. IEEE Eng. Med. Biol. Soc.*, Baltimore, 306, 836, 1994.

86. Hochmair, E., System optimization for improved accuracy in transcutaneous signal and power transmission, *IEEE Trans. Biomed. Eng.*, BME-31(2), 177, 1984.

87. Chandran, A. P., Najafi, K., Wise, K. D., A new DC baseline stabilization scheme for neural recording microprobes. *Proceedings of the First Joint BMES/EMBS Conference*, 386 Oct. 1999 Atlanta GA.

88. Gray, P., Meyer, R., *Analysis and Design of Analog Integrated Circuits*, 3rd edition, Wiley and Sons, New York, 1993.

89. Loloee, A., A Programmable Gain VLSI Circuit and System for Sensing and Stimulating Action Potentials of Neuronal Networks, Ph.D. dissertation, Southern Methodist University, 1996.

90. Rebeschini, M., *Delta Sigma Data Converters: Theory, Design, and Simulation*, IEEE Press, 1996, chap. 6.

91. Morizio, J., Adaptive Gain Sigma Delta Architectures, Ph.D. Dissertation, Duke University, 1995, 19-37.

92. Morizio, J. et al., 14-bit, 2.2MS/s Sigma Delta ADCs, *IEEE J. Solid State Circuits*, 35, 968–976, July 2000.

93. Carden, F., *Telemetry Systems Design*, Artech House, Massachusetts, 1995.

94. Razavi, B., *Monolithic Phase-Lock Loops and Clock Recovery Circuits Theory and Design*, IEEE Press, 1996.

95. Best, R., *Phase-Locked-Loops Theory, Design, and Applications*, 2nd edition, McGraw-Hill, Inc., New York, 1992.

96. Wheeler, B.C., Automatic discrimination of singe units, in *Methods for Neural Ensemble Recordings*, Nicolelis, M.A.L., Ed., CRC Press, Boca Raton, FL, 1999, 61.

97. Mackay, R., *Bio-Medical Telemetry: Sensing and Transmitting Biological Information from Animals and Man*, 2nd ed., IEEE Press, Piscataway, NJ, chap. 6.

98. Schrom, G., Liu, D., Fischer, C., Pichler, C., Svensson, C., Selberherr, S. VLSI performance analysis method for low-voltage circuit operation, *Fourth Int. Conf. on Solid-State and Integrated-Circuit Technology*, 1995, 328.

99. Ferry, D., Akers, L., Greeneich, E., *Ultra Large Scale Integrated Microelectronics*, Prentice-Hall, Princeton, NJ, 1998.

100. Franca, J., Tsividis, Y., *Design of Analog-Digital VLSI Circuits for Telecommunications and Signal Processing*, 2nd ed., Prentice-Hall Inc., Princeton, NJ, 1994.

101. Larson, L., Integrated circuit technology options for RFIC's — present status and future directions, *IEEE J. Solid-State Circuits*, 33, 387, 1998.

102. Vittoz, E., Analog VLSI signal processing: why, where and how?, *Journal of VLSI Signal Processing*, 8, 27, 1994.

103. Vittoz, E., Borel, J., Gentil, P., Noblanc, J., Nouailhat, A., Verdone, M., Design of low-voltage low-power ICs, *Proc. 23rd European Solid State Device Research Conference*, 1993, 927.

104. Meindl, J., Low power microelectronics: retrospect and prospect, *Proc. IEEE*, 83:(4), 619, 1995.

105. Svensson, C., Liu, D., Power estimation tool and prospects of power savings in *CMOS VLSI chips, Proc. 1994 Int. Workshop on Low Power Design*, 171, 1994.

106. Svensson, C., High speed and low power techniques in CMOS and BiCMOS, in *Proc. IV Brazilian Microelectronics School*, Santos, E., Machado, G., Eds., Recife, 1995, 265.

107. King, R., Prasad, S., Sandler, B., Transponder antennas in and near a three-layered body, *IEEE Transactions on Microwave and Techniques*, 41, 6, 1980.

108. Toftgard, J., Hornsleth, S., Anderson, J., Effects on portable antennas of the presence of a person, *IEEE Transactions on Antennas and Propagation*, 41, 26, 1993.

109. Besseling, N., Maaren, D., Kingma, Y., An implanatable biotelemetry for six different signals, Department of Electrical Engineering, Delft University of Technology, The Netherlands, Medical and Biological Engineering, 11, 660, 1976.

110. Bakoglu, H., Baldwin, G., Li, Z., Tsai, C., Zhang, J., *Circuits, Interconnections and Packaging for VLSI*, Addison-Wesley, Boston, 1990.

111. Towe, B., Passive biotelemetry by frequency keying, *IEEE Trans. Biomed. Engin.,* 33, 905, 1986.
112. Crols, J., Steyaert, M., *CMOS Wireless Transceiver Design,* Kluwer Academic Publishers, The Netherlands, 1997.
113. Simon, M., Hinedi, S., Lindsey, W., *Digital Communication Techniques,* PTR Prentice Hall, Englewood Cliffs, NJ, 1994.
114. Powers, R., Batteries for low power electronics, *Proc. IEEE,* 83, 687,1995.
115. Reizenman, M., The search for better batteries, *IEEE Spectrum,* 32, 51, 1995.

7 Dynamic Interplay of Neural Signals during the Emergence of Cursor Related Cortex in a Human Implanted with the Neurotrophic Electrode

P. R. Kennedy and B. King

CONTENTS

7.1   INTRODUCTION

Invasive recording from the human central nervous system in order to provide a communication channel to a computer[1] is a radical alternative to other efforts to unlock paralyzed and mute subjects. Such efforts usually use scalp EEG and EMG signals. These alternatives are safer and less invasive but are limited to providing on/off step signals, though various techniques described in these chapters are pushing back the limitations inherent in such signals. To justify invasion of the brain, however, the resulting communication system needs to be of a magnitude better than non-invasive systems. In order to achieve this goal, an understanding of the neurophysiology of the recorded neural signals is essential. We describe here signal characteristics in a patient implanted with the Neurotrophic Electrode.

The neurophysiology of the primate cortex has advanced in recent years. It is now known that somatotopic representations will undergo plastic modification under the influence of new inputs.[3,4] Thus there is a concern that cortical representations of both paralyzed and normally functioning body parts are remodeled in spinal cord lesioned subjects. Hence, the implantation of electrodes into primary motor areas that once controlled now paralyzed limbs might *not* be functionally useful. The search is on, therefore, for alternative implantation sites that might not have undergone these plastic changes.[5]

These concerns, however, may not be warranted. Our experience with subjects implanted with the Neurotrophic Electrode suggests that *plasticity may be the key* to functional utility. We describe here some features of the recorded neural signals from our present subject which suggest that after much training the signals recorded from the implanted hand area of cortex may be devoted solely to moving the cursor. In other words, the patient may have developed cursor related cortex. We present these data at two levels of analysis: first, multiunit neural activity during acquisition of screen targets and second, single unit data that reveal (a) temporal binding of neural activity and (b) directionality.

7.2   METHODS

Patients are assessed prior to receiving an electrode implant. This pre-operative assessment consists of assessing cognition using questions requiring a "yes" or "no" answer and knowledge of pre-morbid cognitive baseline, including education (reading and writing are essential, computer use is preferred).

7.2.1   IMPLANTATION

Electrode fabrication and implantation have been previously described.[6,7] The patient undergoes a functional MRI to determine if, where and when neural activity exists. These fMRI results guide implantation site selection. A craniectomy is performed over the target area identified at surgery by alignment with the active areas noted on the pre-operative fMRI.

7.2.2   RECORDING

Recording techniques are summarized here and described in detail elsewhere.[6-8] The electrode routinely has two wires spaced 0.5 mm apart and hence two recording sites.

One recording site is less than 0.5 mm from the deep end of the cone and the other is 0.5 mm from the wide superficial end. Differential recording (without ground) using both recording sites in the glass cone offers a number of advantages, such as (1) excluding the possibility of recording from tissue outside either end of the cone (for an example, see Figure 3 of Kennedy[6]); (2) minimizing artifacts due to chewing or teeth grinding. The special features of the recorded data are discussed below.

A simple telemetry system was originally used in the animal recordings because it avoided ground loop problems and there were no wires for the animal to entangle. It has been retained in subject implantations because it also allows complete skin closure, thus minimizing the opportunity for infection, and removes the cosmetic problem inherent in wires protruding from the scalp. The radio telemetry system[9,10] consists of a custom transmitter and a modified commercial FM receiver. The transmitter is constructed with surface-mount components mounted on an IC board. During data acquisition, the subject is supine and the antenna of the FM telemetry receiver is positioned within four inches of his head. Transmission frequency can range from 30 to 50 MHz, bandwidth of the transmitter extends from near 0.1 to 5 KHz (the 3 db point), and the filter cutoffs of the post-transmitter amplifier (BMA 831, CWE Inc.) are set at 0.5 and 5 KHz. If for any reason the transmitter were to fail, it could be replaced after disconnection from the electrode connecting pins. A system gain of around 1,000 is determined for each transmitter before implantation. The neural data is archived on an 8 mm digital video recorder whose bandwidth is DC to 12.5 KHz in the PCM mode, with signal amplitude ranging from 10 mV to 1.25 V. The neural signals are recorded in synch with the video signal from the camera that monitors the subject.

7.2.3 WAVESHAPE SEPARATION

During on-line analysis, the video signal is displayed on a monitor and the neural signals are connected to the signal separator (DataWave Systems Inc., Discovery Software, Boulder, CO). Waveshapes that exceed a user-determined voltage level are digitized at a user-determined frequency, typically 32 KHz. When digitizing waveshapes, the user can allow some overlap of waveshapes, within the 1 ms time bins, so that rapidly recurring waveshapes are minimally missed. Seven different parameters are used: (1) peak to peak amplitude, (2) amplitude above baseline, (3) amplitude below baseline, (4) width, (5) time to peak, (6) time to valley and (7) examination of two sample points (of the 32) along the X axis whose values fall within one standard deviation of the waveshape average, thus excluding points outside these values. In addition, only a minimal section of the waveshape (deter- mined by the user) needs to be included in the analysis, thus saving on-line process- ing time. The seven separation parameters are stored in a dedicated set-up parameter file and used in subsequent sessions for classifying waveshapes at subsequent record- ing sessions, or for retrospective off-line analysis from the videotape.

7.2.4 ACQUIRING CONTROL OF FIRING RATES

The subject is encouraged either to visualize the performance of free-form limb movements that are expected to activate the waveshapes, or to focus on moving the cursor. The neural data are fed back as multiunit activity or separated in real time by the signal separation software as described above. After keyboard selection of one or two waveshapes, they are fed back using auditory signals that indicate the firing rate and visual signals that indicate position of the cursor and entry into a target. A video camera records the cursor movement into the various targets and archives them on 8 mm digital recording tape as raw data (Sony EV-S350, stereo video cassette recorder). *Auditory feedback* consists of brief tones of a specific frequency assigned to each waveshape. Each time a specific waveshape fires, the system produces a brief tone of specific frequency. Repetitive firing is represented by a discontinuous tone whose repetition rate is a function of the firing rate. *Visual feedback* consists of the movement of the cursor over the computer screen containing the iconic training task, virtual keyboard, buttons for environmental controls or Internet interface. The individual or multiunit waveshapes are converted to pulses and sent to the second computer to drive the cursor from left to right across the screen or from top to bottom of the screen.

7.2.4.1　Assessing Learning

The training paradigms are described elsewhere[12] and are not the subject of this chapter.

7.2.4.2　Data Analysis

The data are analyzed using DataWave software or transferred to the NEX, Neuron Explorer software package for ease of generation of peri-event time histograms (PETHs), ISIHs, and so on.

7.2.5　PRESENT SUBJECT

Our subject, JR, is a 53-year-old male who suffered a brainstem stroke in December 1997. He remains paralyzed and mute with partial face movements, some left neck movement, disconjugate eye movements, tracheotomy, decubitus ulcer requiring near continuous analgesia with fentanyl patch, gabapentin and codeine. Recently (July 1999) he has become functionally blind due to cataracts. He blinks twice for "yes" and once for "no." He is cognitively intact.

7.3　RESULTS

7.3.1　NEURAL ACTIVITY DURING ACQUISITION OF TARGETS

Following implantation, neural activity appears at about three weeks as expected and provides robust recordings near three months.[6,7] To activate the signals, the subject is first asked to imagine making hand movements. This does produce some possible changes in firings but these are not reliable. Sensory responses are not seen following manipulation of limbs or stroking of skin with emphasis on the contralateral limbs and trunk. At month four, however, face movements reliably activate the

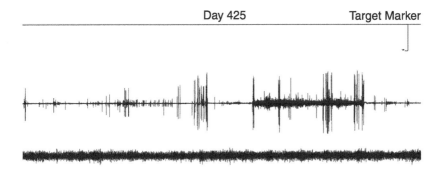

FIGURE 7.1 Neural activity and eyebrow EMG activity from session 425. The cursor is entering the letter space target as shown by the marker on the right. The neural signals are shown in the top trace above the eyebrow EMG. Maximum amplitude is approximately 400 μV. Total time shown is seven seconds.

signals. The patient makes jaw, left neck and tongue movements to the right to produce bursts of activity. At month five, eye movements reliably activate the signals, though there is no preferred direction of eye movements. In addition, eyebrow movements also activate the signals, whereas mouth movements are no longer useful for activation. At month six, he lies quietly with no apparent movements during activation of the signals that are driving the cursor into the target. When queried, he indicates that he is not moving any body part and simply focuses on watching the cursor movement and listening to the signal firings. He then has a period of very poor health and works intermittently. At around fifteen months, the following studies are performed with eyebrow EMG as a control.

The neural activity and eyebrow EMG activity from session 425 (days post implantation) are shown in Figure 7.1 over a time base of seven seconds as the cursor is entering the letter space target as shown by the marker on the right. The neural signals (maximum amplitude is about 400 μV) are shown above the eyebrow EMG. There is neural activity prior to entering target and quiescence on reaching the target. Note that the eyebrow EMG is suppressed, demonstrating that both signals are not co-activated.

The subject could achieve a largely reciprocal relationship between neural activity and eyebrow EMG activity as shown in Figure 7.2 over a time base of 40 seconds. Once again it can be seen that during neural activity that move the cursor across the screen, EMG activity is suppressed, and during EMG activity that move the cursor down the screen, neural activity is largely suppressed. Thus, EMG activity can be suppressed while neural activity drives the cursor.

Repeat attempts to drive the cursor across the row of icons result in improved performance as shown in Figure 7.3 for session 430, where neural activity drives the cursor from left to right across the screen and EMG activity could drive it vertically down. Activation of EMG activity could produce an error. He first takes almost 90 seconds to traverse each of five icons without error. He speeds up and moves accurately across the icons in 50 seconds. On the third attempt he produces

Day 425

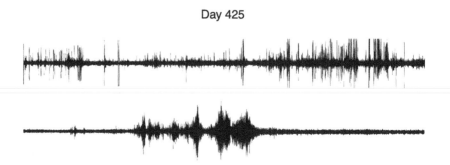

FIGURE 7.2 Neural activity and eyebrow EMG activity used by the patient to control cursor movement. The neural activity moves the cursor across the screen, while the EMG activity is responsible for moving the cursor down the screen.

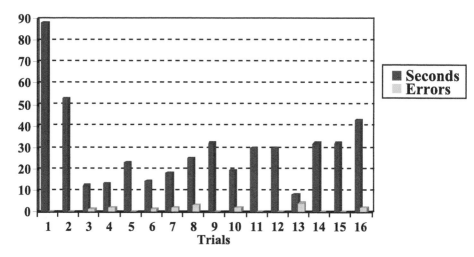

FIGURE 7.3 Neural activity and eyebrow EMG activity from session 430. Neural activity drives the cursor from left to right across the screen, and EMG activity could drive it vertically down.

one error while crossing the icons in less than 20 seconds. Still moving too quickly, he produces further errors until he slows down and takes about 30 seconds to move accurately across the row of icons as shown in trials 9, 11, 12, 14 and 15.

These data demonstrate that the neural signals can be activated without eyebrow EMG signals. They suggest that the subject can control the neural firings alone in order to drive the cursor. These data, along with the subject's denial of thinking about eyebrow, face or other movements during cursor movements, suggest that the implanted cortex may be devoted to moving the cursor rather than moving a body part. Further evidence that there is dynamic interplay between the neural signals is available from single unit data.

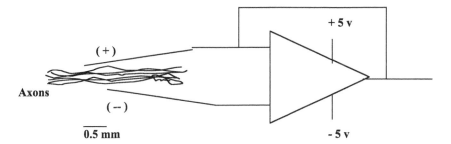

FIGURE 7.4 Special arrangement of the recording setup inside the tip of the electrode. The axons lie inside the glass cone tip between the recording wires that are 500 μm apart. One wire is held in a positive polarity and the other in a negative. There is no independent ground wire.

7.3.2 SINGLE UNIT DATA

Figure 7.4 shows the special arrangement of the recording setup inside the tip of the electrode. The axons lie inside the glass cone tip between the recording wires that are 500 microns apart. One wire is held in a positive polarity and the other in a negative one. There is no independent ground wire. If only one axon traveled between the wires, the depolarization wave would have an initial deflection in one polarity and then a deflection in the opposite polarity. In addition, the interval between each deflection would be constant, but it is not. We know from histological results in rats and monkeys that there are many axons in the tissue inside the cone[8] that contribute to the neural activity.[6-7] It is important to note that the polarity of the initial depolarization wave indicates proximity to one electrode or the other. This is because the external axonal membrane is at a constant polarity (conventionally positive) and membrane depolarization consists of a transient opposite polarity (conventionally negative), whereas the wires are fixed in opposite polarities. Thus a transient negativity of the axonal membrane is registered at each wire as an initial depolarization that is opposite in sign depending on which wire is closest to the axon.

The polarity is illustrated by the up-going waveshapes in Figure 7.5. Waveshapes similar in amplitude and shape can be discerned as shown in Figure 7.5 for three waveshapes, the two largest being close in amplitude. The time base is 1.5 ms, and voltage is about 200 μV for the tallest waveshape. These three waveshapes all have initial depolarizations in the same direction and therefore are close to one wire. In addition, these data indicate that waveshapes of similar shape can be seen from session to session (days 425 and 470).

Many waveshapes are not unidirectional, however, but have dual polarity, as shown in Figure 7.6. On the left are two waveshapes with down-going polarity immediately followed by up-going polarity, with equal amplitudes on each phase. Such a pattern can only be explained by an action potential that is first recorded at one wire and then immediately at the other wire from an axon that is equidistant from each wire. This could only happen, however, if the recording wires have minimal spatial separation in the longitudinal dimension. Instead they are fixed at 500 μm apart. Up-going and down-going (composite) waveshapes therefore ought to be separated by a fixed time and ought to originate near different wires if they

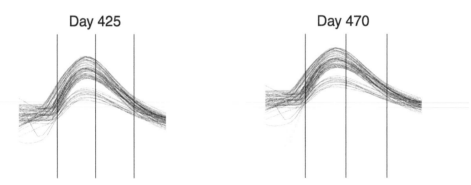

FIGURE 7.5 Examples of three waveshapes recorded from the Neurotrophic Electrode on two different days (Day 425 and Day 470). These three waveshapes all have initial depolarizations in the same direction and therefore are close to one wire. In addition, these data indicate that waveshapes of similar shape can be seen from session to session (days 425 and 470). The time base is 1.5 ms. Voltage is about 200 μV for the tallest waveshape.

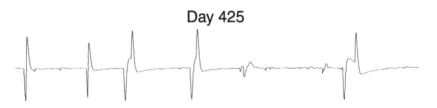

FIGURE 7.6 Example of waveshapes with dual polarity. Peak to peak amplitudes are approximately 200 μV. Time base is 150 ms.

are the same waveshape. Figure 7.6 shows that this does *not* happen. The 150 ms of continuous recording show waveshapes of different amplitudes and polarities with varying latencies between the waveshapes. Peak to peak amplitudes are about 200 μV.

These differing waveshapes raise the possibility that the first and second waveshape to the left of the sequence in Figure 7.6 might be a composite of different waveshapes recorded near different wires. This question was addressed by examining waveshapes of similar amplitudes from different recording sessions.

The waveshapes in Figure 7.7 are chosen randomly and aligned in sequence to illustrate the point that the bidirectional waveshapes on the left of each row appear to be composites of up-going and down-going waveshapes. To the right are shown waveshapes of opposite polarities with several milliseconds between them. There are often 10, 20 or more ms in between (not shown). Tracing the data from right to left, the waveshapes of opposite polarity come closer and closer in time and coincide to form one large spike with an equal up- and down-going phase amplitude. These waveshapes with equal phase amplitudes coinciding in time, then, may well be composites of the different waveshapes recorded at different wires from different axons.

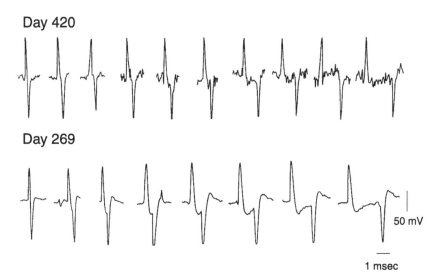

FIGURE 7.7 Waveshapes chosen at random from two different days, Day 420 and Day 269. The waveshapes are arranged from left to right such that the bidirectional waveshapes are on the left of each row and waveshapes of opposite polarities with several milliseconds between them are shown on the right.

These composite waveshapes may be examples of *temporal binding*: waveshapes of opposite polarities may coincide in time, driven by some undetermined *common input*, related perhaps to some *behavioral event*. If there is no temporal binding, then the waveshapes ought to fire in random temporal order. However, this does not happen, as demonstrated by data from sessions 420 and 269 (shown in Figure 7.7). The behavioral event may be the cursor driven horizontally across the screen from left to right under the influence of these waveshapes as described in Section 7.2, Methods. In all those sessions the up phase almost always preceded the down phase as shown above in Figure 7.7. Therefore, we conclude that the firing is not random, at least in these examples. There appears to be temporal binding.

The behavioral event that might drive an input to result in this temporal binding is not immediately evident in our data. It is nearly impossible to determine behavior in a subject who is paralyzed because there can be no *objective* correlation with movement or EMG activity. The only output available, apart from face movements, is the cursor movement on the monitor. In the above examples in Figure 7.7 for sessions 420 and 269 (and many others), the cursor is driven from left to right across the screen. To our surprise, when we introduce three consecutive sessions of driving the cursor in the vertical direction down the screen, the phase relationship switches 180 degrees so that the down phase precedes the up phase, as shown in Figure 7.6 for session 425.

To assess whether or not these phase relationships are present throughout the recording session, PETHs are constructed between the up waveshape and the nearby down waveshape. Only those waveshapes with the same shape and amplitude as those shown in Figures 7.6 and 7.7 are used. The PETHs in Figure 7.8 use *up* waveshapes

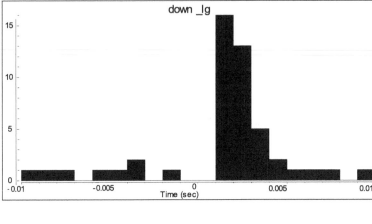

-10ms   0ms   -10ms

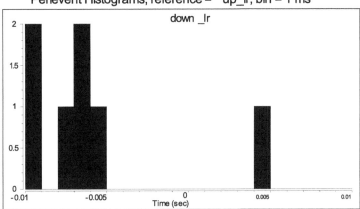

-10ms   0ms   10ms

FIGURE 7.8 Perievent time histograms using *up* waveshapes as the reference node (time zero) and bins are filled when a down waveshape appears. A. On day 420, when the cursor was moved horizontally across the screen, down waveshapes followed up waveshapes (peak to the right of zero time [i.e., time of up waveshape]). B. When the cursor was moved vertically across the computer screen, down waveshapes preceded up waveshapes (peak to the left of 0 time). Zero time is time of up waveform. The counts per bin on the abscissa are 6 (left panel) and 2 (right panel).

as the reference node and have zero time as the centerpoint, with time before to the left and time after to the right. As the PETHs illustrate, there are few exceptions to the observation that when driving the cursor in the horizontal direction (panel A), the up reference waveshapes are followed by the down waveshapes, and when driving the cursor in the vertical direction the phase relationship is reversed (panel B).

Thus, these phase relationships may be associated with *directionality* of cursor movement. *The mental events associated with expressing directionality may be the factors determining the temporal binding in this example.*

7.4 DISCUSSION

7.4.1 SUMMARY

We present several lines of evidence demonstrating that while driving the cursor across the computer screen, the subject, JR, eventually learns to drive the cursor without the use of eyebrow movements that once activated the signals. We interpret these events to indicate that the face representation on motor cortex initially spreads to include the implanted hand area. This occurs because the subject uses his face actively to indicate "yes" and "no" and other facial expressions with much emotional content, whereas the hand is not active due to the paralysis. In addition, we do not encourage the subject to imagine phantom hand movements. The initial plasticity leading to enlargement of the face representation reverses itself as the subject learns to focus on the cursor. The observation that eyebrow EMG activity is quiet (Figures 7.1 and 7.2) while the neural signals are active is evidence in favor of this interpretation. These data support the notion of the development of cursor related cortex.

The unique design of the recording electrode shown in Figure 7.4 provides a means of recording from axons close to each wire as described. This has revealed interesting properties such as *temporal binding* of waveshapes where bidirectional waveshapes separated by variable latencies coincide to produce a composite waveshape, but rarely reverse the phase relationship during the same behavioral event (Figure 7.8). Only when the cursor is required to move in a direction 90 degrees out of phase does the phase relationship reverse. Thus, up-going waveshapes almost always precede down-going waveshapes when the cursor is driven in the horizontal direction and vice versa for the vertical direction. Our analysis supports the hypothesis that *directionality* is encoded by these waveshapes relationships.

7.4.2 IMPLICATIONS

These findings have important practical implications. First, the emergence of cursor related cortex has implications for the field of brain–computer interfaces that use implanted electrodes for recording individual neural signals. The possibility of being able to implant areas almost at random and provide functional control signals from these areas expands enormously the potential areas for implantation. It makes less critical the area implanted. In our work, we may not need to implant the areas that show functional activation on the fMRI in response to imagined movements. The idea of "plug and play" may not be as far in the future as we now think. Furthermore, if almost any area can be implanted the number of implantable electrodes increases.

When training the patient to produce appropriate signals, arrays of electrodes covering large areas of cortex would be unlikely to result in the emergence of cursor related cortex because trying to feed back all these signals for training would be too confusing for the patient. Instead, signals ought to be fed back individually or in small groups. To try to feed back large numbers of signals at once would be analogous to trying to re-train musicians in an orchestra whose conductor is suddenly missing and whose members cannot hear their own playing. By the time their output is restored they would be playing in total confusion or not playing at all. Therefore, trying to re-train all the musicians at once with one feedback channel would probably result in more confusion. It would be better to take individual musicians (analogous with waveshapes) and try to re-train them to play the same melody (analogous with cursor movement). Thus, re-training a few waveshapes will probably be more successful than simultaneous re-training of an immense number of waveshapes spread out over a large expanse of cortex. This remains to be seen of course, and data from subsequent patients are needed to confirm and extend the present data.

Second, the data on directionality suggest that the timing relationships of waveshape firings vary in a systematic way in relationship to direction of cursor movement. These data may provide preliminary evidence for our suspicion that directionality may be encoded in the firing patterns in a way not previously described. Georgopoulus et al.[11] have described cortical neurons in monkeys with preferred directions of firing. These results may not be as incompatible as they seem. Perhaps we are listening to neurons in the *processing layers* of cortex that are "upstream" or "downstream" from the neurons described by Georgopoulus et al.[11]

Third, the data have enormous practical implications *if verified in other patients*. They raise the possibility that *merely thinking* about the desired direction might produce a firing pattern that could be interpreted by an algorithm and used to change the direction of the moving cursor under the subject's mental control alone. This would allow control of direction from one electrode, thereby enhancing the value of each electrode. In monkey behaving studies that used the Neurotrophic Electrode, waveshapes with differing functionalities are described, namely, hand flexions *and* extensions.[7] It should not be surprising, therefore, to find similar multimodalities in human recordings.

7.4.3　Future Directions

Clearly, these results need to be replicated in future patients. The present data impel us to implant multiple electrodes and extract all the information that is available from the waveshape firing patterns. There may be information unique to each electrode. The original idea of separating individual waveshapes and feeding them back to the patient for training may not be necessary. For example, if directionality can be reliably extracted, the patient will drive the cursor using the firing rates of *several* signals and switch direction of the cursor *at will*.

These results also suggest that it may be appropriate to implant areas not necessarily related to imagined hand movements as seen on the fMRI. If cursor related cortex is confirmed to emerge in subsequent patients, one might consider implanting cortical areas other than motor cortex. If so, this would offer hope to those patients who have extensive cortical damage due to trauma or stroke but who still retain adequate cognition to control a cursor. In addition, extracting multiple control signals from each electrode implies that for applications requiring multiple degrees of freedom, such as control of a paralyzed or robotic limb, fewer electrodes may be needed than were once assumed by workers in this field.

ACKNOWLEDGMENTS

We thank Prof. Melody Moore, Mr. Greg Montgomery and Dr. Chris Russell for software programming. We thank Dr. Roy Bakay for his instrumental role in implanting the subjects and Dr. Hui Mao for performing the functional MRI. The cost of implantation of the first subject was borne by Emory University Hospital, the second subject by the Veterans Administration Hospital, Clermont Road, Atlanta, GA, and the third subject by a grant to Neural Signals Inc. from the National Institutes of Health, Neural Prostheses Program, SBIR #1R43 NS 36913-1A1.

POTENTIAL CONFLICT OF INTEREST

The first author, P.R. Kennedy, is entitled to receive income from Neural Signals Inc. if and when it receives income from the sales of the products related to the research described in this paper. The terms of this arrangement have been reviewed and approved by Emory University in accordance with its conflict of interest policies.

REFERENCES

1. Kennedy, P.R. and Bakay, R.A.E. Restoration of neural output from a paralyzed subject by a direct brain connection, *NeuroReport,* 9, 1707, 1998.
2 Kubler, A., et al. The thought translation device: a neurophysiological approach to communication in total motor paralysis, *Exp. Brain Research,* 124, 223, 1999.
3. Merzenich, M.M., et al. Variability in hand surface representation in areas 3b and 1 in adult and squirrel monkeys, *J. Comp. Neurol.,* 258, 281, 1987.
4. Wall, J.T., et al. Functional reorganization in somatosensory cortical areas 3b and 1 of adult monkeys after median nerve repair: possible relationships to sensory recovery in humans, *J. Neuroscience,* 6(1), 218, 1986.
5. Batista, A.P., et al. Reach plans in eye-centered coordination, *Science,* 285, 257, 1999.
6. Kennedy, P.R. The cone electrode: a long-term electrode that records from neurites grown onto its recording surface, *J. Neuroscience Methods,* 29, 181-193, 1989.
7. Kennedy, P.R., Bakay, R.A.E., and Sharpe, S.M. Behavioral correlates of action potentials recorded chronically inside the cone electrode, *NeuroReport,* 3, 605, 1992a.
8. Kennedy, P.R., Mirra, S.S., and Bakay, R.A.E. The cone electrode: ultra-structural studies following long-term recording in rat and monkey cortex, *Neuroscience Letters,* 142, 89, 1992b.
9. Mackay, R.S. Bio-medical telemetry, *NASA Report,* SP-50994, 1970.
10. Motchenbacher, C.D. and Fitchen, F.C. *Low-Noise Electronic Design,* John Wiley and Sons, 1973.
11. Georgopoulus, A.P. *Cold Spring Harbor Symposium Quantitative Biology,* 55, 849, 1990.
12. Kennedy, P.R., et al. Direct control of a computer from the human central nervous system. *IEEE Trans. Rehab. Eng.* 8(2), 198–202, 2000.

8 Brain Control of Sensorimotor Prostheses

John K. Chapin and Miguel A. L. Nicolelis

CONTENTS

0-8493-2225-1/01/$0.00+$.50
© 2001 by CRC Press LLC

8.1   INTRODUCTION

The introduction to this volume outlines the possibilities inherent in utilizing electronic interfaces with the brain to alleviate problems of paralysis, such as that caused by spinal cord injury. The possibility of using electroencephalographic recordings for this purpose has been explored, but the amount and specificity of the information that can be extracted in this way is questionable. As an alternative, it would theoretically be more efficient, and more elegant, to utilize control information extracted from the part of the brain that normally is directly involved in processing "commands" for voluntary movement of the arm. In fact, the information contained in the motor areas of the brain is (by definition) completely sufficient to reproduce all of the normal movements of the body. It has long been considered that paralysis victims might benefit from such an ability to control devices directly from the brain. It would of course be preferable to utilize non-invasive techniques for recording the brain activity necessary to control these devices, but no currently available technique can provide sufficient information for control of external movement in real time. Electroencephalography (EEG), for example, involves recordings from the scalp, and thus can only detect the common electrical dipoles emerging from billions of individual neurons in wide areas of the subjacent brain. Though it has commonly been used for measuring global brain rhythms, it has not been particularly useful for extracting brain information specific to a particular limb movement. As such, when EEG is used as a brain–computer interface, the subjects must normally learn to control the expression of these global rhythms. Such approaches have been successfully utilized for tasks that are not time-critical, such as selecting letters by using EEG recordings to move a cursor across a computer screen.[2]

More recently, the technique of functional magnetic resonance imaging (fMRI) has become extremely popular because of its ability to record functional activity from specific areas of the brain. Unfortunately, MRI scanners are large, cumbersome and expensive, although these problems might be partially mitigated when improved superconductor technology becomes available. A more difficult problem is that the fMRI technique does not directly measure brain activity, but instead measures changes in blood flow that may (or may not) be caused by increased functional activity. Thus, fMRI will never be able to extract truly real-time brain information.

In 1980, Schmidt[3] suggested the possibility of controlling external devices by using control signals obtained from multi-single neuron recordings in the motor cortex. Such a possibility was suggested by the previous findings of Humphrey et al.[4] that the resolution of motor information in the brain could be improved by averaging the discharge of up to seven neurons recorded simultaneously in the motor cortex. Such averaging is required because each neuron broadcasts its output signals through sequences of action potentials ("spikes") that by themselves contain little information.[5,6] Further advances in this field therefore depended on availability of technologies for simultaneously recording large numbers of neurons. Beginning in the 1980s, we and a few other laboratories began to develop such technologies,[7-12] so that it is now possible to record well over 100 neurons simultaneously through chronically implanted electrode arrays (see Section 8.3.1). This has allowed us to demonstrate the feasibility of neurorobotic control in experimental animals (see Section 8.2.1), as more fully described in a recent paper.[13]

8.2 RECENT SUCCESSES IN NEUROROBOTIC CONTROL

8.2.1 Feasibility Demonstrated in the Rat

8.2.1.1 Neurorobotic Paradigm

Six rats were trained to press a lever whose angular position directly controlled proportional movement in one radial dimension of a robot arm (Figure 8.1). The rats learned to move the lever such that the robot arm (RA) was positioned to receive a drop of water from a dropper. Upon releasing the lever the RA moved through a gap in a Plexiglas wall, allowing the rats to drink. These animals were then surgically implanted with 24–32 microelectrodes in the primary motor (MI) cortex, plus the ventrolateral (VL) thalamus. Two animals were also implanted with bipolar electromyographic (EMG) electrodes in appropriate forelimb muscles. All implants were in the forelimb representations of these areas, except two rats that received implants in the MI mouth area. After one week, the rats were then retrained on the original paradigm during parallel recording of up to 48 single neurons from the above brain areas.

In typical experiments, about 90% of recorded single neurons exhibited some statistically significant discharge relating to the timing of the forelimb movement. Among the six animals with MI forelimb area implants, between 23 and 46 neurons (average 33.5) were obtained with such significant task correlation. Various neurophysiological analysis techniques were used to define the "motor tuning" of these recorded neurons, in terms of the range of movements normally utilized in the task.[14] The neurons recorded in the MI and VL were often active during multiple phases of this movement, but most tended to be correlated best with the onset of the downward pressing movement. This neuronal activity preceded onset of wrist-flexor/elbow-extensor EMG activity by up to 80 ms, and onset of detectable downward movement by up to 100 ms. Moreover, for about 75% of our MI neurons, this activity that occurred during the end phase of the reaching movement was much more strongly correlated with the subsequent onset of pressing than with the timing of the reaching movement. In contrast, a smaller group of neurons (12% of total) were found to be best correlated with the onset of reaching.

The next aim was to investigate methods for transforming the activity of these simultaneously recorded neurons into a neuronal population vector capable of accurately controlling the movement of the robot arm. The data analysis quickly demonstrated that the information capacity of individual neurons was insufficient to this task because they cannot produce a signal that approaches the positioning accuracy and smoothness of native limb movements. This is because single cortical unit spike counts cumulated over appropriately fine time bins (e.g., 20 ms) can normally achieve a maximum of about three spikes/bin. Thus, only four "levels" of discharge can be specified (i.e., two bits of binary information). For example, a typical single neuron exhibited its greatest discharge in the 100 ms period around the onset of the

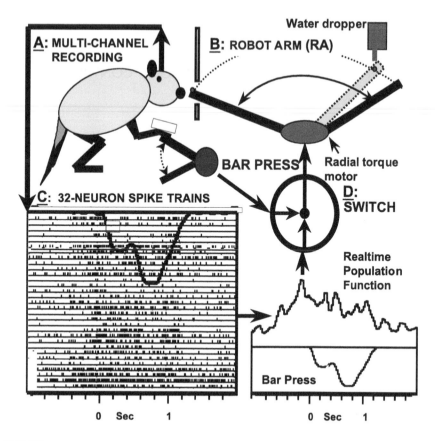

FIGURE 8.1 Neurorobotic experiments in rats. A. *"Bar press->Robot-arm" mode*: Rats were initially trained to press down a spring loaded lever for a water reward. The graded lever displacement was electronically translated to produce proportional displacement of a robot arm from its normal rest position (protruding through a slot in a Plexiglas barrier) B. to a water dropper. Upon lever release the robot arm (with the water drop) moved passively to the rest position where the rat could drink it. C. *"Neuronal-population-function->Robot-arm" mode:* Rats were chronically implanted with multielectrode recording arrays in the MI cortex and VL thalamus, yielding simultaneous recordings of up to 46 discriminated single neurons. Spike trains are shown from a subset of 32 neurons chosen for the neurorobotic control. These spike trains are shown over a 3 s period around onset of bar movement. D. Neuronal-population function (NPF; weighted according to PC1) extracted electronically in real time using a weighted 32-channel integrator network with a 20 ms time constant. A switch determined the source of input signals (i.e., bar press or NPF) for controlling position of the robot arm. In experiments, rats typically began working in the lever-movement->robot arm. Without warning the paradigm was then switched to the NPF>robot arm mode, allowing testing of whether the animal could routinely obtain its water through direct neural control of the robot arm.

pressing movement, yet the mean spike count during this period was about 0.45, ranging from 0–3 over different trials. Overall, such a neuron's correlation (R) with lever movement was about 0.3, and thus did not approach the level of trial-by-trial reliability and temporal specificity that would be required for controlling a motor device.

The resolution and reliability of neural signals (as measured by R) were found to be markedly greater in multi-neuron ensemble averages than in any single neuron recorded. This was true even for raw (unweighted) averages, in which the signal information capacity (i.e., number of levels) increases with the number of neurons, but can still only specify integer values. Weighted averages were still better, as their information capacity varies with the square of the ensemble size. Such averaging is necessary because of the stochasticity of recorded neurons: In our data sample, the total communality (i.e., covariance) between the neurons in such a population generally did not exceed 30%, even when integrated into 25 msec time bins.

Principal components analysis (PCA) was used as a direct, assumption free approach for the first stage of defining weighting functions for this ensemble averaging.[15] It allowed us to condense the salient information within a whole ensemble of 23–46 partially correlated neuronal spike trains into 3–5 uncorrelated components. In other neural systems,[16,17] we have shown that PCA optimally maps the range of tuning functions within a neuronal ensemble, thus providing an ideal reduced variable set for more directed mathematical treatments (such as discriminant analysis or neural networks). Here, the information in the data set was deliberately reduced to one variable (the first principal component, PC1) because accurate electrode placement ensured that most of the recorded neurons discharged around the onset of forelimb lever pressing movement. Though each neuron had a unique but partially overlapping tuning function, averaging the activity of these neurons using weights defined by PC1 collapsed this information into a smooth and distinct neural activity profile with a greater signal/noise than the ensemble average. Such PC1 profiles, as obtained from all animals, tended to peak sharply during the active downward forepaw movement just as it touched the lever.

Further studies demonstrated that this PC1 signal effectively predicted the timing and magnitude of lever movement over the 300–500 msec following lever movement onset. Interestingly, the period during which PC1 predicted these movement parameters was not during the time that the movement was occurring, but during the peak of PC1 activity that occurred during the 100 msec just prior to movement onset (Figure 8.2). As such, this PC1 signal did not encode movement trajectory as a direct real-time image, but instead within a short burst of premovement activity. Thus we experimented with mathematical transformations of the time domain of this PC1 signal. First, the PC1 signal was transformed into a tapped delay line, by dividing it into five parallel signals lagged at successive 100 ms delays. Three multivariate statistical techniques, regression, discriminant and canonical correlation analysis, utilized this tapped delay signal as input to produce predictions of the actual lever movement trajectory (Figure 8.3). All three predictions were much better correlated with the movement than PC1 (Rs ranging from 0.74–0.76, as opposed to 0.51 for PC1). However, the linearity of these techniques did not allow them to reject neural signals relating to non-lever pressing behaviors. In contrast, such inappropriate signals were effectively rejected by a recurrent neural network whose transformation of the unlagged PC1 signal produced an output which was highly correlated (R = 0.86) with the movement trajectory.

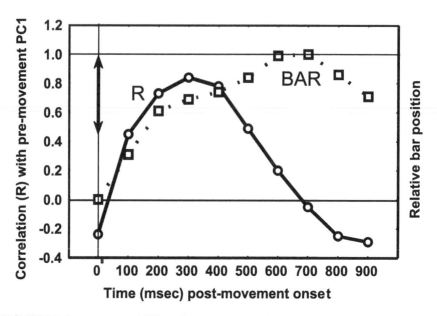

FIGURE 8.2 Pre-movement NPF predicts movement trajectory. The amplitude of bar movement (squares, dotted lines) was correlated over 100 msec intervals with the amplitude of pre-movement NPF activity (i.e., 32 neuron integration weighted according to PC1). The correlation coefficients (circles, solid lines) peaked at R = 0.84 during the interval from 200–300 msec post-movement-onset.

The rANNs (Neural Dimensions, Gainesville, FL) consist of input units, recurrently connected hidden layer units, and output units. Since rANNs use their recurrent connections to encode temporal information, the learning data sets do not require tap-delaying. Dynamic back-propagation learning (using the learning data set) is used to adjust the weightings within this network to optimally transform the input (from the test data set) into an output function that closely matches the MM predictor function (Figure 8.4).

Our analysis of this transformation revealed that the recurrent neural network learned to recognize and use distinct features of the PC1 signal to predict the timing and magnitude of the lever movement. In particular, the onset and termination of lever press was found to be very precisely predicted on the basis of the slope steepness during the rising and falling phases of the PC1. Similar results were found in the other five animals used in this study. Thus, the neural encoding of motor output functions may be related less to the overall magnitude of neural population activity than to highly distinct spatiotemporal patternings of neuronal activity within the population.

To determine whether the PC1 signal could be used to control the robot arm, it was translated in real time into a voltage signal that positioned the robot arm. Initially, a summing amplifier circuit was used for this function. This allowed a determination of whether the animals could use the brain (PC1) signal as a substitute for lever pressing as the operant behavior necessary to obtain the water reward. In order to maintain the animals' association between lever press and robot arm movement, the animals were allowed about five minutes at the beginning of each experimental day to work in the lever press mode. This ensured that the animals would continue to make lever pressing movements when the control of the robot arm was suddenly switched to the brain-derived signal. These movements provided a direct measure of the timing of the animals' motor-behavioral intentions, as specifically defined by the conditioning task.

The results were that all animals were able to move the robot arm immediately after switching its control to the brain (Figure 8.5). Indeed, two of the six rats were able to move the robot arm to obtain water with 100% efficiency (i.e., they obtained water whenever they pressed the lever correctly). These rats plus two more that reached 70–90% efficiency were able to obtain their water for the day using this technique. On the other hand, the final two rats could not reach 50% efficiency, and therefore the behavior was extinguished.

As this training continued over several days, the animals tended to dissociate their lever movements from the brain activity normally observed during those movements. Thus, they made smaller movements and often did not press the lever at all. This result suggests that the cortical coding for limb movement is sufficiently abstract to allow a certain drift in terms of the actual motor output that is made when the cortical neurons discharge in a certain pattern. This and other evidence that the motor cortical output is trainable[18-20] is good news for the prospect of neurorobotic control in paralyzed patients. On the other hand, we found it extremely difficult to train animals *de novo* to exhibit a certain pattern of activity without first training them to perform a physical act such as lever pressing. This suggests that even though the cortex can exhibit some dynamic plasticity, it is still dependent on the overall context of external movement to generate its output.

These results have thus demonstrated the feasibility of controlling external devices from the brain. Moreover, since this control is derived from the brain area that controls the normal limb movement, its timing and trajectory coordination are closely related to the normal processing of movement information in the brain. This made it possible to achieve accurate real-time "neurorobotic" control in a simple animal with a minimum of training.

8.2.2 Neurorobotic Control in Monkeys

Our techniques for simultaneous multisite neural ensemble recordings, originally developed in rats,[7-12,21-22] were next adapted for behaving primates, chiefly owl monkeys.[23-24] More recently, such recordings have been utilized for experiments in neurorobotic control. In this paradigm, the extracellular activity of large populations of cortical motor neurons, located in multiple motor areas of the primate neocortex, was sampled, and processed in real time in order to generate a neuronal population motor control signal suitable for controlling the movements of a robotic arm.

To obtain such a neurorobotic motor control signal, it was first necessary to investigate how neuronal populations in the motor cortex encoded visually guided arm movements in these monkeys. In this paradigm, visual cues were used to prime the animal to choose between two possible directions of arm movement. The monkeys achieved criterion performance in this task in about three to five months, at

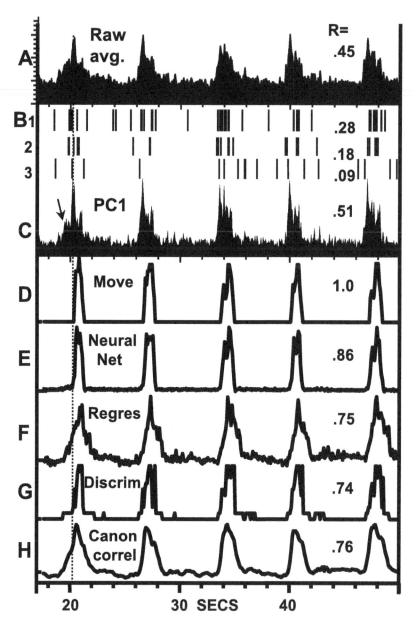

FIGURE 8.3 Bar movements predicted by neuronal population activity, as encoded by different methods for calculating neuronal population functions. All stripcharts cover the same 33 s sequence, during which the rat made five rewarded bar presses. Correlations (R) with bar movement are shown for all NPFs. A. Raw average of 32 neuronal activity over this time. B. Spike train rasters from three neurons representative of low (R = 0.09), middle (R = 0.18) and high (R = 0.25) levels of correlation with forelimb movement. C. Integration of the same neurons weighted by PC1. D. Vertical position of the bar over the same time frame, inverted

which point they were capable of executing from 500–700 trials per day. During the initial stages of training, these monkeys received implants of multiple multielectrode arrays (see Section 8.3.1.2) in different motor and somatosensory cortical areas. Each multielectrode array contained 16 Teflon-coated stainless steel microwires (Nblabs, Dennison TX). In all animals, multielectrode arrays were surgically implanted in the deep layers of two motor cortical areas: the primary motor cortex (M1) and dorsal premotor area (PMd). A week after the surgery, the animal was brought back to the setup and large populations of single motor cortical neurons were isolated and recorded during the execution of the behavioral task for long periods of time. Thus, this approach allowed us not only to study new ways to sample and process motor cortical signals in real time, but also to investigate how these cortical motor signals changed as a function of training and during the initial stages of learning of a visuomotor task.

As expected from previous reports in the literature,[26] cortical motor neurons were found to be broadly tuned to the direction of intended movement. This was true for both naïve and overtrained owl monkeys. Consequently, the same sample of neurons exhibited increased firing prior to movements in two different directions. The first important observation regarding the potential use of this approach for the design of brain-controlled prosthetic devices was that our cortical recordings remained stable and viable for over 18 months. This finding further confirmed our previous data obtained in the somatosensory cortex of owl monkeys and suggested that chronically implanted microwires are still the best electrodes available for long-term single- and multi-unit recordings in freely behaving animals. Indeed, to this date, despite the introduction of very elaborate designs, no other electrode has been able to match the long-term stability and yield of microwires.

The next issue addressed in our primate recordings was to measure how well simultaneously recorded cortical signals could be used, at different stages of training, to predict the behavioral outcomes (direction of movement, or correct vs. incorrect sensory-motor associations) of single trials of a visuomotor task. During the entire training period of this task, the animal's behavioral performance was monitored while patterns of cortical motor activity were recorded from the same chronically implanted microwires. Thus, for the first time in the field of primate neurophysiology, chronic and simultaneous recordings from large populations of single cortical motor neurons were obtained during the period required for naive primates to learn to associate distinct sensory stimuli (visual cue) with different directions of movement.

Over several weeks of training, the behavioral performance of the first monkey employed in this task improved from 62% to 95% correct, meaning that it significantly enhanced its ability to associate a particular spatial visual cue with the

---

FIGURE 8.3 (continued) to show threshold "drinking" position at bottom and "water drop" position at top. E. Results of using a recurrent neural network to transform the PC1-NPF (C) into a prediction of the timing and magnitude of lever-movement . F-H: Results of using regression, discriminant and canonical correlation analysis to transform the PC1-NPF signal (C; 5 tap delays with 100 msec intervals) into a prediction of lever movement (D). All NPF traces were quantized into 20 msec bins and smoothed over 2-bin windows. For all, the vertical scale is arbitrary, and scaled between maximum and minimum.

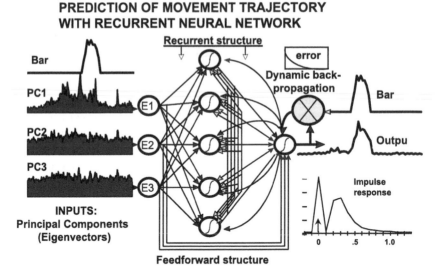

**PREDICTION OF MOVEMENT TRAJECTORY
WITH RECURRENT NEURAL NETWORK**

FIGURE 8.4 Use of recurrent artificial neural networks (rANNs) to transform the time domain of a neuronal population function (NPF). The rANN used here was like a typical three layer perceptron (i.e., with complete feedforward connectivity), but with recurrent connections between neurons in the middle ("hidden") layer. Input signals could cycle several times through these recurrent connections, allowing the hidden layer neurons to encode temporal as well as spatial patterns of input activity. Here, the simultaneously recorded spiking activities of 32 single neurons in the rat motor cortex and thalamus were reduced to three principal components (PC1, PC2 and PC3), which were used as inputs to the rANN. Though none of these signals by themselves came close to matching the time course of the lever movement trajectory, the output of the rANN (right) predicted the trajectory with good accuracy. Inset ("Impulse response") shows the temporally patterned response (in secs) of this ANN to a single input pulse (at arrow). The time course of this response is virtually the same as the correlation function between the pre-movement NPF and bar movement, as shown in Figure 8.3.

direction of arm movement they were supposed to make. During this period, the variance of their reaction-times to the visual stimuli also decreased significantly. As the animal's behavior improved, the direction of arm movement was encoded earlier (from 200 to 400 ms before movement onset) in the trials by motor cortical ensembles. Wavelet discriminant analysis,[27] using an artificial neural network classifier, was then employed to quantify whether fine spatiotemporal patterns of M1 and PMd ensemble activity (i.e., instantaneous firing rate across the neuronal population in 5–10 ms intervals) discriminated between trials with different movement directions (left and right) and different outcomes (correct vs. error). Over several weeks of training, the classification of single trials (left vs. right, correct vs. error) based on spatiotemporal patterns of cortical motor ensemble activity increased from ~80% to 95%. Independent component analysis (ICA)[28] was then used to detect synchronous (zero-lag) correlations between subsets of the neuronal ensembles and to quantify the contribution of neuronal correlated activity over training on a trial-by-trial basis. Classification of single trials based on ICA improved from ~65 to ~85% correct over training. Interestingly, mean neuronal firing rates remained stable over the course of training and provided above chance discrimination for movement direction. Altogether, these data suggest that changes in fine spatiotemporal patterning of cortical ensemble firing and correlated activity in populations of neurons in multiple motor-related areas of the primate cortex occur during the initial acquisition of a visuomotor task.

To further describe the functional alterations that take place in the primate cortex during learning of this visuomotor task, we investigated the single trial predictive value of ensembles of neurons located in either the primary motor or the premotor cortical areas. In these experiments, a simple multi-layer feedforward network (supervised Kohonen network; Learning Vector Quantization, LVQ) was employed to quantify how well spatiotemporal patterns of firing from either cortical area could discriminate between trials with different behavioral outcomes (i.e., left vs. right movements and correct vs. errant trials). Interestingly, both areas independently provided enough information to discriminate outcomes on a single trial basis with over 85% accuracy in an overtrained animal. Once again, independent component analysis (ICA) was employed to detect synchronous (zero-lag) correlations between subsets of the neuronal ensembles and to quantify the contribution of neuronal correlated activity over training on a trial-by-trial basis. Discrimination based on correlated activity in the premotor cortex improved from 70% to over 95% of trials correctly classified over training. However, correlated activity within motor cortex never provided enough information to discriminate single trials with more than 80% accuracy. These results suggest that learning this task had a differential effect on the encoding of trial outcome in each of the two motor areas

The next step was to create, in collaboration with Plexon Inc (Dallas, TX; see Section 8.3.2.1), a real-time interface to control a robotic arm based on the cortical motor signals (see Section 8.3.5.3). The first prototype of this interface allowed us to sample the information provided by large populations of single cortical motor neurons, as well as multiunit activity derived from the same microwire electrodes, and generate, in real time, a neural population function that encoded information about the arm movement direction, velocity, and amplitude. This neural population function was generated by a linear weighted sum of the activity of individual neurons, through a principal component algorithm, and more recently, through different types of artificial neural networks (see Section 8.3.3). This neuronal population function, as calculated in real time by this computer interface, provided an analog electronic output signal that was able to directly control a robotic arm.

This approach has been used to generate continuous neural population functions that depict the pre-motor cortical activity that just precedes single visually guided arm movements. Further analysis revealed that the correlation between this pre-motor activity and manipulandum displacements in primates can reach the values (0.80–0.89) obtained in similar studies carried out in rats.[13] Interestingly, this pre-movement population signal could be correlated with the amplitude, trajectory and direction of the visually guided arm movements. The same signals could also be used to predict correct vs. error trials, i.e., whether the movement was the one indicated by the visual cue. These results are important because they suggest that chronic recordings from small populations of cortical motor neurons (between

MOVEMENT OF ROBOT ARM WITH POPULATION VECTOR ACTIVITY

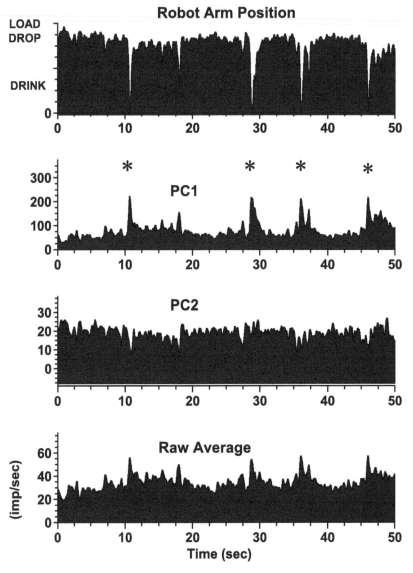

FIGURE 8.5 Neurorobotic mode. "Robot Arm Position" over the first 50 s period after switching to NP-function->robot-arm (NPF->RA) mode. "PC1" NPF over the same time period. This signal was electronically manifested as robot arm movement in real time (inverted in A) by electronically extracting the first principal component of the spiking activity of 32 simultaneously recorded neurons (the best 26 MI and 6 VL neurons out of 46). Asterisks

35 and 65) may be enough to generate the type of basic signals required for controlling a prosthetic robot arm.

Other supervised and unsupervised methods for real-time processing of the neural population function are currently being investigated in order to produce a signal that reproduces in great detail the 3D arm movement produced by the animal in each trial. We should point out, however, that this is the first time in which a signal derived from populations of individual cortical motor neurons in primates was used to control a robotic arm in real time. The implications of this demonstration are multifold and are likely to lead to several technological breakthroughs in the field of artificial neuroprostheses and robotics.

8.3 APPROACHES TO MULTI-NEURON RECORDING AND NEUROROBOTIC ACTUATION

The above results have shown that it is possible to use signals derived from simultaneously recorded neuronal populations to control external devices. This demonstration was made possible by our efforts across a broad spectrum of technological and scientific areas, ranging from electrode design, to computer based multi-spike discriminators, to real-time neurorobotic actuators. Much further progress must be made in all of these areas to reach the goal of restoring movement to paralysis victims. The following sections discuss the current state of such techniques and plans for future improvement.

8.3.1 Electrodes

8.3.1.1 Integrated Multielectrode Devices

Because of the explosion of interest in multi-neuron recording, a variety of multi-electrode recording techniques are currently under development. Chapter 6 contains a detailed discussion of these techniques, particularly our plans for increasing the number and density of electrode contacts in the brain through use of multi-contact ceramic electrodes. The following discussion outlines our techniques for implantation and utilization of microwire arrays, which have been utilized for virtually all published reports involving multi-neuron recordings.

8.3.1.2 Microwire Electrode Arrays

Our technique involves use of arrays of Teflon-coated microwires (25 or 50 μm), which can be implanted in multiple areas of the same brain. Currently, our implantable

---

FIGURE 8.5 (continued) indicate trials in which the rat successfully used this real-time NP signal to move the robot arm to the water drop position (black dotted line labeled "T (NPF)"). All asterisked peaks in the NP function occurred during the "pre-lever-movement" periods of their trials. The signal/noise of this PC1 NPF was superior to functions generated using "PC2" (second principal component) or the "raw average" (bottom).

microwire electrode devices are obtained premanufactured from NBLabs (Denison, TX). The microwires are typically implanted in bundles of 8 microwires apiece or arrays of 16 microwires apiece. These are attached along with ground and stimulating wires to Microtek connectors, or more recently, to Omnetics connectors. These comprise an integrated unit called a "hat," 1–6 of which can be cemented to a rat's head during surgery. The microwire arrays consist of two rows of eight microwires, spaced at exact intervals (usually 250 mm), to yield a total length of 2.0 mm. The rows are usually spaced 0.5 mm apart. The advantage of these arrays is that they can be implanted across cortical subzones, e.g., to cross the forelimb representation in the rat motor cortex. They can also be implanted in deeper brain structures, such as the thalamus and brainstem. However, for implants in smaller structures it is often preferable to utilize the eight microwire arrays, which can be placed so as to sample from a circular area of about 1.0 mm diameter in a deep brain nucleus. Since at least four such hats can be fit on a rat's head, up to 64 microwires can be stereotaxically implanted in appropriate target nuclei. The accuracy of electrode placement during surgery is assured not only by these stereotaxic techniques, but also by recording neural activity through the microwires during the implantation. Further verification of electrode placement can be obtained subsequently through use of a dental X-ray which shows the locations of the metallic microwires relative to skull landmarks in frontal, horizontal and sagittal views. Finally, histological reconstruction is carried out after all recording experiments are finished. The Prussian Blue reaction can be used to mark specific electrode sites.

8.3.1.3 Chronic Implantation of Microwire Arrays

The surgical procedures for implantation of multielectrode arrays are a critical bottleneck for obtaining long-term cortical ensemble recordings in both rodents and primates. Many have reported difficulty in obtaining stable long-term recordings using mirowires in old world monkeys such as macaques. We have found, however, that such difficulties are most often attributable to inadequate surgical methods. Our experience is that good surgical technique yields a high single-unit yield and improved long-term recording stability. In fact, it is much easier to implant chronic electrodes in the cortex of owl or squirrel monkeys than in the rat cortex. In a demanding surgical procedure such as the one required in these experiments, there are no ways to substitute for good planning, experience and patience. Thus, the first thing one needs to remember is that there is no need to finish the procedure at a record pace. Indeed, in more than ten years of experience with these procedures, we have learned that the slower the microelectrode implantation surgery, the higher the single neuron yield. Careful implantation also seems to increase the longevity and stability of neural ensemble recordings in both rodents and primates.

In our primate surgeries, an initial dose of ketamine (15 mg/Kg I.M.) is used to produce a superficial sedation of the animal. Primates are then maintained under deep anesthesia with a 0.5–1.5% isoflurane–air mixture administered through an endotracheal tube with spontaneous respiration. Before intubation, the larynx is sprayed with 2% lidocaine to prevent laryngospasm. The animal's EKG, heart and respiratory rate, and temperature are continuously monitored during the surgery.

Slow I.V. infusion of fluids containing the required electrolytes and glucose is used to avoid dehydration and hypoglycemia during the surgery. Arousal level is determined by continuously monitoring reflexes and the electrical activity of cortical neurons (EEG) through metal skews implanted in the skull.

The implantation of microelectrodes is carried out through a series of steps. First, prior to their implantation in the brain, microwire arrays are sterilized using gas or ultraviolet light to minimize the risk of infection. Several microwire arrays are implanted in each animal through small craniotomies, each of which can accommodate only a single 16 microwire array (2 mm^2 in area). A series of smaller holes are then drilled in the skull for placement of 6 to 12 metal screws which provide support for the microwire arrays and a common reference for our electrophysiological recordings. Once the craniotomies are completed, a small slit is made in the dura mater just prior to slowly (approximately 100 μm/minute) lowering the electrode array through the pia surface into the cortex. Over the years, we have learned that very slow implantation is required to reduce the amount of cortical dimpling produced by the electrode array. Blunt tip microwires can be driven slowly into the primate cortex, unlike the cerebellar cortex, without causing major tissue depression. However, the experimenter has to be aware that cortical dimpling can occur in the early stages of implantation if the electrodes are moved too quickly.

Single and multiunit recordings are performed throughout the implantation procedure to ensure the correct placement of the microwires. The receptive fields of both single neurons and multiunits are qualitatively characterized to define the relative position of each individual microwire. Once the microwire array reaches the lower layers of the cortex, the craniotomy is filled with small pieces of gelfoam and then covered with either bone wax or 4% agar. Once the craniotomy is sealed, the entire microwire array is fixed in position using dental cement. This procedure is repeated several times until multiple microelectrode arrays have been implanted in different brain regions. In our hands, each microwire array implant requires about two hours to be completed. Therefore, 4 to 6 arrays (64 to 96 microwires) can be implanted during an 8–12 hour surgery. After the completion of the chronic experiments, we have found that individual microelectrode tracks can be readily identified in Nissl stained sections months after the initial surgery.

8.3.2   SIGNAL PROCESSING AND TRANSMISSION

8.3.2.1   Recording System

Once electrode arrays have been implanted in the brain, the electrode tips are used to record the very small extracellular action potential currents emitted by nearby neurons. These single neuron action potential currents are amplified first through a head stage (from NbLABS, Dennison, TX), containing 16 field effect transistors (FETs, Motorola MMBF5459) arranged in two rows of eight at the end of insulated cables. The FETs are configured as voltage followers with unity voltage gain, but with a current gain of about 100. This protects the recorded signals from electrical noise associated with wire movement (piezoelectric currents) and other forms of electromagnetic interference. These signals are then transmitted to the Plexon Multi-Neuron Acquisition Package (MNAP). This system was developed over many years of collaboration with Plexon Inc. (Dallas, TX) to develop electronic devices capable of carrying out such single unit waveform discrimination simultaneously on up to 96 recording channels. Up to four different neurons are able to be discriminated from each electrode channel. The device features computer controllable amplifier-filters for each channel and then utilizes multiple digital signal processors (DSPs) to carry out the discrimination.

The MNAP system features two stages of signal amplification, the first consisting of a "front end box" containing differential OP-Amps (gain 100, bandpass 100 Hz to 16 KHz). The OP-Amp output signals are transmitted, through ribbon cables, to A/C coupled differential amplifiers located on modular input boards, each containing 16 amplifiers. Once in the input boards, the analog signals pass through another level of amplification (jumperable gain of 1, 10 or 20), are filtered (bandpass 400 Hz to 8KHz), and reach the final stage of amplification (programmable multiplier stage, ranging from 1 to 30). These boards also include one 12-bit analog-to-digital (A/D) converter per channel, which simultaneously digitizes the waveforms defining extra-cellular action potentials at 40 KHz. After A/D conversion, the signals are routed to DSP boards, each of which contains four digital signal processors (DSP, Motorola 5602) running at 40 MHz (instruction read at 20 MHz). Each DSP handles data from eight input channels and contains 32 K 24-bits of SRAM and 4K 16-bit words of dual port SRAM memory.

8.3.2.2   Spike Waveform Discrimination

One to two weeks after microwire electrodes are implanted, they typically yield excellent single neuron recordings, as manifested by stable action potential wave-form profiles. Each such profile is presumed to represent a distinctive "signature" of a single neuron located somewhere near the electrode tip. Discriminable single units are typically obtained from 50–100% of the implanted microwires, but since many microwires record two to four discriminable units, it may be possible to record up to 128 well discriminated neurons with 64 or fewer microwires. The shape of the action potential waveforms recorded from different neurons are distinct because of normal variations in their size, dendritic geometry and proximity to the electrode. Thus, multiple waveform shapes can typically be found in the same microwire recording, requiring sophisticated computer algorithms to discriminate between them. In the Plexon system, spike discrimination programs are downloaded from the PC host to the DSPs. Single spikes are discriminated in real time by any of the following algorithms: (1) Multiple time-voltage discriminators, in which each of four possible units is discriminated by a threshold crossing, and two time-voltage boxes, all user configurable on the computer screen using drag-and-drop mouse manipulations; (2) A principal component spike waveform clustering algorithm; and (3) A template matching algorithm that, based on user specifications, automatically builds up to four spike templates for real-time discrimination based on user specified thresholds for mean squared error. Each of these techniques tests spikes captured at 40 KHz pre- and post-threshold crossing, and classifies them as units 1–4, or rejects them. During the experiment, the time of occurrence of each of all valid spikes, for all recorded channels, is transferred to the hard disk of the PC host through a parallel

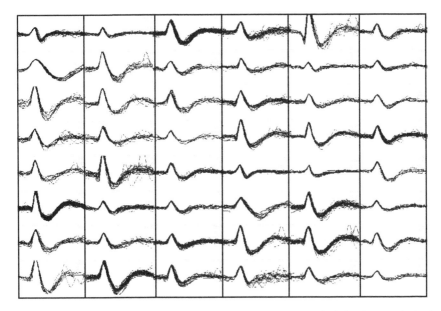

FIGURE 8.6 Superimposed waveforms of 48 single neurons recorded simultaneously in the sensorimotor cortex of an awake, behaving rat.

bus (MXI-Bus, National Instruments, Austin, TX) that is capable of transferring 2 MB of data per second. Digitized samples of the spike waveforms are also recorded periodically and stored for off-line analysis using a visualization program (Spike-Works, Plexon, Dallas, TX). See Figure 8.6.

A MNAP timing board is responsible for distributing timing stamps and synchronizing all input signals to the system. This board also provides a digital time output that can be used to synchronize external devices or to drive a video timer for a video tape recorder employed to document the animal's behavior. The timing board can also receive up to 16 TTL inputs generated by other devices which are responsible for controlling the behavioral cages used in our experiments. A single host Pentium microcomputer (100–200 MHz, with 64 Mbytes of RAM, and 2.1 Gbytes of disk space), running C++ software (SSCP, Spectrum Scientific, Dallas, TX) in the Windows NT Workstation 4.0 operating system (Microsoft, Seattle, WA) is responsible for controlling the MNAP over a serial line.

8.3.3   NEURONAL POPULATION CODING

8.3.3.1   Necessity of Multiple Neuron Recordings for Neuroprosthetic Control

It was revealed about 50 years ago that the information recordable from single neurons using small electrodes was far more specific than that recorded as electro-encephalographic field potentials from the cortical surface or the scalp. Since then, single neuron recordings have provided most of our current knowledge about the coding of information by the brain. The traditional method of revealing the information processed by single neurons has been to average the spiking discharge of single neurons over large numbers of trials. This is because the temporal patterning of neuronal spike trains recorded in single trials (especially in higher brain areas) is highly stochastic.[29] Whether this trial-to-trial variability represents noise or unexplained signal, it is clear that weighted spatial averaging across neurons can increase the specificity of their common information.

Despite these advantages of multi-single neuron recording, multiunit recordings may be preferable in some cases. In topographically mapped cortical areas, for example, multi-neuron cluster recordings may be useful because they are inherently more stable than single neuron recordings, especially over long time periods. Since these neurons exist in the same local cortical region, they may have a number of similar physiological properties. Thus, to the extent that all of the neurons in the cluster share the same information, a major increase in statistical resolution can be achieved by pooling them. On the other hand, if these neurons contain divergent or canceling information, their pooling will *decrease* the statistical resolution of important neural output signals. Our previous experience suggests that use of pooled neuronal clusters will generally increase resolution of significant population codes, especially over a single dimension of movement. This increased resolution may not be seen, however, when finer, multidimensional movements must be encoded. Ultimately, the choice of the number of neurons to pool in a single discrimination could be evaluated automatically as an adjunct to automatized spike discrimination software.

8.3.3.2 Population Vectors

The term population vector has been used since the 1980s[26] to define the direction of hand movement predicted by the combined activity of neuronal populations in the motor cortex. The direction vector was calculated by combining the spike trains from single neurons (serially recorded) in the motor cortex of monkeys trained to move their hands in eight different directions from a central point. First, the "preferred direction" of each neuron was measured through use of traditional multi-trial averaging. Each neuron's "motor field" was defined by assuming that its discharge during movement in any of the eight directions would be equal to the cosine of its angle away from the preferred direction. This cosine rule then provided a tuning function for calculating the direction of any movement, as the "population vector" formed by the average cosine weighted discharge rates of multiple neurons with different preferred directions. Greater predictive accuracy was achieved by increasing the number of neurons included in the population vector calculation. This work was a milestone in that it was the first to demonstrate the importance of distributed coding in the motor cortex. While no one neuron was found able to accurately specify the direction of movement in a single trial, very good accuracy was obtained by decoding and combining the information provided by neuronal populations. Thus, even though these original studies only utilized serially recorded neurons, it predicted that simultaneously recorded neurons should be useful for predicting movement direction in real time.

More recently, simultaneously recorded single neurons in the motor cortex are being used to encode online population vectors. This work, as yet unpublished, is currently being carried out by three groups, including Nicolelis and Chapin (see Section 8.2.2), Donoghue and Hatsopoulos, and Schwartz. Ultimately, however, it is unlikely that the cosine function technique will be used exclusively for online neurorobotic control. First, this approach is difficult to generalize to the majority of neurons whose properties cannot be explained in terms of simple cosine functions. For example, our studies in sensory and motor cortices (in rats and monkeys) have generally found that the shapes of neuronal tuning functions can vary widely. Not only do they range from narrow to broad, but they can also be markedly skewed and can have multiple peaks. Moreover, since motor system neurons tend to have both motor and sensory properties, their encoding of the temporal domain can be quite complex. Some neurons (especially in premotor areas) may begin to discharge well before the onset of the movements they encode. Others may discharge only after the onset of movement. Finally, neuronal coding properties tend to be dependent on context. Thus there is often a highly nonlinear relationship between neurons and the sensory and/or motor events with which they are correlated. For these and other reasons, the use of cosines and other basis functions may prove to be highly inaccurate when used to transform neuronal population activity into complex movement trajectories. Moreover, the time required to measure the complex, conditional and nonlinear topographies of each neuronal basis function will become prohibitive.

8.3.3.3 Statistical Approaches

Multivariate statistical techniques can provide an excellent alternative to the problems of the population vector approach. These can be used to empirically derive functions capable of predicting variables (e.g., movement parameters) from the activities of simultaneously recorded neurons. Because of the diverse nature of these techniques, we have utilized the term "neuronal population function" (NPF) in place of "population vector" to specify the mathematical encoding of output variables by simultaneously recorded neuronal populations. Two general approaches exist: principal and independent components analysis (PCA and ICA) and canonical correlation analysis (CCA) can be used to map the complex multi-neuronal data set into a more parsimonious structure which can be easily manipulated. Techniques such as multiple regression analysis (MRA) and discriminant analysis (DA) can be used for direct prediction of output variables (see Figure 8.3).

8.3.3.4 Principal Components Analysis (PCA)

We have utilized PCA as a pre-processing step to extract underlying informational factors that are distributed across multi-neuron ensembles.[30-31] Such factors are often detected as recurring patterns of covariance among the neuronal activity. PCA provides differential weighting of each neuron's contribution to the population average according to its patterns of correlation with other neurons, thereby concentrating the salient "signal" embedded within the neural activity into a small number of orthogonal principal components (PCs) and effectively separating them from the stochastic activity of individual neurons. This is accomplished by calculating a correlation matrix between the variables (neurons) over the range of samples (time bins) in selected periods of the recording. Eigenvalue rotations of this matrix yield a sequence of factors or "components" that explain successively smaller amounts of the total variance. Each component can be reconstructed as a weighted population average of the recorded neurons, using the rotational coefficients as weightings. The first component (PC1), which typically explain from 10–30% of the total variance, resembles the population average with the removal of the "background" activity. The main use of PCA here is to use PC1 as a first approximation of an NFP and as an ideal information base for more directed mathematical treatments.

8.3.3.5 Independent Components Analysis (ICA)

Like PCA, we have used ICA to extract salient informational factors that are distributed across neuronal ensembles.[28] Unlike PCA, which defines rules for distinguishing between dependent factors in the data (e.g., different dimensions of a topographical map), ICA detects completely independent factors (e.g., different stimulus frequencies). Whereas principal components define linear subspaces of the data, independent components define nonlinear subspaces.

8.3.3.6 Multiple Regression Analysis (MRA)

MRA generates linear models capable of predicting one or more dependent (output) variables from a set of independent (input) variables.[30] As an example, we have utilized the weighted discharge of multiple neurons to predict movement, i.e., the angular movement of the lever (and hence, the robot arm). To fine-tune such a model, a variety of nonlinear estimation techniques are available in commercially available statistics packages (see below). These will identify nonlinearities in the neural encoding of the movement, allowing them to be incorporated in the model. The purpose will be to optimize the performance of the predictive model by using all available techniques for minimization of a loss (error) function. Next, more elaborate statistical techniques, such as CART, may be implemented. Finally, a number of neural network techniques (especially three-layer back-propagation nets) are available in commercial packages (e.g., Neural Dimensions, Gainesville, FL) for use in developing regression models.

   MRA is the statistical technique of choice for defining the linear mathematical relationship between a set of independent predictor variables (e.g., a population of motor system neurons) and a dependent variable (e.g., a motor output function). In its simplest form, MRA will yield a linear equation of the form:

$$Y = a + b_1 X_1 + b_2 X_2 + \ldots + b_p X_p$$

in which Y = the dependent variable (e.g., lever angle), X_n = the independent variables (e.g., neuronal population activity), a = an offset, and b_n = coefficients for weighting the independent variables. This relation is easily calculated from a sample data set by *linear least squares estimation*, which computes a line through the observed data points such that the squared deviations of the points through that line are minimized.

This calculation also yields the coefficient of determination (R^2) which quantitates the predictive reliability of the model (in a range from 0–1). Beyond this, it is very important to test the model through application of a number of residual analysis procedures, which can identify outliers, and gross violations of the assumptions of MRA.

Traditional MRA depends on the assumption that predictor variables are normally distributed and have a linear relation to the dependent variable. Fortunately, linear MRA is generally used as a starting point for a variety of *nonlinear estimation* procedures which can define virtually any significant relation between independent and dependent variables. The first step in identifying such nonlinearities is to examine scatterplots between the independent and dependent variables, and then utilize various nonlinear estimation techniques to best identify any nonlinearities involved. These techniques often use a weighted least squares method to estimate a nonlinear "loss function" (i.e., the error). This procedure generally includes a linear least squares component and also a nonlinear component. Depending on the types of nonlinearities revealed by these procedures, any of several nonlinear regression models can be implemented, including (1) linearizing mathematical transformations, in which the observed nonlinearities are corrected by simple mathematical transformations; (2) polynomial regression, as used when relations between independent and dependent variables are curvilinear, and thus can be optimally fitted to particular polynomial expressions; (3) piecewise linear regression, used when neurons have different linear relationships with motor output over different segments of the output function's range; and (4) breakpoint regression, used when there are sudden steps in the regression line.

8.3.3.7 Discriminant Analysis (DA)

DA is performed by randomly dividing the data from the neuronal population into a learning set and a test set. Data recorded during identifiable movements (e.g., flexion vs. extension) are group categorized for the discriminant analysis. A discriminant function is then derived (using a package such as Statsoft-Statistica, Tulsa, OK), providing a matrix of neuronal weightings to optimally discriminate between these two movements. These weights were used to construct the NPF in Figure 8.1 using the test data set. *Temporal transformations using tapped delay lines*: Multivariate statistical procedures (regression, canonical correlation and discriminant analysis) can be performed on learning data sets of tap-delayed input variables, e.g., five successively time shifted (100 ms) replicas of the PC1 signal. This "tapped delay line" thus allows the temporal information in the signal to be linearly recombined, producing a closer approximation of the target motor output signal

8.3.3.8 Artificial Neural Networks (ANNs)

ANNs consist of networks of processing elements that simplistically mimic the structure of neuronal networks in the brain. ANNs are generally composed of arrays of "neurons," each of which receives "synaptic input" from sets of other neurons or input vectors. The input/output transformation of each neuron is usually defined by a simple squashing function (e.g., a sigmoid) which confers nonlinearity on the network's processing. This allows the network to learn more complex problem spaces than linear statistical methods. Learning occurs iteratively by modifying to adapt the strengths of the synaptic connections through successive trials. We mainly use variants of the supervised "back-propagation" learning rule because it is relatively fast and effective.

As discussed in Section 8.2.1 (see Figures 8.3 and 8.4), and also in Section 8.2.2, we have found that ANNs are often the most accurate method for transforming neuronal activity into predictions of motor outputs. Since the structures of ANNs are neurally inspired, it is not surprising that they might be useful for encoding information in real neuronal networks. Moreover, ANNs are one of the best methods currently available for mapping complex and nonlinear problem spaces. On the other hand, ANNs often seem to solve difficult problems by "black magic," leaving the investigator with little new knowledge about the system being studied. Instead of forcing the experimental system into a generalized linear predictive model, an ANN may instead reach a narrow "local" solution by focusing on statistical outliers characteristic of a specific learning data set. Moreover, ANNs can be tricky to use because a user is required to guess at many of the parameters, including the number of neurons, their transfer function, the learning algorithm, the rate of learning, and the criterion for success. Our belief, however, is that ANNs may represent the best overall solution for the problem of neurorobotic control.

8.3.4 MOTOR SIGNALS OBTAINABLE FROM DIFFERENT BRAIN AREAS

The aim in brain-controlled neuroprosthetics is to translate brain activity into normal movement. While laboratory investigations will inevitably employ highly simplified tasks, it will ultimately be necessary to utilize decoded brain signals to control a more natural range of body-like movements. This will require obtaining information from a large number of neurons, because distributed coding is the method used by the brain to handle complex, multidimensional information. It is known, for example, that different subcomponents of the motor system are involved in different movements. While the frontal cortices have a major role in planning and execution of voluntary movement, the brainstem and spinal cord are more importantly involved in moment-to-moment regulation of posture and locomotion. Therefore, if we wish to ultimately utilize a neuroprosthetic technique to restore locomotion, we must either record signals from such lower CNS regions, or computerize that function.

8.3.5 NEUROROBOTIC ACTUATION

8.3.5.1 Hardware Devices

The neurorobotic feasibility study in rats (described in Section 8.2.1) utilized a hardware device designed to electronically actuate a user defined neuronal population vector. It consisted of four operational amplifier stages which transformed multiple channels of simultaneously recorded neural spiking inputs into a single analog control output that controlled the moment-to-moment position of a one dimensional robot arm. The inputs to this device came from a multichannel neuronal spike discriminator (from Plexon Inc), whose real-time outputs were 5 V electrical pulses (of 1.0 msec duration) on each of 32 channels. The first stage was a summing amplifier that encoded the population vector by individually weighting the pulses from the 32 channel discriminator. The second stage was a first order integrator set for a time constant of 25 msec. The third stage was a low pass filter that functioned to remove the high frequency components (>100 Hz) of the integrated input spikes. The last stage allowed the gain and DC level of this signal to be adjusted to roughly match the voltage baseline and range of the analog signal coming from the angular transducer on the lever (see Figure 8.1). This hardware was useful in that it was effective and fully real-time. It was problematic, however, in that all adjustments, including neuronal weightings, were accomplished by turning knobs. Also, it was limited to actuating a linear one layer matrix transform.

8.3.5.2 Server-Client Structure of Multichannel Neuronal Acquisition Processor (MNAP)

In collaboration with Plexon Inc (Dallas, TX), we have more recently implemented a software-based neurorobotic interface which allows a neuronal population function to be derived in real time from the simultaneously recorded single unit data. This interface is the key link between the animal's brain and the robotic arm since it allows massive neuronal activity to be transformed into a real-time control signal that can be used to control a robotic arm. The neurorobotics software runs as part of the Plexon multichannel neuronal acquisition processor. In its current version, the MNAP software is implemented as a distributed client-server architecture. This permits the development of separate client application programs that can tap into the real-time spike event data stream that is supplied by the server program. The server handles all timing and communications with the MAP hardware box and with other data acquisition peripherals. Client programs may either run in the same host computer as the server or run in other computers distributed over a local area network (LAN). Ideally, the RoboticsClient runs in a dual Pentium processor Windows NT workstation, which also runs the server.

8.3.5.3 RoboticsClient

After spike sorting has been set up using the SortClient program the RoboticsClient picks up the real-time spike event data stream from the server. The RoboticsClient is thus a separate program with its own user interface windows. The server schedules the RoboticsClient to run every 15 ms, at which time it performs computations on the event data stream and updates the position of the servo-controlled robot arm. The first version of the RoboticsClient program performs a weighted-sum and integration algorithm to control the arm position. The user interface window shows scrolling spike event activity displays for selected units. A signed weight factor is assigned to each unit. The weighted activity streams are then summed by a leaky integrator which has a time-constant parameter to control the overall response of the arm. The output of the integrator drives the robot arm servomotor. The Robot-icsClient provides for entry of parameters controlling scaling, limiting, and default positioning of the arm position. A rotating arm position object indicates on the screen the motion of the actual arm. In the next funding period, the RoboticsClient will be further developed to incorporate a variety of control algorithms, including a series of unsupervised artificial neural network architectures, and semi-automatic means of setting the parameters.

8.3.5.4   Adaptive Maintenance of Neurorobotic Control

Current efforts are addressing the problem of adaptively maintaining stable neuro-botic control during nonstationary changes in the recorded neuronal population. These nonstationarities normally come from two different sources: (1) loss and gain of single neuron recordings, and (2) conditioning-induced adaptive changes in neural coding properties. The loss and gain of neurons is relatively easy to deal with because it is predictable and occurs slowly over days and weeks of recording neuronal populations from the brain. Often the waveform of a particular single neuron gets too small to discriminate, or another neuron appears in a recording and must be discriminated and added to the population function. When the total number of recorded neurons is large (>30), the immediate changes in the neurorobotic control signal are minimal. Over the long term, however, the newly appearing neurons need to be discriminated and included in the population function. For this reason, the population function must sometimes be redefined. In our case, this involves matching it against the recorded kinematic output of the limb. In the future, however, when such recordings are obtained in a paralyzed individual, there will be no kinematic output to match. Our plan for this situation is to weight newly discriminated neurons according to their covariance with other neurons already being recorded.

A further issue will be that of dealing with adaptive changes in the actual encoding of information by the recorded neurons. For example, we have observed changes (see Section 8.2.1) in the correlation of motor cortical neurons with limb movement kinematics after several days of neurorobotic training. One might predict that such conditioning-induced adaptation of neuronal discharge would facilitate the experimenter's covariance based recoding of the population function. In fact, several investigations have suggested that cortical neurons can be conditioned to arbitrary output functions.[32-33] Although this is likely to be much more difficult than using one's cortical neurons in their normal manner, it will likely be essential to the coding of neurorobotic conrol functions in paralysis victims (see Chapter 7).

8.4   FUTURE DIRECTIONS

It appears certain that successful neurorobotic control will soon be demonstrated in monkeys, which will later set the stage for human studies. Initially, this will involve brain-control of movement in one dimension, either in Cartesian or polar coordinates. Next, neurorobotic control of movement will be attempted in two and then three dimensions. At this point, a number of decisions must be made. Should we, for example, attempt to use brain activity to completely specify the robot arm movements, as though the robot is a virtual extension of the real arm? Alternatively, the brain activity might be used to provide much more abstracted information, such as the final position of a robot arm movement. Whatever the strategy, it appears certain that larger numbers of neuronal recordings will be needed for performance of more complicated tasks. Given the recent pace of technological innovation, however, one can assume that the next few years will see development of appropriate technologies for recording and handling information from several hundred neurons simultaneously. If so, we predict that new categories of brain–machine interfaces might soon emerge.

ACKNOWLEDGMENTS

This work was supported by NIH contract NS62352, NIH grants NS26722 and NS40543, and DARPA/ONR grant N00014-98-1-0679 to JKC and DARPA/ONR N00014-98-1-0676 to MALN.

REFERENCES

1. Bhadra, N. and Peckham, P.H. Peripheral nerve stimulation for restoration of motor function. *J. Clin. Neurophys.* 14, 378-93 (1997).
2. Wolpaw, J.R., Ramoser, H., McFarland, D.J., and Pfurtscheller, G. EEG-based communication: improved accuracy by response verification. *IEEE Trans Rehabil Eng 1998* 6(3), 326-33 (1998).
3. Schmidt, E.M. Single neuron recording from motor cortex as a possible source of signals for control of external devices. *Ann. Biomed. Eng.* 8, 339-49 (1980).
4. Humphrey, D.R., Schmidt, E.M., and Thompson, W.D. Predicting measures of motor performance from multiple cortical spike trains. *Science.* 170, 758-62 (1970).
5. Erickson, R. Stimulus coding in topographic and nontopographic afferent modalities: on the significance of the activity of individual sensory neurons. *Psychol. Rev.* 75, 447 (1988).
6. Dormont, J.F., Schmied, A., and Condé, H. Motor command in the ventrolateral thalamic nucleus: neuronal variability can be overcome by ensemble average. *Exp. Brain Res.* 48, 315-322 (1982).
7. Chapin, J.K. and Patel, I.M. Effects of ethanol on arrays of single cortical neurons recorded for several days. *Soc. Neursci. Abst.* (1987)
8. Chapin, J.K., Shin, H.-C., and Patel, I.M. Many-neuron recordings define synchronous activity and behavioral gating in thalamic and cortical networks. *Abst. in Ann. Mtg. of Soc. for Neurosci.* (1988).
9. Shin, H.-C. and Chapin, J.K. Modulation of afferent transmission to single neurons in the ventroposterior thalamus during movement in rats. *Neurosci. Lett.* 108, 116-120 (1990).
10. Shin, H.-C. and Chapin, J.K. Movement-induced modulation of afferent transmission to single neurons in the ventroposterior thalamus and somatosensory cortex in rat. *Exp. Brain Res.* 81, 515-522 (1990).
11. Nicolelis, M.A.L., Lin, C.-S., Woodward, D.J., and Chapin, J.K. Peripheral block of ascending cutaneous information induces immediate spatio-temporal changes in thalamic networks. *Nature* 361, 533-536 (1993).

12. Nicolelis, M.A.L., Lin, C.-S., Woodward, D.J., and Chapin, J.K. Distributed processing of somatic information by networks of thalamic cells induces time-dependent shifts of their receptive fields. *Proc. Natl. Acad. Sci.* 90, 2212-2216 (1993).

13. Chapin, J.K., Markowitz, R.A., Moxon, K.A., and Nicolelis, M.A.L. Direct real-time control of a robot arm using signals derived from neuronal population recordings in motor cortex. *Nature Neurosci.* 2, 664-670 (1999).

14. Chapin, J. K. and Woodward, D.J. Distribution of somatic sensory and active-movement neuronal discharge properties in the MI-SI cortical border area in the rat. *Exp. Neurol.* 91, 502-523 (1986).

15. Chapin, J.K. Population-level analysis of multi-single neuron recording data: multivariate statistical methods. In: *Neuronal Population Recording.* Ed. M.A.L. Nicolelis, CRC Press, Boca Raton, FL (1998).

16. Chapin, J.K and Nicolelis, M.A.L. Population coding in simultaneously recorded neuronal ensembles in ventral posteromedial (VPM) thalamus: multidimensional sensory representations and population vectors. *J. Neurosci. Meth.* 94(1), 121-140 (1999).

17. Nicolelis, M.A.L., Baccala, L.A., Lin, R.C.S., and Chapin, J.K. Synchronous neuronal ensemble activity at multiple levels of the rat somatosensory system anticipates onset and frequency of tactile exploratory movements. *Science* 268, 1353-1358. (1995).

18. Fetz, E.E. Operant conditioning of cortical unit activity. *Science* 28, 955-8 (1969).

19. Fetz, E.E. and Finocchio, D.V. Operant conditioning of specific patterns of neural and muscular activity. *Science* 174, 431-5 (1971).

20. Fetz, E.E. and Finocchio, D.V. Correlations between activity of motor cortex cells and arm muscles during operantly conditioned response patterns. *Exp. Brain Res.* 23, 217-40 (1975).

21. Nicolelis, M.A.L. and Chapin, J.K. The spatiotemporal structure of somatosensory responses of many-neuron ensembles in the rat ventral posterior medial nucleus of the thalamus. *J. Neurosci.* 14(6), 3511-3532. (1994).

22. Nicolelis, M.A.L., Lin, C.-S., and Chapin, J.K. Neonatal whisker removal reduces the discrimination of tactile stimuli by thalamic ensembles in adult rats. *J. Neurophys.* 78(3), 1691-706 (1997).

23. Nicolelis, M.A.L., Ghazanfar, A.A., Stambaugh, C.R., Oliveira, L.M., Laubach, M., Chapin, J.K., Nelson, R.J., and Kaas, J.H. Simultaneous representation of tactile information by distinct primate cortical areas rely on different encoding strategies. *Nature Neurosci.* 1, 621-630 (1998).

24. Nicolelis, M.A.L., Ghazanfar, A.A., Stambaugh, C.H., Oliveira, L.M.O., Laubach, M., Chapin, J.K., Nelson, R., and Kaas, J.H. Simultaneous encoding of tactile information by three primate cortical areas. *Nature Neurosci.* 1, 621-630 (1998).

25. Nicolelis, M.A.L. Neuronal ensemble recordings in behaving primates. In: *Methods for Simultaneous Neuronal Ensemble Recordings.* Nicolelis, M.A.L., Ed., CRC Press, Boca Raton, FL, 121-156 (1999).

26. Georgopoulos, A.P., Kettner, R.E., and Schwartz, A.B. Neuronal population coding of movement direction. *Science* 233, 1416-1419 (1986).

27. Laubach, M. and Nicolelis, M.A.L. Cortical ensemble activity increasingly predicts behavior outcomes during learning of a motor task. *Nature* 405, 567–571 (2000).

28. Laubach, M., Shuler, M., and Nicolelis, M. A. L. Independent component analyses for quantifying neuronal ensemble interactions. *J. Neurosci. Meth.* 94, 141-154 (1999).

29. Moore, G.P., Perkel, D.H., and Segundo, J.P. Statistical analysis and functional interpretation of neuronal spike data. *Ann. Rev. Physiol.* 28, 493-522 (1966).

30. Chapin, J.K. Population-level analysis of multi-single neuron recording data: multi-variate statistical methods. In: *Many-neuron recording*. Nicolelis, M., Ed., CRC Press, Boca Raton, FL, 193-228 (1999).

31. Chapin, J.K and Nicolelis, M.A.L. Population coding in simultaneously recorded neuronal ensembles in ventral posteromedial (VPM) thalamus: multidimensional sensory representations and population vectors. *J. Neurosci. Meth.* 94(1), 121-140 (1999).

32. Fetz, E.E. and Finocchio, D.V. Operant conditioning of specific patterns of neural and muscular activity. *Science* 174, 431-5 (1971).

33. Kennedy, P.R. and Bakay, R.A. Restoration of neural output from a paralyzed patient by a direct brain connection. *Neuroreport* 9, 1707-11 (1998).

9 Drug Deliveries into the Microenvironment of Electrophysiologically Monitored Neurons in the Brain of Behaving Rats and Monkeys

Nandor Ludvig

CONTENTS

9.1   INTRODUCTION

When this book chapter was written, a century was about to close. It was in this era that neuroimaging techniques were introduced into the repertoire of diagnostic tools, antibiotic and psychotropic drugs were developed for the management of neurological and psychiatric diseases, and the advances in neuroscience and engineering laid down the foundations of modern neurosurgery. These triumphs revolutionized the medicine of the brain. Yet, at the dawn of year 2000, effective therapy is still unavailable for the treatment of Alzheimer's disease[1] and other degenerative disorders, stroke is still the third leading cause of death in the United States after heart disease and cancer,[2] and the currently marketed drugs are still ineffective in about 60% of patients with complex partial seizures.[3] These few examples demonstrate that the challenges for neurology and psychiatry in the new century will be as enormous as the achievements of these medical fields in the past one. New therapeutic strategies are needed, which capitalize on the progress in drug research, molecular biology, computer technology and electronics. Construction of neuroprosthetic devices will contribute greatly to this effort.

The first generation of successful neuroprostheses included diaphragm pacing devices to stimulate the phrenic nerve in patients with respiratory paralysis,[4] the Neurocybernetic Prosthesis for seizure control via vagus nerve stimulation[5] and cochlear implants for acoustic nerve stimulation in individuals with hearing loss.[6] A new generation of neuroprostheses, designed to stimulate the central and peripheral nervous system to restore sensory and motor functions, is the subject of the present book. Indeed, neuroprosthesis research focuses on the electrical stimulation of neural tissue. This chapter suggests that the viability of hybrid, molecular–electrical neuroprostheses may also be worth exploring. This idea is based on the fact that neurons are not merely living electronic machines. Rather, we emphasize[7] that they "receive, process, store and transmit information by continuous interconversion of analog neurochemical and digital electrical signals." Therefore, neuronal functions can be manipulated not only by electrical means, but by pharmacological means as well. This can be exploited in future neuroprostheses.

The present chapter will first examine the functional units of the brain, emphasizing the dual, molecular–electrical nature of neuronal activity. Understanding the operation of these units as they form larger networks and produce behavior is critical for designing neuroprostheses. Therefore, a method will be described which has been tailored to determine the molecular regulation of neuronal firing in the brains of behaving rats and monkeys. Finally, these theoretical and methodological analyses will be put in a clinical perspective, and the concept of hybrid neuroprostheses will be evaluated.

9.1.1   Neurons as Molecular–Electronic Computers

9.1.1.1   The Operation of the Functional Units of the Brain

While the functions of the human brain are based on the intercommunication of approximately 10^{11} neurons, each of these neurons has intimate relations with neighboring glial cells. Each neuron also receives oxygen and nutrients from the local

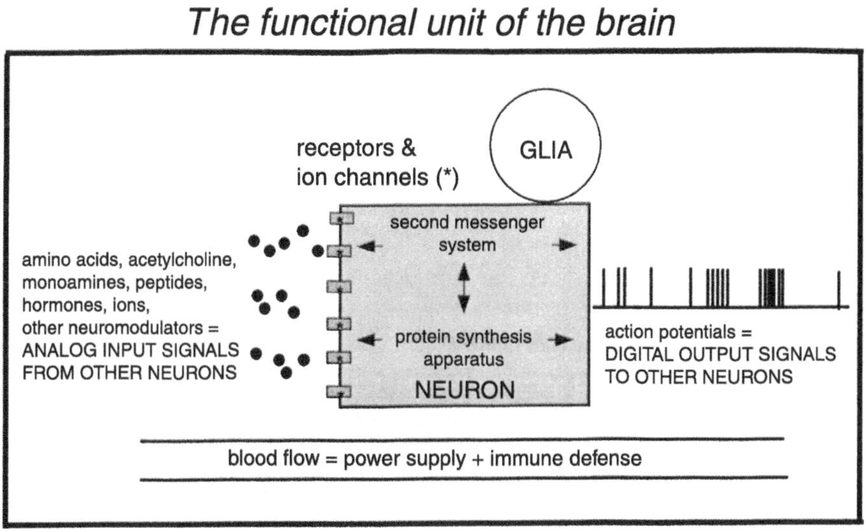

The functional unit of the brain

FIGURE 9.1 A schematic depiction of the functional unit of the brain. The electrical output of the neuron is determined by molecular mechanisms which operate on diverse neurochemical inputs. This fundamental feature of the neuron can be exploited in future neuroprostheses, which can manipulate neuronal activity by both electrical and pharmacological means.

microcirculation, which serves as an interface between the neuron and the immune system. Thus, it is appropriate to view each neuron in the brain as the central component of an extended functional unit, as illustrated in Figure 9.1. While neurons transmit information to other neurons through purely electric tools, the action potentials, the generation and the spacing of these action potentials are regulated by molecular mechanisms. In fact, these mechanisms are sophisticated interplays of a large number of intra- and extracellular molecular systems. Thus the neurons work as molecular–electronic computers as they process analog, neurochemical inputs with molecular mechanisms in order to generate digital, electrical outputs. Pharmacological agents can alter the efficacy of neurochemical inputs (e.g., neurotransmitter uptake blockers), can interfere with the molecular input-processing (e.g., second messenger modulators), and can modify the generation of electrical outputs (e.g., local anesthetics). Importantly, these neuronal computers are powered, protected, and maintained by glial cells and the local blood flow, both also sensitive to pharmacological agents. As a consequence, stimulation of brain tissue via neuroprosthetic devices can be accomplished by electrical stimulation, pharmacological stimulation, or both.

9.1.1.2 Abnormal Neuronal Firing and Brain Disorders

There is an agreement among neuroscientists that the flow of information in the brain is principally coded in the sequence of action potentials generated by the neurons, although this code is not yet deciphered.[8] Indeed, throughout life each neuron seems to produce a characteristic stream of action potentials, which is called the cell's firing pattern. We think that when "aberrant firing patterns occur in a large number of neurons and remain uncompensated, neurological and psychiatric disorders develop."[9]

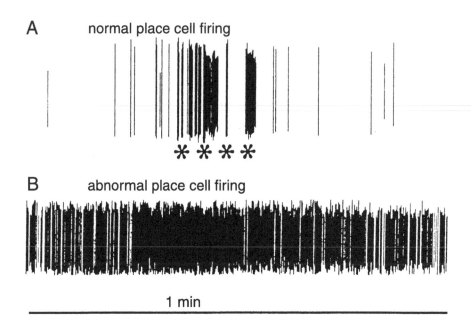

FIGURE 9.2 Normal and pathophysiological firing patterns obtained in freely moving rats. One min traces of discriminated action potentials (140–280 μV) are shown. Panel A: The normal firing pattern of a hippocampal place cell. Note the marked firing rate increase (asterisks), preceded and followed by low frequency discharges. In this cell type, such firing rate increases are displayed at specific spatial locations. Panel B: Pathophysiological firing pattern in another place cell, following microdialysis with 500 μM NMDA in its environment. Note the abnormally high frequency discharges, which eventually leads to epileptiform electrographic seizure. Identifying the key firing pattern abnormalities in various brain disease models will help design future neuroprostheses.

These aberrant firing patterns have not been elaborated for most brain disorders, partly because controlled cellular studies cannot be performed in humans, and partly because many diseases do not even have proper animal models. Nevertheless, for some diseases, typical firing pattern abnormalities have been identified. For example, seizures emanating from limbic structures have been shown to be preceded by or associated with abnormally high cellular firing rates in animals,[10] as well as in patients who subsequently underwent temporal lobectomy.[11] Figure 9.2 provides examples of normal and abnormal neuronal firing patterns in the rat hippocampus.

Since signal transmission within the network of the functional units of brain is manifested in neuronal firing patterns, understanding the laws that govern these firing patterns is critical for neuroprosthesis research. These laws include the principles of the molecular regulation of action potential generation in the brain. The question is: What methodological approaches are appropriate for uncovering these principles? Single-cell recording, at least presently, is an invasive technique. This excludes controlled human studies. Thus, animal experimentation is necessary. Non-primates, such as rats, are essential to gather the fundamental data. At the same time, the concepts that emerge from these data must ultimately be tested at the proper phylogenetic level. Thus, carefully designed monkey studies are also important. As Figure 9.1 illustrates, the firing of neurons in the brain, that is, their electrical output, is driven by the unceasing waves of diverse neurochemical inputs. Substantial numbers of these inputs originate from distant, synaptically connected circuitry. Since such inputs are mostly disconnected in tissue cultures and slice preparations, these may not be the best approaches for studying spontaneous neuronal firing. Experiments on anaesthetized subjects are also problematic because of the profound modulatory effects of the anesthetics themselves.[12,13] Therefore, the molecular regulation of neuronal firing in the rat and monkey brain are best studied in natural circumstances, during behavior. With the combined single-cell recording–intracerebral microdialysis method, such studies can be performed.

9.2 MAPPING THE MOLECULAR MECHANISMS OF NEURONAL FIRING IN THE BRAIN OF BEHAVING ANIMALS

The most popular approach to gaining insight into the neurochemical machinery of the neurons in the brain has been monitoring single-cell activity following systemic drug administration. But in this case the drug affects the cells of the whole body and often drastically changes the subject's behavior. These problems obscure the specific effects of the drug on the recorded neurons. Intracerebral methods are needed to eliminate these problems. However, no single technique has been able to capture the neurochemical/molecular mechanisms that control the firing of neurons in the brain of behaving animals. Therefore, for this purpose integrative techniques have been developed. In 1993, Glynn and Yamamoto[14] constructed special electrodes for simultaneously detecting cellular electrical activity and local dopamine release. In 1994, West and Woodward[15] successfully combined iontophoresis and single-cell recording in freely moving animals. These ingenious methods have proved to be too difficult for routine use. As the author learned from personal communications, several groups have attempted to record single neurons and to inject drugs into the extracellular space via microcannulas and push–pull cannulas in behaving rats. However, to date these attempts have not yielded publishable results.

In the meantime, it has become apparent that the intracerebral microdialysis method, invented by Bito et al.[16] and refined by Delgado et al.[17] and Ungerstedt and his colleagues,[18] is suitable not only for collecting neurochemical compounds from the extracellular space. It is also suitable for delivering drugs, via the implanted microdialysis probes, into the brain.[19] This process, often called "reverse dialysis," is demonstrated in Figure 9.3. The illustration shows that a radiolabelled compound, perfused through an intracerebrally implanted microdialysis probe, diffuses out to the surrounding brain tissue.

In another area of neuroscience, single-cell recording techniques underwent significant progress,[20] culminating in the ability to simultaneously record the firing of more than 100 neurons in the moving rat.[21] These advances made it possible to combine single-cell recording and microdialysis techniques, in order to perfuse drugs in the environment of electrophysiologically monitored neurons in freely behaving animals.[7]

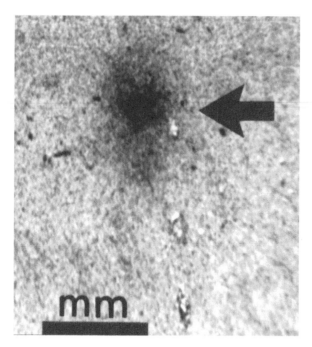

FIGURE 9.3 Autoradiogram showing the distribution of (^{3}H)cAMP molecules in the rat hippocampal tissue (arrow), following local (30 min) microdialysis with a mixture of unlabeled dibutyryl cAMP (1 mM) and the labeled compound. Note that the isotope diffused into a restricted, approximately 1 mm diameter area. Thus, microdialysis is an effective method for localized intracerebral drug delivery.

9.2.1 NEURONAL RECORDINGS COUPLED WITH LOCALIZED DRUG PERFUSIONS

9.2.1.1 The Combined Single-Cell Recording–Microdialysis Method

The concept of integrating extracellular recording and microdialysis techniques to pharmacologically manipulate neurons is illustrated in Figure 9.4. The advantages of this method are the following: (1) It affects only a discrete brain site; (2) It may not cause behavioral alterations; (3) It can be readily performed in freely behaving animals; (4) It allows continuous recording and dialysis for as long as 4–5 days; (5) It makes it possible to remotely alternate the perfused control and drug solutions, without disturbing the animal; (6) It is suitable for multiple, successive drug deliveries into the recording site; and (7) It offers the determination of local neurotransmitter release by neurochemical analysis of the collected dialysates. The disadvantages of the method are that: (1) It allows only a small fraction (5–30%) of the perfused drugs to diffuse through the dialysis membrane into the extracellular space, and (2) As a consequence, it makes it difficult to know the actual extracellular drug concentrations in these experiments. Nevertheless, the technique has made it possible for the first time to pharmacologically manipulate neurons in the brain during

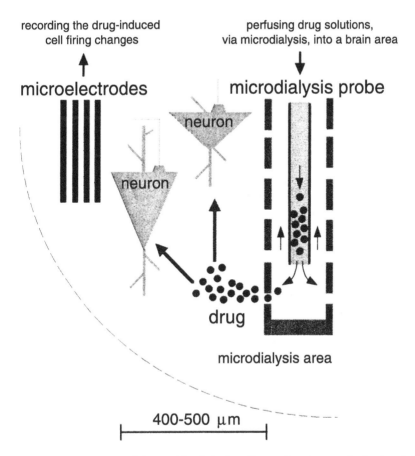

FIGURE 9.4 The concept of the combined single-cell recording–intracerebral microdialysis method. Drugs, perfused through an intracerebrally implanted microdialysis probe, diffuse into the environment of extracellularly recorded neurons. The drug-induced neuronal firing pattern changes are recorded and analyzed. From this analysis, information can be obtained on the neurochemical/molecular machinery of the exposed neurons. The method proved to be suitable, for the first time, to pharmacologically manipulate electrophysiologically recorded neurons in the brain of freely behaving animals.

behavior. Indeed, it has been successfully used by other investigators,[22] and its application was recently proposed for "research on the human brain in an epilepsy surgery setting."[23]

9.2.1.2 The Experimental Apparatus

The experimental apparatus consists of two major blocks: the electrophysiological recording system and the microdialysis system. The electrophysiological system includes the test chamber, the actual recording apparatus, and the computerized data acquisition and analysis system. The test chamber is an electrically shielded, sound-attenuating, illuminated and ventilated chamber.

The electrophysiological system is partly inside and partly outside this chamber. An electric commutator and a recording cable are inside. The recording cable, equipped with operational amplifiers, connects to the driveable microelectrode assembly implanted in the brain. Outside the chamber are differential AC amplifiers, oscilloscopes, the computer interface and the computer, and a patch panel that distributes the electrical signals from the recording cable to these components. The patch panel also connects the power supply to the operational amplifiers. This basic setup can be extended with speakers (to "listen" to the neuronal discharges) and a video apparatus to view and tape the behavior of the examined animal. These are very useful, although optional items.

The microdialysis system consists of syringe pumps, a liquid swivel, a traditional valve or the newly developed minivalve, and the tubing that connects these components, and runs, along the recording cable, to the microdialysis probe. The microdialysis probe is inserted in the brain through a probe guide, which is integrated into the driveable microelectrode assembly. The liquid swivel and the electric commutator are also integrated into a single unit, which rotates together as the animal moves around in the test chamber. It is optional to extend this system with an HPLC or other neurochemical analyzer to measure neurotransmitters from the dialysates.

The above listed items are either available commercially or can be manufactured at any machine shop. However, there are three special devices in the experimental apparatus, which are described in the next section. The construction of these devices made it possible to develop the combined single-cell recording–intracerebral microdialysis method.

9.2.1.3   Special Devices

The first special device is the electric commutator–liquid swivel unit. Separately, each device consists of a stationary and a rotating part. In the unit, the stationary and the rotating part of the commutator are secured to the corresponding part of the swivel. In this way, the rotation of the commutator drives the rotation of the swivel. This unit has been described previously.[7] It is absolutely necessary for rat studies, since the rats are moving around in a test chamber. For monkey studies, where the animal is seated in a primate chair, the use of this device is optional.

The second special device is the driveable microelectrode/microdialysis probe guide assembly (Figure 9.5). The tips of the electrodes and the probe-guide are implanted in the brain; the rest of the device is anchored to the skull with dental cement. The microdialysis probe is inserted in the brain through the probe guide. This assembly has several advantages. Its microelectrode-array is made of 10–13 nichrome wires, 25 μm diameter each. These wires have proved to yield excellent extracellular recordings.

The assembly is anchored to the skull with three nylon cuffs, which include the driving screws. This arrangement provides a structural stability for the device. Furthermore, the electrode/probe unit can be advanced in the brain simultaneously, in a well-controlled fashion, by turning the driving screws. The integration of the microdialysis probe guide into the assembly assures the critical 400–500 μm distance between the tips of the microelectrodes and the probe. Certainly, other designs can

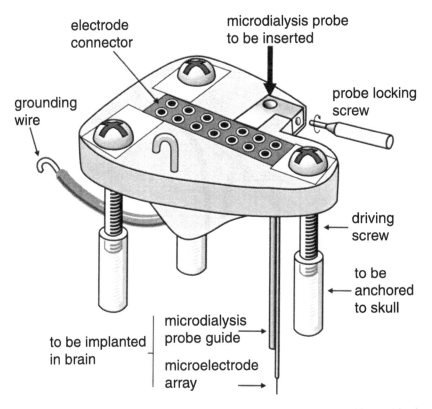

FIGURE 9.5 The driveable microelectrode/microdialysis probe guide assembly used in single-cell recording–microdialysis studies. The figure shows the assembly for rats. The assembly for monkeys differs only in the length of the microdialysis probe guide/microelectrode array unit. The device weighs only 4 g and is easily carried by monkeys and rats on their heads.

also be used, but the presented one and its previous versions have been tested in many experiments.[7,9,10]

The third special device is a minivalve (Figure 9.6). It should be noted that single-cell recording/microdialysis studies can be conducted with traditional valves and liquid switches (Figure 9.7). Furthermore, in monkey studies the solutions can be alternated manually by the experimenter from behind the animal's chair (Figure 9.8). However, with the use of the minivalve, faster and more reliable intracerebral drug deliveries can be achieved, and the subject's behavior is not disturbed (Figure 9.9). This device has just recently been developed. It is a lightweight (2 g) valve, portable by the animals on their heads. Due to the closeness of the device to the recording/dialysis site, the drug solutions reach the brain quickly. As a consequence, at a 10 µl/min flow rate the drug-induced cell firing changes can be detected within 1–3 min after switching the minivalve. In contrast, the heavy and large traditional valves and switches are placed on or out of the test chamber. Therefore, the drug solutions reach the brain slowly, and thus it often takes 20–40 min to achieve drug-induced cell firing changes. This lengthens the experiments and

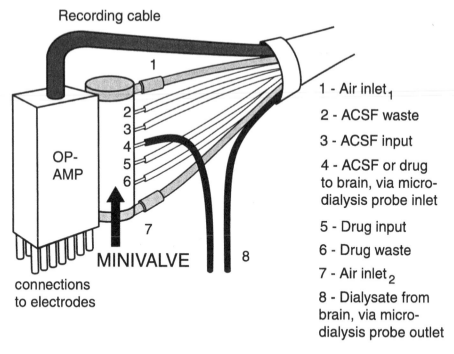

FIGURE 9.6 The minivalve, as it is attached to the operational amplifier (op-amp) of the recording cable. This cable connects to the "electrode connector" shown in Figure 9.5. The valve inside the device is switched by air pressure, regulated remotely by a microprocessor-based controller. The device directs either ACSF or a drug solution into the brain via the implanted microdialysis probe. Due to the closeness of the device to the head of the animal, the drugs are delivered into the recording/dialysis site rapidly, and the drug-induced cell firing changes can be recorded instantly. This is an important advantage in single-cell recording/microdialysis studies.

obscures the onset of drug effects. The minivalve eliminates these difficulties. The valve inside this miniature device is activated by air, remotely, by a microprocessor-based controller. The air is transmitted from the controller to the minivalve via two designated channels of the liquid swivel. Thus, when the minivalve is used, these swivels are, in fact, functioning as air/liquid swivels.

9.2.1.4 Surgical Procedures

While the same experimental apparatus and special devices can be used for both rats and monkeys, the protocol for the surgical implantation of the microelectrode/microdialysis probe guide assembly substantially differs in the two species. The monkey surgery requires sterile conditions, the involvement of an anesthesiologist, a special protective cap to prevent infection around the implant, and an elaborate postoperative monitoring. Therefore, the surgical procedures will be described separately for the rat and the monkey (squirrel monkey; *Saimiri sciureus*).

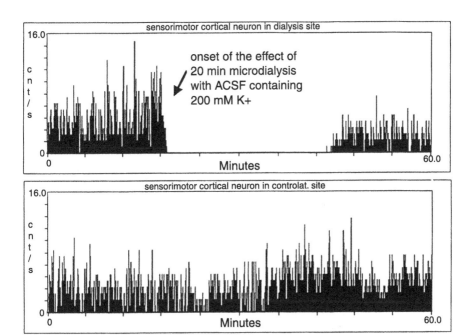

FIGURE 9.7 Demonstration of the ability of the single-cell recording/microdialysis method to manipulate the firing of an electrophysiologically monitored neuron in the rat sensorimotor cortex (upper panel), without causing similar effects in the contralateral site (lower panel). Firing rate histograms are shown; X axis: recording time (min); Y axis: firing rate (spike counts per second). Note that intracortical microdialysis with a high K^+ concentration solution reversibly silences the recorded neuron, an effect that is absent in the contralateral site. This experiment was conducted with a traditional valve placed on the ceiling of the test chamber. Because of the distance between the valve and the animal, 20 min elapsed between the activation of the valve and the onset of the drug action.

The rat is anaesthetized with 50 mg/kg pentobarbital, i.p., and placed in a stereotaxic apparatus. The skull is exposed, and a 2 mm diameter craniotomy is made above the targeted brain site. With the use of the manipulator of the stereotaxic apparatus, the microelectrode/microdialysis probe guide assembly is positioned on the skull in such a way that the tip of the electrode array penetrates into the brain, just above the target site, while the tip of the probe guide is 5 mm above that of the electrode array. Then the assembly is anchored to the skull by applying dental cement around the nylon cuffs of the driving screws, as well as around four anchoring screws placed in the skull. One of these screws will also serve for grounding the rat. The craniotomy is closed with sterile bone wax, and the assembly is covered with sterile tape. The rats are introduced in the experiments 3–4 days later.

The monkey receives i.m. injections of atropine (0.05 mg/kg) and Bicillin (100,000 units/kg), followed by a mixture of ketamine (11 mg/kg) and xylazine (0.5 mg/kg). In the sterile operating room, endotracheal intubation is performed, and the anesthesia is maintained with 1.2%–2% isoflurane in oxygen. The femoral vein

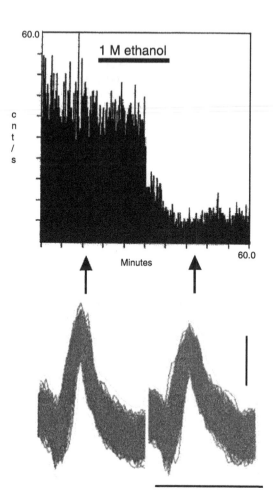

FIGURE 9.8 Demonstration of the ability of the single-cell recording/microdialysis method to manipulate the firing of an electrophysiologically monitored neuron in the monkey hippocampus. Firing rate histogram is shown as in Figure 9.7. In this experiment the control solution (ACSF) and the drug solution (1 M ethanol) were alternated manually from behind the animal's chair. The drug suppressed, but did not silence, the firing of the cell. The overlaid action potentials of the neuron, collected before and after the ethanol exposure, are also shown. The similarity of the waveforms proves that the recording was made from the same single cell during the entire 60 min period. Calibrations: horizontal bar: 0.8 msec; vertical bar 100 µV.

and artery are cannulated for fluid administration and blood pressure measurement. Besides blood pressure, ECG, SpO_2, respiratory rate, end tidal CO_2, respiratory rate, and rectal temperature are monitored. The monkey is placed in the stereotaxic apparatus, and the microelectrode/microdialysis probe guide assembly is implanted in the brain in essentially the same fashion as in the rat. However, in the monkey,

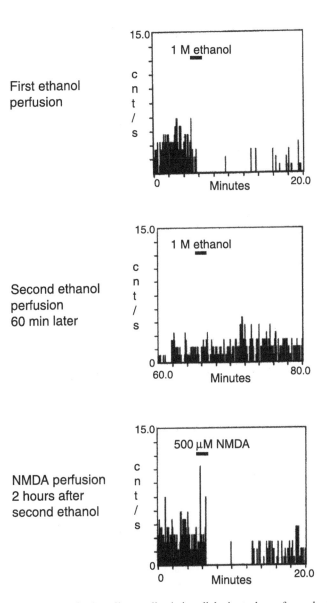

FIGURE 9.9 Data from a single-cell recording/microdialysis study performed with the minivalve. The recordings were made from the same single neuron in the hippocampus of a freely moving rat. Firing rate histograms are shown, as in Figure 9.7. Note that the first ethanol perfusion, delivered via the minivalve into the hippocampal recording site, suppressed the firing of the recorded neuron as quickly as two minutes after valve-activation. However, the second ethanol delivery was ineffective, indicating the development of rapid cellular alcohol tolerance. A subsequent N-methyl-D-aspartate (NMDA) application through the probe induced an initial firing rate increase, followed by electrical silence. This proved that the inefficacy of the second ethanol perfusion was not due to microdialysis probe dysfunction.

a protective cap is placed around the assembly. This protective cap consists of a base, which is anchored to the skull with dental cement, and a cover, which can be removed during the experiments to provide access to the assembly. Betadine is extensively applied both inside and outside this cap. Finally, the skin is approximated, the anesthesia is reversed, and the monkey returns to the home cage. After a week of postoperative monitoring, which includes neurological examinations, as well as antibiotic (Bicillin) and analgesic (Buprenex, 0.01 mg/kg, i.m) treatments, the animal is ready for the experiments.

For both animals, the electrode/probe guide is cleaned with 100% ethanol, and for the monkeys, the surgical tools are steam-sterilized. We use the stereotaxic brain atlas of Paxinos and Watson[24] for rats, and that of Gergen and MacLean[25] for the squirrel monkeys. More detailed descriptions of the surgical procedures are available in our previous publications.[7,26]

9.2.1.5 Single Cell Recording/Drug Perfusion Sessions in Behaving Rats and Monkeys

These sessions start with transferring the animal to the test chamber. Here the rats can move freely, but the monkeys are seated in a primate chair without head restraint. The monkey is trained for about two weeks to sit in the chair calmly and consume food. The microdialysis probe, already perfused with artificial cerebrospinal fluid (ACSF), is inserted into the brain through the probe guide, and the recording cable is plugged into the microelectrode connector. This critical, 3–4 min procedure can be readily done in the seated, awake squirrel monkey. The rats, however, should be trained to be immobile during probe insertion and cable connection. The microdialysis is maintained with artificial cerebrospinal fluid. We use a 10 µl/min flow rate, which does not interfere with normal cell firing and allows sufficient drug delivery through the probe. However, if the microdialysis procedure is also used for neurotransmitter release measurements, the flow rate should be reduced to 1–3 µl/min. After probe insertion, the brain tissue is allowed to recover from the microtrauma for 1–2 hours in the monkey and for 2–14 hours in the rat. Then the electrode/probe unit is advanced into the targeted brain area to record cellular electrical activity.

The extracellular electrical signals, recorded between microelectrode-pairs, are amplified (10,000×), filtered (between 300 Hz and 10,000 Hz), and displayed on oscilloscopes. They are also digitized at a 40,0000 Hz sampling rate and collected on the hard disk of a computer. For this purpose, we use the Discovery data acquisition system from DataWave Technologies, Inc. It is convenient to also record, between one of the microelectrodes and the grounding screw, the local EEG activity. These recordings are important indicators of pathophysiological electrical changes, such as epileptiform discharges, etc. Movement artifacts are mostly eliminated from the recordings by the operational amplifiers built in the recording cable. Therefore, the use of these impedance-lowering devices is necessary. When clear action potentials with amplitudes at least 2.5× higher than the 60–80 µV background noise are recorded from identified neurons over at least 30–60 min, and the microdialysis fluid flow is constant, data storage and drug perfusions can start. The Discovery program can also be used for data storage. The stored data files are analyzed later.

Each data file must comprise recordings before, during, and after the perfusion of a drug solution. The recordings before the drug application, when ACSF is perfused, provide data for the control (pre-drug) cell firing patterns. The recordings during the drug perfusion provide data for the immediate/short-term cell firing modulatory effects of the delivered drugs. The recordings after the drug perfusion provide data that characterizes both the recovery of the control (pre-drug) cell firing patterns and the delayed, long-term effects of the drug. The duration of the drug perfusions can vary from 2 min to 60 min, depending on the mechanism of action of the delivered compound. A given solution can contain one drug, or it can be a mixture of several drugs. In one animal, multiple, successive drug perfusions can be readily conducted. In rats, each recording/dialysis session can be conducted continuously for as long as 3–4 days, while the animal moves, sleeps, eats and drinks. Of course, the actual data storage periods are restricted to several selected hours. The monkeys are seated in a primate chair, requiring a different experimentation strategy. Here, each recording/dialysis session is terminated after 5 hours. However, with 1–3 days of intersession intervals, the drug perfusions can be repeated. We performed a single, 3- to 4-day session on each rat, and ten to twelve 5-hour sessions on each monkey. The stored electrophysiological data are analyzed off-line, for example with the CP Analysis software of DataWave Technologies, Inc. These analyses include the discrimination of the action potentials and the generation of firing rate histograms.

Figure 9.7 shows the effect of delivering excess potassium into the extracellular space of a neocortical neuron, while another neuron in the contralateral site is recorded simultaneously, in a freely moving rat. In this experiment, a traditional valve was used to alternate the solutions in the microdialysis probe. The figure demonstrates (1) the clear firing suppressant action of the perfused potassium solution, (2) the long, 20-min period that elapses between the start of the potassium perfusion and its cellular effect, (3) the lack of this pharmacological effect in the contralateral neocortex, and (4) the recovery of the firing pattern of the exposed neuron after washing out the excess potassium from the extracellular space.

Figure 9.8 shows the effect of delivering ethanol into the extracellular space of a fast-firing hippocampal neuron in a squirrel monkey seated in a primate chair. Head restraint was not used, as this is not necessary with the described method. The control and ethanol solution were alternated manually by the experimenter, without the use of a valve. The figure demonstrates (1) the firing suppressant action of ethanol and (2) the similar action potential waveforms at the beginning and end of the session. This proves that the recording was obtained from the same neuron throughout the recording/dialysis period. However, a slight amplitude decrease can also be recognized following the drug exposure, which indicates that the perfused ethanol affected not only the frequency of the action potentials but also the process of their cellular generation.

Figure 9.9 shows the effects of two consecutive ethanol perfusions and a subsequent NMDA (N-methyl-D-aspartate) perfusion on the firing of a hippocampal pyramidal cell in a freely moving rat. In this experiment, the minivalve was used to alternate the solutions in the microdialysis probe. The figure demonstrates (1) the firing suppressant action of the ethanol perfusion, (2) the instant development of this pharmacological effect, (3) the lack of ethanol effect during and after its second perfusion due to the induction of rapid cellular alcohol tolerance, (4) the biphasic, excitatory/inhibitory action of NMDA on the electrical activity of the neuron, which proves that (5) the lack of effect of the second ethanol perfusion could not be due to a dysfunction of the microdialysis probe.

9.2.1.6 Histological Examination of the Implanted Brain Sites

After the termination of the recording/dialysis sessions, the brains are processed for histological studies. These are important, because they determine the localization of the track of the microelectrode-array, as well as that of the microdialysis probe. The histological studies are also needed to assess the extent of tissue damage around the electrode/probe unit. Figure 9.10 shows the result of such a histological study.

9.3 CLINICAL PERSPECTIVES

Mapping the electrophysiological responses of single neurons to intracerebrally perfused drugs in behaving animals serves two goals. First, it serves to elucidate the molecular regulation of neuronal firing in the brain. Second, it serves to develop effective strategies for the treatment of intractable neurological and psychiatric disorders. These two goals are inseparable, as solid basic scientific data on the operation of brain cells is the prerequisite of pioneering new pharmacotherapeutic strategies. Certainly, these strategies include the administration of new drugs or drug combinations via traditional systemic routes. However, systemically administered drugs will always cause adverse reactions, as they act not only on the targeted, pathophysiologically functioning neuron-population in the brain, but also the rest of the nervous system and in the whole body. Furthermore, the blood–brain barrier prevents the penetration of most systemically administered peptides, proteins, oligonucleotides, or vectors into the brain, which, at present, renders many of these potentially powerful therapeutic agents clinically irrelevant. Therefore, intracerebral drug administration strategies should be considered. Indeed, the use of an intracerebroventricular pump system for delivering neurotrophic factors into the brain was already proposed.[27] An alternative method is to load the drugs into ethylene vinyl acetate copolymer (EVAc) rods and implant these rods chronically in the brain. The drugs then diffuse from the polymer matrix.[28] The problem with these approaches is that the drug delivery is continuous, with no feed-back from the neural tissue. This may lead to the administration of drug solutions whose concentration is too low and, therefore, ineffective. It can also lead to the administration of drug solutions whose concentration is too high, resulting in pathophysiological reactions. Furthermore, traditional drug delivery designs do not allow periodic drug deliveries, applying the drug solutions only when they are needed. Hybrid neuroprosthetic devices can offer regulated intracerebral drug delivery.

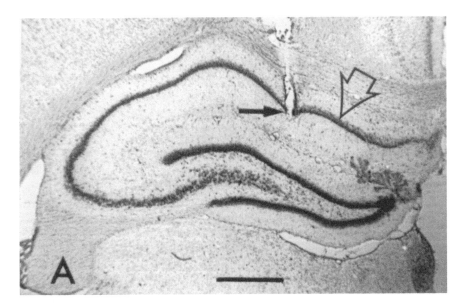

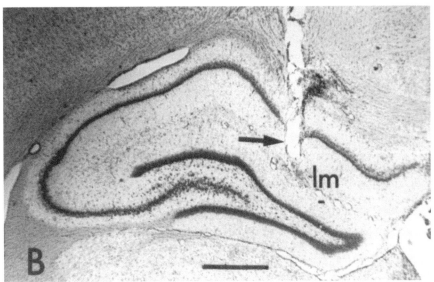

FIGURE 9.10 Results of a histological study in a rat implanted chronically with a microelectrode/microdialysis probe unit in the CA1 region of the left hippocampus. Coronal sections, at the level of the tip of the microelectrode (A) and that of the microdialysis probe (B), are shown. Nissl staining of formaldehyde-fixed tissue. Arrow in A points to the end of the electrode-track in the stratum pyramidale (open arrowhead); arrow in B points to the end of the probe-track in the stratum lacunosum-moleculare (lm). Calibration bars: 1 mm.

9.3.1   THE FUTURE OF INTRACEREBRAL PHARMACOTHERAPIES

9.3.1.1   Hybrid Neuroprosthetic Devices

Our term "hybrid neuroprosthesis" refers to a microcomputer-controlled, intracerebrally implanted drug delivery device, in which the timing and duration of the drug deliveries are determined by the implanted brain tissue's own electrical activity. Thus, the device is a "hybrid" of pharmacological and electrophysiological instruments. The pharmacological components are: (1) a cannula or catheter chronically implanted in the ventricular system or in the brain tissue and (2) a miniature, subcutaneously placed drug reservoir/pump unit, which can be periodically refilled. The electrophysiological components are (1) a recording electrode chronically implanted in the brain and (2) a miniature, subcutaneously placed data recorder/conditioner. A microcomputer, placed in close proximity to these components, both analyzes the electrophysiological data and controls the drug reservoir/pump. All of these components are powered by a nearby battery and sealed in a biocompatible case.

The concept of hybrid neuroprostheses has already been mentioned in a recent article,[10] but the construction of such a device has yet to be accomplished. Nevertheless, recording and stimulating electrodes chronically implanted in the brain are widely used,[29,30] just as are intraventricularly implanted catheters for the treatment of hydrocephalus.[31,32] Essentially, a hybrid neuroprosthesis is a combination of these two elements, coupled with a microprocessor-controlled pump.

Hybrid neuroprostheses may be used for the management of intractable temporal lobe epilepsies, which are currently treated by surgical removal of the epileptogenic tissue. The recording electrodes can monitor the electrophysiological activity of the epileptogenic focus, and from their signals the subcutaneously implanted microcomputer can recognize the initiation of an EEG seizure and activate the pump to deliver an antiepileptic drug solution directly into the pathophysiological tissue. Another application may be the intracerebroventricular administration of drug combinations, continuously adjusted by simultaneous electrophysiological monitoring, in patients with Alzheimer's disease. In rats, chronic intracerebroventricular infusion of phosphoprotein phosphatase inhibitor okadaic acid induces histopathological changes in the hippocampus and neocortex that resemble those that occur in Alzheimer's disease.[33] This indicates that the neural circuitries involved in Alzheimer's disease can be affected by drugs administered into the ventricles. If so, beneficial effects may also possibly be induced in this disease via ventricular drug administrations, for example, with the use of hybrid neuroprostheses. These devices may also be useful in the management of Parkinson's disease. Lindvall and colleagues[34] reported clinical improvements in patients with Parkinson's disease following the grafting of human embryonic dopamine-rich mesencephalic tissue unilaterally into the putamen. Thus, the treatment was achieved with implanting the dopamine-rich tissue at a single occasion and without monitoring the electrical activity of the grafted area. Hybrid neuroprostheses could offer multiple drug administrations into the putamen, regulated by local electrophysiological recordings. The viability of the hybrid neuroprosthesis strategy in the management of other brain disorders, especially in that of stroke, remains to be explored.

9.3.1.2 Theoretical Challenges and Technical Obstacles

The outlined hybrid neuroprostheses would deliver drugs into a specific, pathophysiologically functioning brain site. However, brain disorders usually involve not one but several interconnected regions that function abnormally. Therefore, localized drug deliveries may not be able to correct or reverse diffuse pathophysiological processes. This problem can be partly resolved with multiport cannulas, which can release drugs along their axis into many brain sites, or intraventricular cannulas that can provide widespread drug administrations into the brain. But in these cases normal tissue can also be exposed to the drugs, leading to adverse pharmacological effects. Thus the key question becomes: To what extent can the symptoms of a brain disorder be ameliorated by drug administrations into a specific brain site?

Even if Nature allows improvement of brain functions by administering drugs intracerebrally into a restricted region, necrosis, as well as infection, inflammation, gliosis, vascular reactions, and other adverse effects around the implants[32] can prevent the proper function of hybrid neuroprostheses and can cause severe complications. Thus, understanding the mechanisms of these unwanted tissue reactions is critical, because it can lead to the development of surgical and pharmacological strategies for their suppression. The other major difficulty is related to the subcutaneous placement of the electrophysiological data recorder/conditioner, the drug reservoir/pump, the microcomputer, and the battery. Whether they will be tolerated by the patients cannot be foreseen. Clearly, careful and comprehensive studies in non-human primates will be essential to assess the viability of future hybrid neuroprostheses.

9.4 SUMMARY

The neurons of brain generate electrical outputs, also called firing patterns, in response to neurochemical inputs. These inputs are processed by sophisticated molecular mechanisms. Combination of extracellular recording and intracerebral microdialysis techniques in behaving rats and monkeys allows the study of the molecular mechanisms that regulate the firing pattern of neurons in brain in natural circumstances. Mapping these molecular mechanisms in both normal animals and in animal models for brain diseases helps to devise rational, intracerebral drug delivery strategies for the treatment of neurological/psychiatric disorders. Such intracerebral drug deliveries can be coupled with electrophysiological monitoring of the drug-exposed brain site. This idea is at the core of the hybrid neuroprosthesis concept. Specifically, this concept suggests the use of intracerebrally implanted electrode–cannula units, connected to a subcutaneously implanted microcomputer that analyzes the recorded electrophysiological signals and activates a nearby minipump to deliver drugs through the cannula. In this way, finely controlled intracerebral drug deliveries into a malfunctioning brain area can be achieved. Such devices can be useful additions to the repertoire of future neuroprostheses.

The combined single-cell recording–intracerebral microdialysis method has recently been utilized by two other groups: Thakkar, M. M. et al. (*J. Neurosci.*, 18, 5490, 1998) and Alam, M. N. et al. (*J. Physiol.*, 521, 679, 1999). Patent application for the hybrid neuroprosthesis was filed in 2000.

ACKNOWLEDGMENTS

I am grateful to Lorant Kovacs, President of ESCO, Mt. Kisco, NY, for helping to elaborate the engineering aspects of the hybrid neuroprosthesis. He was also instrumental, together with Laszlo Kando and Geza Medvecky, in the development of the mini-valve. I am indebted to Drs. Paolo Bolognese and John G. Kral for many inspiring discussions on the role of basic neuroscience in advancing neurosurgery. The guidance and advice of Dr. Leonard A. Rosenblum were critical in the monkey studies. I express my gratitude to Drs. Gyongyi Gaal and John K. Chapin for introducing me into the field of neuroprosthesis research. The works presented in this chapter were supported by NIH Grants AA10814 and MH56800, and by a Research Investment Initiative Grant from SUNY at Brooklyn.

REFERENCES

1. Delagarza, V. W., New drugs for Alzheimer's disease, *Am. Family Physician*, 58, 1175, 1998.
2. Wolf, P. A. and D'Agostino, R. B., Epidemiology of stroke, in *Stroke: Pathophysiology, Diagnosis, and Management*, Barnett, H. J. M., Mohr, J. P., Stein, B. M., and Yatsu, F. M., Eds., Churchill-Livingstone, New York, 1998, chap. 1.
3. Williamson, P. D. and Engel, J. Jr., Complex partial seizures, in *Epilepsy: A Comprehensive Textbook*, Engel, J. Jr. and Pedley, T. A., Eds., Lippincott-Raven Publishers, Philadelphia, 1997, chap. 49.
4. Chervin, R. D. and Guilleminault, C., Diaphragm pacing for respiratory insufficiency, *J. Clin. Neurophysiol.*, 14, 369, 1997.
5. Schachter, S. C. and Saper, C. B., Vagus nerve stimulation, *Epilepsia*, 39, 677, 1998.
6. Hambrecht, F. T., The history of neural stimulation and its relevance to future neural prostheses, in *Neural Prostheses: Fundamental Studies*, Prentice Hall, Englewood Cliffs, NJ, 1990, chap. 1.
7. Ludvig, N., Potter, P. E. and Fox, S. E., Simultaneous single-cell recording and microdialysis within the same brain site in freely behaving rats: a novel neurobiological method, *J. Neurosci. Methods*, 55, 31, 1994.
8. Ferster, D. and Spruston, N., Cracking the neuronal code, *Science*, 270, 756, 1995.
9. Ludvig, N., Fox, S. E., Kubie, J. L., Altura, B. M., and Altura, B. T., Application of the combined single-cell recording/intracerebral microdialysis method to alcohol research in freely behaving animals, *Alcoholism: Clin. Exp. Res.*, 22, 41, 1998.
10. Ludvig, N. and Tang, H. M., Cellular electrophysiological changes in the hippocampus of freely behaving rats during local microdialysis with epileptogenic concentration of NMDA, *Brain Res. Bull.*, 51, 233, 2000.
11. Babb, T. L., Wilson, C. L., and Isokawa-Akesson, M., Firing patterns of human limbic neurons during stereoencephalography (SEEG) and clinical temporal lobe seizures, *Electroenceph. Clin. Neurophysiol.*, 66, 467, 1987.
12. Patel, I. M. and Chapin, J. K., Ketamine effects on somatosensory cortical single neurons and on behavior in rats, *Anesth. Analg.* 70, 635, 1990.
13. Aida, S., Fujiwara, N., and Shimoji, K., Differential regional effects of ketamine on spontaneous and glutamate-induced activities of single CNS neurons in rats, *Br. J. Anesth.* 73, 388, 1994.
14. Glynn, G. E. and Yamamoto, B. K., A system for measuring electrophysiological multiple unit activity and extracellular dopamine concentration at single electrodes, *J. Neurosci. Methods*, 47, 235, 1993.

15. West, M. O. and Woodward, D. J., A technique for microiontophoretic study of single neurons in the freely moving rat, *J.Neurosci Methods*, 11, 179, 1984.
16. Bito, L., Davson, H., Levin, E., Murray, M., and Snider, N., The concentrations of free amino acids and other electrolytes in cerebrospinal fluid, *in vivo* dialysate of brain, and blood plasma of the dog, *J. Neurochem.* 13, 1057, 1967.
17. Delgado, J. M. R., Lerma, J, Martin del Rio, R. and Solis, J. M., Dialytrode technology and local profiles of amino acids in the awake cat brain, *J. Neurochem.*, 42, 1218, 1984.
18. Ungerstedt, U. and Hallstrom, A., *In vivo* microdialysis — a new approach to the analysis of neurotransmitters in the brain, *Life Sci.*, 41, 861, 1987.
19. Benveniste, H., Brain microdialysis, *J. Neurochem.*, 52, 1667, 1989.
20. Fox, S. E. and Ranck, J. B., Jr., Electrophysiological characteristics of hippocampal complex-spike cells and theta cells, *Exp. Brain Res.,* 41, 399, 1981.
21. Wilson, M. A. and McNaughton, B. L., Dynamics of the hippocampal ensemble code for space, *Science*, 261, 1055, 1993.
22. Dudkin, K. N., Kruchinin, V. K. and Chueva, I. V., Effect of NMDA on the activity of cortical glutaminergic structures in delayed visual differentation in monkeys, *Neurosci. Behav. Physiol.* 27, 153, 1997.
23. Engel, J., Jr., Research on the human brain in an epilepsy surgery setting, *Epilepsy Res.*, 32, 1, 1998.
24. Paxinos, G. and Watson, C. *The Rat Brain in Stereotaxic Coordinates*, 2nd ed., Academic Press, New York, 1986.
25. Gergen, J.A. and MacLean, P.D., *A Stereotaxic Atlas of the Squirrel Monkey's Brain*, Public Health Service Publication No. 933, Bethesda, 1962.
26. Ludvig, N., Nguyen, M. C., Botero, J. M., Tang, H. M., Scalia, F., Scharf, B. A., and Kral, J. G., Delivering drugs, via microdialysis, into the environment of extracellularly recorded hippocampal neurons in behaving primates, *Brain Res. Protocols*, 5, 75, 2000.
27. Schatzl, H. M., Neurotrophic factors: ready to go?, *Trends Neurosci.*, 18, 463, 1995.
28. Hoffman, D., Wahlberg, L., and Aebischer, P., NGF released from a polymer matrix prevents loss of ChAT expression in basal forebrain neurons following a fimbria-fornix lesion, *Exp. Neurol.*, 110, 39, 1990.
29. Fisher, R. S., Mirski, M., and Krauss, G. L., Brain stimulation, in *Epilepsy: A Comprehensive Textbook*, Engel, J. Jr. and Pedley, T. A., Eds., Lippincott-Raven Publishers, Philadelphia, 1997, chap. 176.
30. Spencer, S. S., Sperling, M. R., and Shewmon, D. A., Intracranial electrodes, in *Epilepsy: A Comprehensive Textbook*, Engel, J. Jr. and Pedley, T. A., Eds., Lippincott-Raven Publishers, Philadelphia, 1997, chap. 164.
31. Pudenz, R. H., The surgical treatment of hydrocephalus: an historical review, *Surg. Neurol.* 15, 15, 1981.
32. Del Bigio, M. R., Biological reactions to cerebrospinal fluid shunt devices: a review of the cellular pathology, *Neurosurgery*, 42, 319, 1998.
33. Arendt, T., Holzer, M., Fruth, R., Bruckner, M. K., and Gartner, U., Paired helical filament-like phosphorylation of tau, deposition of β/A4-amyloid and memory impairment in rat induced by chronic inhibition of phosphatase 1 and 2A, *Neuroscience,* 69, 691, 1995.
34. Lindvall, O., Sawle, G., Widner, H., Rothwell, J. C., Bjorklund, A., Brooks, D., Brundin, P., Frackowiak, R., Marsden, C. D., Odin, P., and Rehncrona, S., Evidence for long-term survival and function of dopaminergic grafts on progressive Parkinson's disease, *Ann. Neurol.*, 35, 172, 1994.

Index